# AGRICULTURAL GROWTH IN INDIA

# AGRICULTURAL GROWTH IN INDIA

Subrata Kumar Ray

SERIALS PUBLICATIONS
NEW DELHI (INDIA)

First Published - 2010

ISBN: 978-81-8387-319-2



*Published by*

**SERIALS PUBLICATIONS**
4830/24, Prahlad Street, Ansari Road
Darya Ganj, New Delhi-110002 (India)
Phone : 23245225 Fax : 91-11-23272135
E-mail: serials@satyam.net.in

Dedicated to my

Grandfather and Grandmother

# CONTENTS

# FOREWORD

Keith Griffin in an essay on the Green Revolution writes: "If one wishes to improve the standard of living of the mass of the population, one must achieve rapid growth not just of GNP in general but of agricultural output in particular. It is in rural areas that the most numerous cases of poverty exist, and it is in these areas that the assault on poverty should begin". (The Political Economy of Agrarian Change, Second Edition, Macmillan, 1979).

The so called Green Revolution, initially introduced by the high yielding varieties of wheat and later extended to rice and other cereals, came as a boon to India at about the same time as the great run of drought for three consecutive years from 1965. Poverty sharply declined between 1967-68 and 1971-72. It then went up when the first flush of Green Revolution had run its course.

The Progress of Green Revolution was tardy until the mid 1970s when the new technology led to changes in crop pattern and in methods of production. It did accelerate the development of market oriented capitalist agriculture in all parts of our countryside aided by subsidized credit provided by state banks. The decade of 1980s witnessed remarkable growth of agriculture production for India as a whole. The state of West Bengal , in particular, had achieved a marked acceleration in the trend rate of growth of agricultural production that was much higher than the all India average.

Of the several issues involved in this context, the author has concentrated on two important issues, viz , flow of institutional credit , especially bank credit, to the agricultural sector and change in the cropping pattern ( as also crop pattern) during the decade of 1980s and early 1990s. The role of the former in influencing the latter is the main theme of the present study, keeping in view that the all important matter is overall growth of agricultural output and that change should always be for the better.

The author has used secondary data at the all India level, the state level(for twelve sample states)and at the district level ( for fifteen districts of West Bengal ) for detailed analysis with statistical tools and econometric models. He has also collected household data from some villages in the district of Midnapore, West Bengal to examine the impact of change in cropping pattern on the profitability ( and employment generation) of the sample households.

Dr. Ray has studied separately the compound annual rate of growth of total agricultural output for the period under survey (1965-95) divided into different sub-periods, the nature of change in crop pattern (e.g, from traditional to modern varieties, from foodgrains to non-foodgrains, etc.), and the trend in institutional credit for the said period (s). He has also examined the relationship among them and has reached some highly interesting and thought provoking conclusions. This book, I am sure, will be found highly useful to all students of Indian Economics and particularly to the researchers in the field of Indian agriculture.

**Hiron Kumar Banerjee**
**Former Pro Vice Chancellor**
**(Business Affairs and Finance)**
**University of Calcutta**

# PREFACE

Indian agriculture has witnessed remarkable changes in the post Independence decades. The most striking aspect of the transformation has been in the production of foodgrains where we have converted a deficit prone country with concern of widespread starvation to a situation of plenty. The experience of Green Revolution has also shown that besides technological advancements and applications, supporting institutions like credit, land reforms, etc. as well as incentives like prices are of paramount importance for technology-led growth in agriculture. The switchover from a food deficit to a food surplus economy was plainly the outcome of this technology-institutional-human interface. Apart from natural factors like rainfall, growth in agricultural production depends on insfrastructural facilities. The Government's intervention in creating infrastructural facilities-coupled with the farmers ability to use modern productive technologies, thus play crucial roles. For the latter to occur the flow of timely and need-based credit to the farmers is an essential prerequisite.

The development of agriculture or crop economy of India is characterised by the process of crop diversification or changes in cropping pattern over time. The change has been associated with the increasing transformation of agriculture from subsistence to commercial farming

Thus it is held that the change in cropping pattern can accelerate the agricultural productivity and growth. The change in cropping pattern is influenced by several natural, technological, economic and institutional factors (viz., availability of irrigation, fertiliser, credit, etc.). It is obvious that the change in cropping pattern, particularly towards larger valued crops, is highly associated with the rate of adoption of new production technology which, in turn, put a greater emphasis on assured supply of sufficient finance. But the financial capability of the majority of the Indian farmers is considerably below minimum level required for the adoption of modern technology. The planned injection of technical change requires that the farmers should have a source of finance through a right agency which can supply the needed capital in the right quantity and at the right time. But credit must be cheap, accessible, safe and productive. Hence the change in cropping pattern rests largely on the cheap external sources of finance. On the other hand, a provision of larger amount of institutional credit to the farmers in India is being given priority from the period of nationalisation of commercial banks.

It is obvious that the change in cropping pattern through the larger application of HYV technology has only been possible due to greater provision of institutional credit to poor Indian farmers, particularly small and marginal farmers. It is evident from a brief review of existing literature that no in-depth study has yet been undertaken to examine the impact of credit availability on change in cropping pattern/ agricultural growth in India. An attempt is made in this book to make up these gaps in the earlier studies as far as practicable

Apart from rigorous statistical examination of the nature and pattern of agricultural growth and the relative role of different components including cropping pattern that influence agricultural growth , the nature of change in cropping pattern, trend, direction/composition of credit and the growth of different factors influencing change in cropping pattern, the book covers some new issues like (I) statistical examination of (a) the relationship between the growth rates of index of change in cropping pattern and the factors influencing change in cropping pattern, (b) the role of credit on change in cropping pattern/agricultural growth, (c) the impact of institutional credit on different factors influencing change in cropping pattern and the impact of these factors on change in cropping pattern, (d) the impact of other factors like PCI or PCSDP, size of land holdings, etc. on change in cropping pattern and (e) the impact of credit availability on commercialisation of agriculture, (II) economics of change in cropping pattern in relation to credit. The findings of this study are quite interesting and the issues covered in the volume have a practical and policy significance in India.

It is hoped that the book will prove very useful to the researchers and students of Economics, Commerce, Management, Agriculture, etc. government executives concerned with the formulation and execution of policies for agricultural/rural development, parliamentarians and legislators, and the common readers interested in agricultural economics/rural development.

**Subrata Kumar Ray**

## ACKNOWLEDGEMENTS

This book is a revised version of my Ph. D. thesis worked out at the Department of Economics with Rural Development, Vidyasagar University, Midnapore, West Bengal. In the course of carrying out this study I have been helped by many persons, both directly and indirectly. My greatest debt to my respected teacher and supervisor Prof. Purnendu Sekhar Das, Department of Economics with Rural Development, Vidyasagar University, Midnapore, West Bengal, who stimulated my interest in the subject and guided me at almost every step of my research work. My greatest debt is to my respected teacher and supervisor Prof. Sachinandan Sau, Department of Economics with Rural Development, Vidyasagar University, Midnapore, West Bengal, who stimulated my interest in the subject. He guided me at various steps of my research work. In spite of his heavy pressure of work, he spared his valuable time to go through my research work meticulously.

I am highly indebted to my teacher Prof. Hiron Kumar Banerjee, University of Calcutta, for his valuable suggestions and deep insights into the present work.

I am thankful to Prof. P.N. Ray, Prof. Ishita Mukhopadhyay, Prof. A.K. Banerjee, Prof. Biplab Dasgupta, Prof. Ashim Dasgupta of University of Calcutta, Prof. Sankar Majumdar, Prof. Debasis Mandal, Prof. Mihir Kumar Pal of Vidyasagar University, Prof. Sudin Kumar Mukhopadhyay of University of Kalyani, Dr. Sudipta Bhattacharya of Visva Bharati University, Prof. Biswajit Chatterjee, Prof. H.N. Sur , Dr. Gautam Gupta of Jadavpur University, Prof. Raj Kumar Sen of Rabindra Bharati University, Dr. Atanu Sengupta , Dr. Arup Chattopadhyay of Burdwan University, Mr. Amitabha Sardar, Mr. Sadhan Chattopadhyay of Reserve Bank of India, Prof. Nripen Bandyopadhyay of Centre for Studies in Social Sciences, Kolkata, Prof. Utsa Patnaik, Prof. Abhijit Sen, Prof. Ramprasad Sengupta , Prof. Jayati Ghosh, Prof. G.S. Bhalla of Jawaharlal Nehru University, Prof. Peter Nijkamp of The Free University, Amsterdam, The Netherlands, Prof. Debdas Banerjee, Prof. Uttam Bhattacharya of Institute of Development Studies, Kolkata, Prof. Kausik Basu of Cornell University for many valuable suggestions made in the course of preparation of this research work. I express my special thanks to Dr. Debidas Ghosh of Vidyasagar University for his advice and assistance during the study period.

I am thankful to the authorities of different institutions, libraries and Government offices particularly Indian Statistical Institute, Kolkata, Centre for Studies in Social Sciences, Kolkata, Fertiliser Association of India( Government of India), Kolkata, Socio-Economic Evaluation Branch (S.P. Mukherjee Road, Kolkata), Government of West Bengal, Reserve Bank of India, Kolkata, NABARD, Kolkata, Bureau of Applied Economics and Statistics, Government of West Bengal, Co-Operative Statistical Cell, Government of West Bengal, Ministry of Agriculture, Government of West Bengal, Ministry of Rural Development, Government of West Bengal, P.A.O., Midnapur(Purba and Paschim), United Bank of India for extending all possible help in consulting the literature.

I must also thank my students who helped me in data collection and processing of data. I shall be failing in my duty, if I do not express my heartiest thanks to my informants and respondants who supplied me the primary data required for the study.

I am thankful to Prof. Jaganmaya Basu, Dr. P. K. Panigrahi of University of Kalyani, Prof. Rajat Acharya, Prof. Ajitava Roychowdhury, Prof. Jayashree Roy, Prof. Arpita Dhar, Prof.

Asim Kumar Karmakar of Jadavpur University, Dr. Anjan Chakraborty, Prof. Sarmila Banerjee, Prof. Soumyen Sikdar, Dr. Sankar Kumar Bhowmick, Dr. Sarbajit Chowdhury , Prof. Pabitra Giri of University of Calcutta, Prof. Dipankar Coondoo, Prof. Abhirup Sarkar, Prof. Pradip Maity, Prof. Pulakesh Maity, Dr. Chiranjib Neogi of Indian Statistical Institute, Kolkata, Prof. J. Debnath of Vidyasagar University, Dr. Prankrishna Pal of Rabindra Bharati University, Mr. Prabal Sen of United Bank of India, Prof. Santanu Ghosh , Prof. Debnarayan Sarkar of Presidency College, Kolkata for extending support and help.

I am thankful to Mr. Chayan Kumar Mandal, Mantu, Anupam, Tarun, Papan, Debu, Mr. Acharya, S. Manna for their neat and timely secretarial assistance.

I must put on records the support and encouragement that I received from my grandfather, grandmother, parents, sisters (particularly Bula) and few relatives in completion of this work. I cannot acknowledge by name many employees of Government of India and Government of West Bengal, students, friends and relatives who have been generous in extending me the facilities and encouragement in completing this work, but I wish to record my debt to them.

Finally, I express my heartfelt thanks to my wife, Malaya, and my son, Soumyadip, who have shown utmost patience and perseverance throughout my busy engagement in this work.

I wish to express my appreciation to Serials Publications, New Delhi (India) for undertaking this publication work in a meticulous manner.

Of course, I am alone responsible for the errors remaining in this study.

**Subrata Kumar Ray**

## LIST OF ABBREVIATIONS

| | |
|---|---|
| IC | Institutional Credit |
| CP | Cropping Pattern |
| NIC | Non-Institutional Credit |
| OF | Own Fund |
| NFA | Percentage of Area Under Non-Foodgrains in Gross Cropped Area |
| GCA | Gross Cropped Area |
| NF | Non-Foodgrains |
| FHs | Farm Households |
| HYV | -High Yielding Varieties |
| GP | Gross Profit |
| NFA/FA | Ratio of Area Under Non- Foodgrains to the Area under Foodgrains |
| FA | Area under Foodgrains |
| HYVA/FA | Ratio of Area under HYV Paddy to the Total Area under Foodgrains |
| LH | Size of Land Holdings |
| A.P. | Andhra Pradesh |
| M.P. | Madhya Pradesh |
| U.P. | Uttar Pradesh |
| W.B. | West Bengal |
| H.P. | Himachal Pradesh |

| | |
|---|---|
| CPI | Cropping Pattern Index |
| PCI | Per Capita Income |
| PCSDP | Per Capita State Domestic Product |
| GIA | Gross Irrigated Area |
| CMIE | Center for Monitoring Indian Economy |
| PACS | Primary Agricultural Credit Societies |
| CV | Coefficient of Variation |
| FC | Fertilizer Consumption |
| LDB | Land Development Bank |
| NI | National Income |
| VA | Value Added |
| GDP | Gross Domestic Product |
| GCFA | Gross Capital Formation in Agriculture |
| AIDIS | All India Debt and Investment Survey |
| CBC | Commercial Bank Credit |
| COC | Co-operative Credit |
| PCBC | Percentage Share of a Size Class in C.B.C. out of Total C.B.C |
| PPACC | Percentage of a Size Class in PACS Credit out of Total PACS Credit |
| PLDBC | Percentage Share of Different Land Holding Size Classes in LDB Credit to total LDB Credit |
| CAGRs | Compound Annual Growth Rates |
| SDP | State Domestic Product |
| NDP | Net Domestic Product |
| GOI | Government of India |

# CHAPTER–1

## INTRODUCTION

India recorded an impressive growth in agricultural output since the mid-1960s, and a noticeable feature of the 1980s was that agricultural growth permeated to all the regions in India. The most remarkable development was the notable acceleration of growth in the eastern region. Specifically creditable was the performance of West Bengal. The most striking aspect of the transformation has been in the production of foodgrains where we have converted a deficit prone country with concerns of widespread starvation, to a situation of plenty. The massive investment in irrigation, rural infrastructure and institutions, research and extension has helped to attain impressive growth in agriculture leading to self-sufficiency in food production. The development of agriculture or crop economy of India is characterised by the process of crop diversification or changes in cropping pattern (CP) over time. The change has been associated with the increasing transformation of agriculture from subsistence to commercial farming. On the other hand, the amount of credit from various institutional sources to rural people has increased considerably since the early 1970s. In this context, the present study is an attempt to highlight (1) the importance of agriculture in general and change in CP in particular in the economy of India, and the inadequacy of the existing literature to give answers to some issues, viz. (a) what has been the role of credit in changing CP and agricultural growth?, (b) what is the mechanism through which credit affects change in CP?, (c) what has been the role of credit in agricultural productivity, employment and income generation, etc.? and (2) also to seek answers to these and other related questions.

### Importance of Agriculture in India's Economic Development

The agrarian question has remained one of the central and abiding issues of classical political economy at least from the time of the physiocrats. That agriculture should have formed the centre piece of political economic enquiry in these early years of the discipline is not really surprising [Breman and Mundle (1991]. In the physiocratic systems, agriculture was given a dominant place because of its strategic role in the economic development of a country. The physiocrats looked upon land as the unique source of wealth. Agriculture was considered to be the only genuinely productive sector in which surplus wealth (or not product) was generated and it was upon this surplus wealth that the growth of the economy depends. Adam Smith also argued, "the most opulent nations, indeed, generally excel all their neighbours in agriculture as well as in manufactures" [Smith]. Again, in focussing on the primary importance of the domestic market (rather than the overseas market) he was closer to the French physiocrats than to the English mercantilists, for in a pre-industrial economy emphasis on the domestic market meant an emphasis on agriculture as a major source of, though not necessarily the motive power for economic growth.

Today also development economists are less sanguine about the desirability of placing heavy emphasis on rapid industrialisation. Perhaps more importantly, they have come to realise that far from playing a passive supporting role in the process of economic development, the agricultural sector in particular and rural economy in general need to be the dynamic and leading elements in any overall strategy at least for the vast majority of contemporary Third World Countries [Todaro (1985)]. Gunner Myrdal was of the opinion that "it is in the agricultural sector that the battle for long term economic development will be won or lost" [Todaro (1985)]. Francis

Blanchard, Ex-Director General, International Labour Organisation, observed that "the main burden of development and employment creation will have to be borne by the part of the economy in which agriculture is the predominant activity" [Todaro (1985)]. The 'grass roots' school of development, which came into fashion as a reaction against the emphasis on industrialisation at any cost, lays stress on policies to raise the level of productivity in agriculture as the best long run development strategy [Thirlwall (1980)]. According to Prof. Kuznets there are three ways in which agriculture can contribute towards the development of the economy [Kuznets]. They are : product contribution, market contribution and factor contribution. Besides these three contributions, Johnston and Mellor (1961) suggested that there are five different ways in which agricultural output and productivity contribute to overall economic growth.

(1) During the process of economic development there will be a substantial increase in the demand for agricultural products. If the agricultural sector fails to increase the volume of production in pace with the growth of demand, economic growth will be retarded. Hence agricultural sector must be expanded to meet the increased demand during the process of development.

(2) Expansion of agricultural exports is likely to be one of the most promising means of earning foreign exchange which is highly needed for importing machinery from abroad.

(3) Transfer of manpower from agriculture to non-agricultural sectors will increase the productivity of agriculture, the volume of industrial output and employment in industries. It follows that manpower for manufacturing and other rapidly expanding sectors can be drawn easily from the agricultural sector.

(4) Agriculture, as the dominant sector of an under-developed economy, can and should make a net

contribution to the capital required for overhead investment and expansion of secondary industry.

(5) Rising cash income of the farm population may be important as a stimulus to industrial expansion.

Traditionally, the role of agriculture in economic development has been viewed as largely passive and supportive to a large extent. Therefore, 1970s witnessed a remarkable transition in development in which agriculture and rural development came to be seen by many as the sine qua non of national economic development. Without such agricultural and rural development the industrial growth either would be stultified or if it succeeded, would create such severe internal imbalances in the economy, that the problem of widespread poverty, inequality and unemployment could become more pronounced.

Five main questions, therefore, need to be asked about Thirld World agricultural and rural development as these relate to overall national development:

(i) How can total agricultural output and productivity per capita be substantially increased in a manner that will directly benefit the average small farmers and the landless- rural people while providing sufficient food surplus to support a growing industrial sector?

(ii) What is the process by which traditional low productivity subsistence farms are transformed into high productivity commercial enterprises?

(iii) When the traditional small farmers and peasants resist change, in their stubborn, and irrational behaviour are they acting rationally within the context of their particular economic environment?

(iv) Are economic incentives sufficient to elicit output increases among peasant agriculturalist or are institutional and structural changes in rural farming "Systems" also required?

(v) Is rising agricultural productivity sufficient to improve rural life or must there be concomitant improvements in education, medical and other social services. In other words, what do we mean by "rural development" and how can it be achieved? (Todaro, 1985).

It is now recognised that in developing economy like India, the importance of agriculture is both crucial and pervasive. Agriculture forms the backbone of the Indian economy. Despite concerted industrialisation in the last five decades, agriculture occupies a place of pride in the Indian economy. The importance of agriculture in Indian economy can be best explained by considering the role of agriculture under different heads.

## Share of Agriculture in National Income (NI)

It is found that the agriculture contributed 48.6%of GDP in 1950-51, 39.7% in 1970-71, 34.7% in 1980-81, 27.7% in 1993-94, 27.43% in 1994-95, 25.0% in 1995-96 and 24.1% in 2003-04. On the other hand, the value added in agriculture as a percentage of the value of output at constant prices of 1980-81 was 79.98% in 1970-71 and decreased to 75.87% in 1994-95. Taking the 35 years period (1960-61 to 1993-94) the annual rate of growth of value added (VA) in agriculture proper was 2.68%. The share of the primary sector which includes agriculture, mining, forestry and fishery has gone down from 55.3%of GDP in 1950-51 to 44.5% in 1970-71 and to still a lower figure of 26.8% in 1995-96 and 21.9% in 2002-03. This only underlines the fact that the agriculture is the most important determinant of the share of the primary sectors in national output. Despite there exist sharp rise in the percentage shares of the other two sectors of industry and services in the economy in the GDP, which is a natural outcome of economic growth, the role of agriculture still may be considered as an important determinant of the growth of national income in Indian economy having characterisrtic of underdevelopment. It is to note that the rural sector generally contributes more in the net domestic product generated in agriculture.

## Role of Agriculture in Employment Generation

Sectoral trends of employment reveal that agriculture which accounted for two thirds of total employment generated, experienced a sharp declining growth rate of employment from

2.32% during 1973-1978 to a low figure of 0.65% during 1983-1988. For the entire period, the growth rate of employment in agriculture was just 1.37% per annum. Agriculture dominates the economy to such an extent that in 1990-91, 80.2%of workers in rural India were engaged in agriculture as against 81% in 1981.It is observed that the percentage of male agricultural workers to total male workers in rural development activites has declined in 1983 as compared to 1971 for most of the states and India as a whole. These states are Andhra Pradesh (A.P.), Gujarat, Haryana, Madhya Pradesh(M.P.), Maharasthra, Orissa, Punjab and West Bengal (W.B.). It is observed that the percentage of female workers is larger than that of male workers both in rural and urban areas of India during 1972-73 to 1987-88. From the National Sample Survey (NSS) data, it is observed that the percentage of male workers employed in agriculture in the rural sector of India has declined from 74.5 in 1987-88 to 74.1 in 1993-94 and that for females remained constant (i.e., 84.7) during the same period. On the other hand, the declining trend has also found for both the male (from 9.1% in 1987-88 to 9.0% in 1993-94) and the female workers (from 29.4% in 1987-88 to 4.7% in 1993-94) employed in agriculture in the urban sector of India. Data provided by the Census of India reveals that in absolute terms, agriculture provided employment to 97 million persons in 1951, the number of people working on land (cultivators and agricultural labourers) increased to 235 million in 2001. Therefore, it can be concluded that in the primary sector agriculture alone not only generates a considerable amount of employment and employs largely the weaker section of society, namely, females.

### Importance of Agriculture for Industrial Develpoment

In India agriculture plays an important role in industrial development. Agriculture provides raw material to the cotton textiles, jute, sugar and vanaspati industries which are of basic importance to the national economy. Besides, all workers engaged in industries depend for their essential consumption requirements on agriculture. Agriculture also provides a market for industrial products. Since the level of income of

Indian farmers and landless labours is very low in India, market for industrial products in the agricultural sector is very limited. This acts as a constraint on industrial development [Samuelson *et. al.* (1975)]. As agriculture develops in India and as the incomes accruing to the rural people increase, the size of market for industrial products in rural areas also increases.

The per capita income in India is very low. Consequently, a large part of this income is spent on fulfilling the basic consumption requirements of the people. It has been estimated that in India approximately 60 per cent of household consumption and 85 per cent of household commodity consumption is of agricultural products (or products based on agriculture). According to NSS, income elasticity of demand for foodgrains is 0.8 in India. Keeping this income elasticity and the possibilities of increase in population and per capita consumer expenditure in view, the planning commision has estimated that demand for agricultural commodities is likely to increase at the rate of 4.7 per cent per annum. Accordingly, the nation can remain self-reliant in agriculture only if agricultural otuput rises at least at the rate of 5 per cent per annum.

## Role of Agriculture in the Field of International Trade

Agriculture and allied products contributed 31.7% of total exports in 1970-71. But the share of agriculture and allied products in total exports has declined to 16.5% in 1992-93 and 15.2% in 1999-2000. If we look at the exports of agricultural commodities alone, it is observed that the percentage share of exports of agricultural commodities in total exports was 29.8% in 1950-51 and declined to 27.8% in 1995-96 and 26.6% in1997-98.The Tenth Plan estimates that agriculture contributes 14.7% of total export earnings. On the other hand, the imports of foodgrains were necessitated by the partition of the country and the growing demand for food for the rising population. The average annual imports of foodgrains which were about Rs.120 crores during the First Plan, rose to Rs.161 crores during the Second Plan, further increased to an average rate of Rs. 241 crores during the Third Plan. During 1992-93 the imports rose to Rs. 1240 crores. Again , the average annual imports of foodgrains rose to Rs. 2170 crores during 2000-01and 2004-05.It is seen that the percentage share of agricultural imports accounted for 15.6% in 1950-51 and declined to 12.5% in 1995-96 and 11.5% in 1997-98.

This dependence on the agricultural imports can be lessened if there is higher agricultural growth and diversification through the application of modern technology and if this is done, the country's crisis of foreign exchange reserves may be solved to some extent. There is no doubt that the export of agricultural commodities has great significance for economic development. For increased exports help the country to pay for the increased imports of machinary and raw materials. As a whole, the higher pace of agricultural growth and diversification is very essential for solving the balance of payment problems of the country.

## Agricultural Growth and Gross Capital Formation in Agriculture

In this perspective of agricultural growth, let us examine the gross capital formation in agriculture (GCFA).GCFA (at 1980-81 prices)has shown a rising trend over years during 1960-61 to 1994-95. The annual growth rate (%) of GCFA as a whole increased from 4.5 during 1969-70 to 1980-81 to 6.4 during 1990-91 to 1994-95. It is to note that the amount of GCFA (at constant price) in the public sector has been lower than that in the private sector during 1960-61 to 1994-95. But the GCFA as per cent of total gross capital formation (GDCF) has declined over years during 1960-61 to 1994-95.

So, it can be commented that the impressive rate of agricultural growth has contributed positively to the capital formation in the economy, which is very urgently needed for expanding the industrial and infrastructural base of our counrty.

## Growth of Agriculture in India and Decomposition of Growth Rates

Higher agricultural growth rate can be achieved through (i) change in area,(ii) change in yield and (iii) change in CP.

But it is true that the supply of land is inelastic in India where the land-man ratio has declined overtime. In this context, it can be argued that the change in CPcan considerably accelerate the agricultural growth through more cultivation of High Yielding Variety (HYV) crops by the application of modern technology in Indian agriculture. Again in view of the rapid decrease in per capita land holdings, changing composition of demand due to faster industrialisation and urbanisation, ecological imbalances, resource chrunch, chronic unemployment problem, massive poverty, high population growth and actue crisis in foreign exchange reserves, higher agricultural growth through change in CP as one of the crucial rural diversification measures would be an urgent need in recent times. Thus, a desirable CP would be one which favours crops which are labour intensive and have greater second round employmernt generating effects. Simultaneously, it should favour crops which are either efficient import substitutes or exportables to save (earn) foreign exchange in balance. Again, given water as the binding constraint, the CP should have the objective to maximise returns to this input and favour crops that can accelerate the agricultural productivity which may assist the process of globalisation of Indian agriculture and in this way, the change in CP can contribute largely to the process of economic reforms in India.

Let us highlight the trends of production and CP in India. It is observed that both the index of production and percentage of area under non-foodgrains have rising trends over years, during 1969-70 to 1996-97 and 1969-70 to 1994-95 respectively. On the other hand, the percentage of area under foodgrains shows a declining trend during the same period. The index of CP has risen over years during 1969-70 to 1981-82, but remained stagnant or has slightly risen thereafter. In the period (1969-1993) the growth rate of production of all crops combined was 2.79% per annum, which is significant at 5% level. The growth rate of area under foodgrains was 0.07% per annum and the growth rate of area under non-foodgrains was 0.50% during the same period. It is also observed that the growth rate of area under foodgrains is not significant, but that under non-foodgrains is significant at 1% level. It is also observed that there is no statistically significant change in the growth of area under all crops combined at all India level. On the other hand, the deceleration in area growth for foodgrains and acceleration for non-foodgrains have been observed. That is to say, inter-crop-shifts favoured non-foodgrain crops at the cost of foodgrains, particularly in the later part of the green revolution, while very slow expansion in aggregate cropped area continued without any perceptible change in its growth rate [Sawant and Achuthan (1995)]. It is also true that yield component has a larger impact on increase in production of foodgrains (as compared to area component), but the area component has greater effect on increase in non-foodgrains production, as compared to yield component.This difference is the result of the larger rate of application of HYV technology for foodgrains production as compared to non-foodgrains production.

The CP has changed significantly during the period from 1962-65 to 1980-83 and from 1980-83 to 1992-95. It is observed from a study by Bhalla and Singh (1997) that in India there has been a steady decline in the percentage of area under foodgrains from 77.85 in 1964-65 to 76.63 in 1980-83 and to 72.21 in 1992-95. But the percentage of area under non-foodgrains has steadily increased from 22.15 in 1962-65 to 23.37 in 1980-83 to 27.79 in 1992-95. The same pattern of changes in the percentages of area under foodgrains and non-foodgrains is observed in the regions, namely, eastern, central and southern regions. But in north-western region there has been steady increase in the percentage of area under foodgrains and the percentage of area under non-foodgrains has continuously declined from 20.53 in 1962-65 to 20.32 in 1980-83 to 17.52 in 1992-95.

If we consider the share of foodgrains and non-foodgrains in total output of the regions, it is seen that the pattern of change in the share of foodgrains and non-foodgrains in the total output of the regions is the same as that of CP. This similarity in the pattern indicates that the change in the value of crops might have induced the farmers to grow more of those crops that bring the farmers higher value by expanding the area under these crops and to grow less of those crops that yield lower value by reducing the area under these crops.

For India as a whole it is seen that during 1962-65 and 1970-73 the annual growth rate of yield of foodgrains was found to be 1.82 and it increased to 1.86 during the period from 1970-73 to 1980-83 and to 3.22 during the period from 1980-83 to 1992-95. Thus while there has been only 0.04 percentage point increase in annual growth rate of yield from the first sub-period to the second sub-period, it was 1.36 percentage point higher in the third sub-period of 1980-83 to 1992-95 over the second sub-period of 1970-73 to 1980-83.

In the case of non-foodgrains, the annual growth rate of yield has steadily gone up from 1.39 during the first sub-period to 1.40 during the second sub-period and to 2.38 during the third sub-period. Thus, increase in the yield growth rate along with the increase in the value of output has been responsible for the change in the CP. The crops with high productivities and /or the high value crops have obviously been preferred to the crops with low productivity and/or the low value crops, and this preference has brought about the change in the CP in India. This change has ,sometimes, taken place among the various foodgrains , sometimes among the non-foodgrains and sometimes between foodgrains and non-foodgrains.

Thus, change in CP has been the cause as well as effect of change in productivity and the value of crops. A two-way relationship between CP, on the one hand and the yield and the value of the product, on the other hand, can be found to exist in the process of agricultural development.

## Factors Influencing Change in CP in India

The change in CP is influenced by several natural, technological, economic and institutional factors. An acceleration in the rate of growth of agricultural output in the regions lagging behind others should come primarily from the increase in the yield levels of individual crops and an increase in cropping intensity and, to some extent, from changes in CP in favour of high value crops. The most important factor contributing to such changes would be the availability of assured irrigation, followed by provision of modern inputs and appropriate price policies. Irrigation, besides leading to increase in yield and cropping intensity, also facilitates change in CP in favour of high value crops, most of which also happen to be more labour using. Such shifts in the CP are important from the view point of employment particularly in the long run to the point where yield increases in staple crops beyond a stage involve technologies using less labour. "At the macro level, diversification was significantly and adversely affected by fertiliser consumption, inter crop value productivity variability, tractor density, skewness in the distribution of land holding, and percentage of cropped area irrigated. It is positively affected by market density. And at micro level, diversification is inversely related to the size of farm, distance from the market and asset per hectare and directly by income" [Singh *et.al.* (1985)]. A study of inter crop price parities undertaken by the Ministry of Food and Agriculture shows how price variations exert an important influence on shifts in CP. "It seems prices influence the acreage under the crops in two ways. One is that the variation in the inter-crop price parities leads to shift in acreage as between the crops. Another is that the maintenance of a stable level of prices for a crop—— provides a better incentive to the producers to increase the output than what a very high level of price does, if there is no uncertainty about this level being maintained over number of years" [Datt and Sundharam (1995)]. Fixed procurement price of wheat and rice and other government controls have induced farmers to shift to cash crops like sugarcane.

According to some authorities, income maximisation objective exercises greater influences in changing the CP. That is, the farmer would favour that combination of crops which would give him maximum income. Dr. Raj Krishna, however, argues that relative profitability is the main consideration which influences the crop pattern. In any case, the farmer is influnced in the choice of his crop by such considerations of price differences between different commodities, or maximisation of income or relative profitability per acre.

**A Schematic Presention of Factors Affecting CP is given below**

**Factors Affecting C.P**

↓

| ↓ | ↓ | ↓ | ↓ | ↓ |
|---|---|---|---|---|
| Natural (e.g. Nature of soil, type of climate, extent of rainfall, etc.) | Historical factors (e.g., historical pattern of land tenure, etc.) | Economic factors (e.g., price and income maximisation, insurance, farm size, availability of agricultural inputs, nature of land tenure, etc.) | Social factors (e.g., social environment, customs, traditions, outlook towards material things, etc.) | Government policy (e.g., availability of credit, supplies of inputs, tax policy,export against risk import policy, etc.) |

## Research Problem

It is obvious that the change in CP, particularly, towards larger valued crops, is highly associated with the rate of adoption of new production technologies which, in turn, put a greater emphasis on assured supply of sufficient finance because of the higher capital intensity of the new production technologies. But the financial capability of the majority of the Indian farmers is considerably below the minimum level required for the adoption of modern technology. The green revolution raised the stature of Indian agricultutre by a few notches and proved that Indian farmers too can succeed like their counterpatrs in more developed countries if they are provided with the required amount of finance due to the higher capital intensity of the modern technology. The timely availability of credit facilities to the cultivator is most significant for agricultural development, especially in an economy like India where majority of the cultivators belongs to the category of small and marginal farmers. The increased importance of capital in making agriculture a powerful engine for growth has been stated by Schultz(1964) as follows. "Once there are investment opportunities and efficient incentives, farmers will turn land into gold". According to Prof. John D. Black(1955) of Harvard University, "If we are all concerned about increasing total agricultural output in the shorter time we must provide credit first and foremost. This will enable them to use more labour saving equipment; more seeds and fertilizer". In the publication of the Union Ministry of Agriculture ,"Agricultural Development: Progress and Prospects", this fact is duly emphasized. It states, "For an agricultural revolution which has to be brought about, an appropriate and organisational structure has to be created. It is one of the most important needs so far as the arrangements are lagging much behind the requirement.The planned injection of technical change requires that farmers have a source of finance through a right agency which can supply the needed capital in the right form, in right quantity and at the right time. But credit must be accessible, cheap, safe and productive without which the sustained growth of agriculturte is almost impossible since the opposite phenomenon may create the 'debt trap' and hardly provides any incentive to effect improvement".

Therefore, Indian farmers should be provided with sufficient amount of institutional credit (IC) due to its lesser detrimental effects as compared to that of non-institutional funds. "In fact, recent technological breakthrough in agriculture have changed the perspective profoundly by emphasizing the relative importance of chemical and biological innovations in today's extent. This has profound significance from our point of view. This is primarily because these innovations can be made scale neutral by providing cheaper credit facilities and a package of agricultural extension facilities" [Chakraborty (1997)]. "The role of formal credit in promoting agricultural growth is to provide the wherewithal for private investment in inputs and fixed capital to enable farmers to switch over to a superior production function. A production oriented lending system implies providing credit for such investment on the basis of its techno-economic feasibility rather than loan on collateral and focusing particularly on those farmers whose low income-saving base precludes their undertaking investment without credit. The role of credit in growth is thus indirect and positive and its productivity is not different from that of the investment financed by it" [Pany (1982)]. It is often assumed that the decision of Indian farmers are not affected or modified in response to price incentives. In other words, the Indian farmers continue to produce the same agricultural output even on more attractive prices. However, the facts are different. The studies of Raj Krishna, Hooper and Stern clearly point out that the Indian farmer reacts rationally to his economic environment. Frequently, on account of lack of marketing facilities, non-availability of loan at fair rate of interest, the cultivators are not able to invest in agriculture. This keeps the level of production per acre of land and per cultivator low. If the government can revitalize the credit co-operative socoeties and the regional rural banks to grant more credit to the small farmers, the level of productivity will undoubtedly increase. The most important consideration affecting CP is economic consideration. The real difficulty in adopting a better CP is that the farmer may not have the requiste capital to invest now or possess the knowhow that may be necessary for changing the crops. It is here that the government may come to his help. Against this perspective, the change in CP through greater application of HYV technology to Indian agriculture rests largely on the cheap external source of finance i.e., instutional finance and hence the change in CP can be considered as an implicit function of IC influencing the farmer's decision to achieve higher agricultural growth. A schematic represention of the need of IC for agricultural growth through change in CP is presented below.

The nationalisation of banks has made a structural change in the banking operations- a change from private banking into mass banking. For meeting the gap in agricultural credit, the potentialities of commercial banks are to be fully utilized. Following social control, commercial banks had shown an increasing interest in the agricultural sector[ Fourth Five Year Plan(1969-74)]. "Consequent on nationalisation, it is expected that the nationalised banks will take further measures to increase their finance for agricultural production and investment. Certain steps have already been taken in this regard. Each district in the country has been alloted to one

**A Schematic Representation of the Need for IC for Agricultural Growth**

**Agricultural Growth**

| Area | Yield | CP | |
|---|---|---|---|
| ↑ | ↑ | ↑ | |
| Technology (Land augmenting) | Technology (HYV, etc.) | Technology (Diversification of crop cultivation) | |
| | ↑ | ↓ | ↓ |
| | Need for larger amount Of capital investment | Rising cropping Intensity | More commercial and superior Cereals cultivation |
| | ↑ | ↑ | ↑ |
| | Need for IC | More labour absorption | Need for larger amount Of capital investment for purchase of technological inputs |
| | | ↑ | ↑ |
| | | Higher requirement of Wage fund | Requirement of IC |
| | | ↑ | |
| | | Need for IC | |

bank called lead bank" [Fourth Five Year Plan (1969-74)]. Various scheduled banks are now embarking on a massive programme of branch expansion, particularly in the unbanked areas. This tempo will be kept up throughout the plan so that the banking facility is brought as near to the villager as possible. The banks will thus attempt to cater to the needs of the individual farmers" [Fourth Five Year Plan(1969-74)]. In the Eighth Plan instutional finance is to be ensured for the thrust area activities to be undertaken. Some of the thrust areas are: (i) livestock improvement, soil and water conservation measures, reclamation of water–logged, saline, fallow land, etc., (ii) minor irrigation and development of shallow tubewells, dug wells, digging of tanks/ponds, (iii) agro-processing of horticultural produce, and (iv) promotion of high-yielding crops in traditional and non-traditional areas" [Eighth Five Year Plan (1992-97)]. Therefore, it can be concluded that starting from the period of nationalisation of banks during the Fourth Plan upto the period of the Eighth Plan, the private sectors, in particular, agriculture have been given a special emphasis by sanctioning more provision of cheap credit to the farmers. In other words, a provision for larger amount of IC to the farmers is being given priority in India from the Fourth Plan upto the recent years. Again, the green revolution in Indian agriculture through the application of HYV technology has led to acceleration of agricultural growth through change in CP. It is obvious that the larger applications of higher capital intensive HYV technology has only been possible due to the greater provision of IC to poor Indian farmers whose own financial abilities do not permit them to take the advantage of the modern technology. Against this backdrop, a research may be made for studying the impact of IC on change in CP.

**Scope and Objectives of the Present Study**

The objectives of the study are stated below:

(i) To examine the role of credit, particularly IC, on change in CP both at macro and micro level.

(ii) The change in CP is dependent on the growth of state of different economic, social, climatic, geographical, etc. factors. Here, the credit may play the role of a catalyst for the expansion or application of some of those factors such as, irrigation and fertilizer, needed for the change in CP. Therefore, our another objective is to examine the impact of ICon the several factors influencing change in CP, particularly irrigation, and fertilizer.

(iii) The factors like per capita income (PCI) or per capita state domestic product (PCSDP), profitability, employment generating capacity, etc. may also influence change in CP. But, there exist no such study which has examined the impact of these factors viz.,PCI, PCSDP, profitability, etc. on the change in CP. This aspect of the problem has been discussed in detail in the survey of literature in Chapter 2. So there is another gap in the existing literature that needs to be filled.Hence, our third objective is to study the impact of these factors like PCI, PCSDP, profitability, etc. on the change in CP both at macro level and micro level.

Within the broad frame of the objectives stated above, we would like to pursue the following things.

(i) To examine the nature and pattern of agricultural growth and the relative role of different components, including CP, that influence agricultural growth.

(ii) To examine the nature of change in CP.

(iii) To examine the growth of different factors influencing change in CP viz., irrigation, IC, fertilizer consumption, PCI, PCSDP, etc.

(iv) To examine whether there exist significant relationship between the growth rates of index of change in CP and the factors influencing change in CP or not.

(v) To examine the role of credit, particularly IC, on the change in CP.

(vi) To examine the impact of IC or credit as a whole, on different factors influencing change in CP ,viz., percent of gross irrigated area (GIA) to gross cropped area (GCA), fertilizer consumption per hectare, etc. and then to examine the impact of these factors on change in CP.

(vii) To examine the impact of the other factors (influencing change in CP) like PCI, PCSDP, size of landholdings, etc. on the change in CP.

(viii) To examine the effect of change in CP on profitability, employment generating capacity, etc.

## Hypotheses

The hypotheses of the present study are as follows:

(i) The larger the availability of agricultural credit, more diversified is the CP and higher the growth of agricultural production.

(ii) The larger the availability of agricultural credit, greater is the use of factors like irrigation, fertilizers ,etc. and larger is the change in CP.

(iii) The larger the availability of agricultural credit, greater is the percentage of HYV area in GCA and larger the change in CP and higher is the agricultural growth.

(iv) The larger the availability of agricultural credit, larger is the percentage of area under non-foodgrains and greater is the profitability.

(v) The larger the availability of agricultural credit, higher is the prospect of employment generation in agriculture.

(vi) The smaller the size of land holdings, higher is the percentage of area under non-foodgrains.

(vii) The higher the PCI or PCSDP, larger is the percentage of area under non-foodgrains.

(viii) The larger the percentage of GIA in GCA, more diversified is the CP,

(ix) The larger the proportion of IC to total credit, larger is the profitability.

(x) The larger the use of fertilizer, the larger the change in CP.

### Plan of the work

The plan of the rest of the work is as follows:

- Chapter-II makes a survey of literature in the study area.
- Chapter-III discuss the sample design, data collection and methodology.

- Chapter-IV highlights the nature and pattern of agricultural growth in India, sample states, districts of W. B. and blocks of Midnapore.
- Chapter-V examines the relative importance of CP in agricultural growth in the sample states of India and the districts of W.B.
- Chapter-VI examines the nature of change in CP in the sample states as well as India.
- Chapter-VII focuses on the nature of change in CP in the districts of W.B. and blocks of Midnapore.
- Chapter-VIII highlights nature, trend and structure of the agricultural credit in the sample states of India, districts of W.B. and blocks of Midnapore.
- Chapter-IX examines the impact of IC and the related factors such as irrigation, fertilizer consumption, PCI and PCSDP on the change in CP in the sample states of India, the districts of W.B. and blocks of Midnapore.
- Chapter-X analyses the relationship between credit and the change in CP in some selected villages in Midnapore district.
- Chapter-XI gives a summary of work and makes concluding observations.

# CHAPTER–2

## REVIEW OF LITERATURE

In this chapter an attempt is made to make a review of the existing literature on changes in CP and on IC.The existing literature on changes in CP and on IC in India is fairly developed. This literature may, broadly, be classified into the following major groups.

### Impact of Change in Cropping Pattern on Agricultural Growth

The decomposition of growth of agricultural output was attempted by Minhas and Vaidyanathan (1965). The contribution of area, yield and CP and the interaction of the latter two factors to the increase in output were estimated by using an additive scheme of decomposition. Dharam Narain's (1977) work is an extension of the study of Minhas and Vaidyanathan. He decomposed the growth of productivity in agriculture into changes in CP, locational shifts of area under individual crops and pure yield effect. His study concluded that the growth of productivity during the fifties was mainly due to the first two factors, i.e., CP effect and locational shift effect. During the 60's the major factor influencing growth of productivity was identified as pure yield effect, which was the result of technological change.

The above studies, however, do not examine the effect of CP, irrigation and fertilizer simultaneously on agricultiral production. But Ranade (1980)examined the effect of these factors upon agricultural output per hectare across 54 agroclimatic regions covering 16 major states for the pre-green revolution period from 1962-65 and then for the post -green revolution period from 1970-1973. However, Ranade too has not paid any attention to extend the same analysis at the disaggregated district level. This gap has been covered by P.Devasena Naidu (1989). His study shows that the variations in infrastructural facilities have caused wide variations in CP as well as yield in various districts of A.P.

L.S. Venkataraman and M.Prahladachar (1980) have made an estimate and analysis of the growth rates in area, yield and output of major crops in six states, viz., Punjab (including Haryana), Rajasthan, U.P., Bihar, Maharasthra and A.P. during the period 1950-51 to 1974-75 and analysed the impact of growth rates of crops on CP in these states. It also attempts to study the area, yield and CP effects on crop output growth in these states in the period under reference.

The pioneering work in this field done by S.R. Sen (1967) took agricultural production data for undivided India for 48 years from 1900-01 to 1947-48 and for Indian union for 30 years from 1936-37 to 1965-66. He further sub-divided each of the above mentioned two periods into two equal parts using a simple statistical method. He measured annual average growth rate and also measured fluctuations by fitting trend on peak points and through points seperately. Further, he considered the amplitude and frequency of fluctuations by analysing the changes in the peak production. However, from this analysis he concluded that an increase in the rate of growth is accompanied by an increase in instability. He gave emphasis on irrigation that will effect changes in C.P. which will, in turn, reduce instability.

Nilkantha Rath (1980) examined the performance of agricultural production in India, during 1955-78, on the basis of summary measure that have been arrived at by fitting exponential trend functions to the index numbers of production, area and yield rate of the major crops. He improved the fitting of trend line by arbitrarily excluding the large deviant observations. To get the idea about production potential he also fitted a trend line by taking only the recorded peak levels. By estimating growth rates he came to the conclusion that except for wheat and ragi either there was no significant change or if changes took place that was in downward direction in the period 1964-65 to 1970-71. However, during the last five years (1973-74 to 1977-78), the indices for many crops not only persisted at high level rarely recorded in the previous ten years or so, but also some crops recorded rapid increase in it.

Contrary to the belief that the period since 1965-66 or 1967-68 is regarded as the post green revolution period for agricultural growth analysis, Y.K. Alagh and P.S. Sharma (1985) accept the cut off year as 1969-70 and fit the log -linear function seperately for the two sub-periods (1) 1960-61 to 1969-70 and (2) 1969-70 to 1978-79. The following test was applied to examine if the growth rates in the two periods are significantly different from each other. The variation of output has been measured by taking the absolute deviation from trend. From the study they came to the following conclusions:

1. Green revolution started having an appreciable effect on the Indian economy only since 1969-70,
2. The estimated growth rates in second period are generally higher than those for first period,
3. Growth is more evenly spread in period II as compared to period I and,
4. The variation around the growth trend is still large.

S. K. Ray (1983) went a little deeper and analysed the nature and causes for growth and instability in Indian agriculture during the period 1950 through 1980. He used a production function approach of the from, logYt = log A+B log Pt +C log Wt+Ut, where Yt is output, t is time, Pt is prices and Wt is rainfall. He estimated this equation stepwise:

First, only with time trend; then with time trend and rainfall and finally, with time trend, rainfall and price variables. This procedure was followed in order to examine seperately the impact of the systematic changes (if any)in rainfall and prices on the estimated growth rates. The actual production of foodgrains was compared against their estimated values and the impact of rainfall and price variations on production was estimated. The findings of this investigation suggests that the major causes for change in the pattern of growth and instability in production were due to (1) an increase in the variabilities of rainfall and prices., and (2) an increase in the sensitivity of production to variations in rainfall and prices. He was of the opinion that with appropriate irrigation and price policy, growth could be established with stability.

P. S. Das (1978) using sophisticated statistics, examines whether there was acceleration, deceleration or stagnancy of growth rates in crop output in Eastern India during 1950-51 to 1973-74. He first considers the problem of auto correlation, then estimated three trend equations namely linear, loglinear and Gompertz curve for declining, constant and rising growth rates. He finds that exponential curve i.e., constant growth rate is best fitted for Assam, Orrisa and W.B., whereas for Bihar,the straight line (or falling growth rate) is best fitted. For the measurement of instability he fitted straight lines to the peak points and the through points and by examining the regression coefficients he statistically measures not only the instability but also the nature of the instability i.e., whether that is diverging or converging. From this he comes to the conclusion that in Assam,Bihar and Orissa the total agricultural production indices showed tendencies of increasing instability whereas in W. B. they had the tendency of declining instability.

S. Mahendradev (1987) deals with an inter-state time series analysis of foodgrains production for the period from 1960-61 to 1984-85 to examine: (a) unadjusted and weather adjusted growth rates; (b)trends in instability; (c) relationship between growth and instability. The new thing in this paper is that (i) in examining growth performance, it takes explicit account of the impact of rainfall as a proxy for weather on crop production; (ii)trends in instability apart from measures of instability for one or two time points (as done by Mehra, Hazell and Ray) are estimated; (iii) it examines the relationship between growth and instability across the states.

The entire period is divided into three sub-periods: (a) 1960-61 to 1969-70, (b) 1970-71 to 1979-80 and (c) 1980-81 to

1984-85 and the trend rates of growth for the three periods are estimated by introducing appropriate intercept and slope dummies in a loglinear function and to obtain weather adjusted and unadjusted growth rates equation is estimated with and without variables relating to rainfalls. Apart from the summary measures, an alternative approach of measuring instability is used by fitting semilog function to the nine year moving standard deviation.

The normal procedures in growth analysis is to choose a growth function and to estimate its parameter. But V.N. Reddy and N. Krishnaji are very critical about these procedures. Reddy explains the limitations of criteria like minimum residual sum of squares and highest $R^2$. Reddy asserts that in some cases the alternative criterion of $R^2$ (viz., minimum possible standard errors) also falls and in these cases, criterion of accuracy of prediction of growth rates may be the right one. N. Krishnaji argues that these procedures of measurement of growth and fluctuation have no theoretical basis, either economic or statistical, and that , consequently, the inferences of patterns of growth or magnitude of fluctuations drawn from fitted trends are not valid. Taking a hard look at the assumptions of the statistical model and these validity in agriculture, Krishnaji recommends that the best one can do in measuring agriculturtal growth is to describe the year to year changes. Krishnaji is of the opinion that since observed variation is attributed to two sets of factors, viz., rates of input, etc., on the one hand and rainfall, etc., on the other, the seemingly modest aim of measuring growth in output, purged of the weather induced component can not be realised without an adequate knowledge of the working of at least one set of factors. That can be done either by (1) estimating production function or by (2) estimating rainfall output relationship. However, both of these approaches are crippled by theoretical and practical difficulties.

An alternative approach to eliminate or at least reduce the influence of any time-trend in weather on the time trend in output is to eliminate extreme observations showing exceptionally high or exceptionally low output. Since exceptionally high or low output may reasonably be attributed to exceptionally good or bad weather and omitted from the series in order to determine a 'weather free'time trend in output. This is what Rao, Nadkarni, and Deshpande (1980) did. The decision to omit extreme observations should be governed by two considerations:

1. the extreme observations must be judged as exceptional by some objective criteria (using probabilities and level of significance);
2. the omission of the observations should significantly alter the time-trend of output. In this objective criteria also some element of subjective judgement is unavoidabale so far as the choice of the level of significance and decision regarding the significant difference in the time-trend of output are concerned.

S.D. Sawant and C.V. Achuthan (1995) measured growth rates in agriculture across crops and regions in India for the period, 1968-69 to1991-92. They distinguished two sub-periods, viz., (i) 1968-69 to 1981-82 and (ii) 1981-82 to 1991-92. From this they came to the conclusion that there must be an upsurge, a significant one,in the growth of aggregate production and productivity in Indian agriculture and it can not be attributed merely to a favourable weather. The fact that the role played by yield improvement in inducing higher output growth has been far more important than that of expansion in area indicates that the process of growth has been technologically more dynamic too.

Bhalla and Singh (1997) have done a study on recent developments in Indian agriculture (a state level analysis). It reveals that there was a marked acceleration in the growth rate of agricultural output in India during 1980-83 to 1992-95 as compared to that in the earlier periods, and that agricultural growth had become regionally much more diversified. The period 1980-83 to 1992-95 was also characterised by important CPchanges away from coarse cereals towards rice and wheat cultivation, on the one hand, and towards oilseeds, on the other. Finally, the 1980s also witnessed acceleration in per capita agricultural worker productivity in many Indian states.

If sustained, higher productivity growth is likely not only to result in higher wages but also to trigger growth in the non - agricultural sector through input-output and consumption linkages.

The study made by A. Chattopadhyay and P.S. Das (1997) on the trends in crop output in W.B.reveals that the performance of W.B.'s agriculture during left front government (LFG) rule has not been properly estimated. There has been upward bias in the growth rates as estimated by R.B.I. studies. W.B.'s agriculture during LFG regime has improved marginally and that too has been possible because of growth in production of rice and oilseeds (others are more or less stagnant or declining). One of the possible reasons of this marginal improvement may be the lower base in the pre-LFG regime. Perhaps, this has also happened in case of Mizoram and Nagaland, where estimated growth rates in foodgrains production are respectively 9.3% and 6.7% during 1981 to1991 compared to 6.5% in W.B. as per the report of the C.M.I.E. (1993).

Nadkarni (1996) concluded that there is considerable evidence of an accelerated commercialisation of Indian agriculture since 1980s. In all possibility, the process has further accelerated in 1990s and has induced growth, making agriculture much more dynamic than ever in the past. Yet there exists the paradox of dynamic agriculture along with stagnating peasants. Later Sau and Ray (1998) have criticised the study made by Nadkarni (1996). Their study has concluded that there exists constrained accleration of commercialisation of agriculture in India and identified the constraints on commercialisation as supply, and demand constraints where supply side constraints are infrastructures, viz., irrigation and finance and demand side constraints are economic factors,viz., per capita income, structural changes consisting of industrial economy and agricultural exports and social factors, viz. ,poverty and food security. Despite these bottlenecks the sample marginal peasant households are seen to carry forward commercialisation (i.e., the farmers are not static but dynamic).

J. K. Mallik (1997) has analysed the growth performance of agriculture in independent India focusing on the slow down in the growth rate during the 90s. The deceleration in agricultural growth during the 90s owes primarily to the stagnation of crop area attributable to the reduced pace of coverage of irrigation. An alarming decline in the foodgrains area has been observed. But this study does not find much merit in the generally held perception that foodgrains production suffered because of the expanded cultivation of non-foodgrains like oilseeds. For further growth, the need for increasing the coverage of irrigation and the greater attention of R&D underscore the urgency of rising agricultural investment substantially.

There are many more works on the estimation of agricultural growth in India and her states and on the identification of factors contributing to the growth or decay in agricultural production and yield. Worth mentioning are the works of Vidyasagar (1977, 1980), Rao (1985), Dayal (1966), Sharma (1974), Minhas (1966), Saha (1994), Vani and Vysulu (1996), Rao (1998), etc.

### Impact of C. P., Farming Practices on the Income and Employment

Pal and Pal (1985) have examined the causes and impacts of crop diversification on the farm economy of W.B. They found that, the model of farming, the degree of mechanisation and the farm size have been effective in enhancing diversification. Regression of farm business income (FBI) entails that both crop diversifications (CD) and variety diversification (VD) have direct significant impact on FBI. FBI is more influenced by CD than by VD.

Thakur, Kapila and Moorti (1985) have conducted a study in Sarproon Valley of district Solan in Himachal Pradesh (H.P.) and concluded that the farmers can earn very high income annualy by diversification towards vegetables growing.

The study made by Saini and Singh (1985) shows that the diversification of crop farming with high yielding milch animals can play an important role in increasing income and employment in small farms.

Singh, Singh and Sharma (1991) have shown that the optimum combination of enterprises in a farming system is helpful in augmenting the levels of employment tremendously.

Singh and Grover (1991) have made a study in Punjab and concluded that the rapid increases in crop productivity, changes in crop patterns and rise in cropping intensity in the cotton belt and central districts were together responsible for the higher increase in the value of output per hectare. The green revolution period witnessed significant shifts in the value productivity which were influenced significantly by the C.P.

One can mention a large body of literature dealing with the impact of C.P./farming practices on the income and employment. Specific mention may be made of the works done by Giri, Sashtri and Somayajulu (1966), Chatterjee (1966), Kaul (1966), Mathur (1963), Sharma and Sirohi (1980), Joshi and Alshi (1985), Chand, Sidhu and Kaul (1985), Throve and Gadgilkar (1985), Satheesh, Sharma and Sharma (1985), Singh, Bal and Kumar (1985), Dangat and Patil (1985), Swamy (1985), Sharma and Singh (1985), Kalita (1985), Chand and Singh (1985), Reddy (1985), Singh, Singh and Sengar (1985) , Gil and Patel (1985), Balistar, Singh and Singh (1985), Yadav, Rana, Singh and Nigam (1985), Chawla and Chahal (1985), Prahladachar and Mohanasundaram (1985), Mukundan (1985), Singh and Sharma (1985), Mani and Varadarayan (1985), Joshi (1991), Dangat, Sate and Nawadkar (1991), Singh, Singh Gupta, Kadian and Singh (1991), Waghmare, Tak and Hedgire (1991), Chawla and Chahal (1991), Singh and Pandey (1991), Verma, Dhaka and Gupta (1991), Thakur, Sharma and Shaini(1991), Singh and Singh (1991), Gupta (1963), Deoghare, Sharma and Goel (1991), Ramasubhan (1963) , Kymal and Ramakrishna (1973), Singh (1989), Singh, Verma and Gangwar (1991), Iyyampillai and Swamikanan (1991), Bhosale and Gawade (1991), Upadhyay , Singh and Upadhyay (1985), Sharma (1985), Azad, Prasad, Yadav and Singh (1985), etc.

### Impact of change in C. P. on Ecology

The study made by Hussain (1991)has concluded that the change in C.P. has accelerated the growth of market oriented economy resulting into the destruction of the traditional agriculture. It has not only increased the power of land owners resulting in the labourers and landowners polarisation, intensifying class conflict, but also affected adversely the health of the soils, damaging their nutrients, making the ecosystems more fragile.

Joshi and Tyagi (1991) have examined the sustainability of existing farming system in the Punjab and Haryana and calculated that the rate of changes in production and crop yields of important crops, viz. rice, wheat, etc. has slowed down in the Punjab and Haryana. In good water quality regions the existing production levels are maintained by over exploitation of ground water. Contrary to this, the poor quality water regions are underutilising ground water and deteriorating the soil health by salinity and waterlogging. Both the phenomena are undesirable and threaten the production and ecological sustainability.

Sharma,Bhati and Singh (1991)have highlighted that the post -green revolution changes in Himachal Pradesh have led to non-sustainable use of land resources.

Chaudhary and Aneja (1991) have concluded that the green revolution successes have led to breakthrough in foodgrains production in Haryana. But the unscientific use of modern technology has resulted in a number of problems. Over exploitation of land and water resources during the last two decades has led to deterioration of soil health, created nutritional imbalances and distributed the natural hydrology, particularly in intensively irrigated areas.

Singh and Sankhayan (1991) found that the CP in the Punjab has been shifting towards wheat-paddy and wheat-cotton rotations depending upon the suitability of the area for paddy and cotton. The requirements for irrigation water are increasing there also but not the supply and therefore shortages are being felt over time.

Khatkar, Chamola and Manocha (1991) have examined the post green revolution scenario in Haryana agriculture and concluded that the existing CP mainly rice-wheat rotation and indiscriminate use of water have resulted in creating ecological imbalance.

One can mention a large body of literature dealing with the impact of change in CP on ecology.

### Optimum C.P. Plan

Mehta, Sharma and Singh (1982) have conducted a study to show the impact of mixed farming on economic efficiency of different farm sizes in Punjab. Nadda, Sharma, Bhaduria and Swarup (1978) showed that the existing CP was sub-optimal, more so in the low and mid-hills in different agro-climatic zones of H. P. where diversification of agriculture was observed. Singh and Rahim (1978) have developed a few alternative optimal cropping system for a typical sub-watershed in the hill region of U.P. Rajagopalan and Varadrajan (1978) have concluded that the best policy to minimise risks in adopting modern technology is to choose the mixed farming and mixed cropping and improved market practices to help price stabilisation.

Sharma and Morti (1991) have evaluated different risk efficient plans showing different levels of risk and income associated with them in Kangra district of H. P. Sain (1991) has made a comparative study of wheat-paddy versus wheat-cotton farming system in the South-Western Punjab. Deshmukh, Hange, Kab and Pawar (1991) have made an attempt to discuss the costs and returns structure and production elasticities of different resources in the crop production of different farming systems with a view to indicating resources use efficiency/inefficiency of different farming systems in the drought-prone area of Western Maharashtra.

### Study on I.C.

### Assessment of Credit Requirements

A brief review of the growing literature on rural credit in India reveals that prior to the publication of the All India Rural Credit Survey Report (1954), the crushing weight of 'debt'burden was widely discussed. With the advent of planning, the debt burden has visibly yielded place to 'credit requirements', and this shift in the conceptual framework is very relevant as the present day focus is on the 'credit needs'.

As regards the approach and methodology used for assessment and sometimes for projections of credit-requirements, mainly the approaches, viz., 'borrowing approach' and 'expenditure approach'have been adopted (Choubey,1977).There has been a shift towards 'the expenditure approach' or 'cost of production approach' with the adoption of the New Agricultural Strategy in India. A.C.Shah's study probably was the first research to give a projection of the demand for agricultural credit at the end of Third Plan (ICSSR, 1975).

The working group set up by Agricultural Production Board (1965), the pannel of Economists headed by Dantwala (1996), the fertiliser committee (1968), the study group under the chairmanship of Gadgil (1969), the All India Rural Credit Review Committee (1969), the Sub-Group on Agricultural Credit and the Working Group on Co-Operation of the Fifth Plan (1974) and the National Commision on Agriculture (1976) have mainly adopted 'the cost of production' method in their assessment, projection of production credit requirments for different times in India. However, a close examination of all these macro estimate shows that these have been made on the basis of different assumptions, with differences in the components of farm expenses. For example, the working group (1965) had estimated the anticipated total cash requirements for 1970-71 for various current inputs both under traditional and improved methods of cultivation of the major crops in the country (AIRCRC).

However,the methodology and assumption were changed when the above estimate was revised by the pannel of economists. The panel adopted two approaches in such revision. Under the first, credit requirments were assumed to bear a certain relationship to the value of agricultural produce. Under the second method, per acre borrowings were multiplied by the estimated net acreage under cultivation in 1966-67 and 1970-71 and the estimates so obtained were then inflated by 25% to allow for the increase in price level between 1961-62 and 1966-67. Under each of these methods, again, two sets of estimates were made. The economists' panel presented, accordingly, four estimates in regard to production credit requirements of agriculturists in India during 1970-71.

Although the working group's estimate and the panel estimate were made for the same years, i. e., 1970-71 ,yet these are not comparable because they are based on different appioaches, separate sets of data and assumptions.

With some modifications in the method of the panel, the study group estimated production credit requirments at Rs. 1200 crores for the year 1967-68 (National Credit Council).

The Review Committee (1969) taking into account the technological change in agriculture and the psychological change in the attitude of agriculturists towards the adoption of the new technology, made an attempt to project production credit requirments in 1973-74. Compared to the earlier estimates of credit requirments, the Review Committee has made an improvement in its estimate by projecting credit requirements area-wise and crop-wise but such requirements in respect of different size groups of cultivators have, however,not been made.

The projection of production credit requirements by the National Commision on Agriculture (NCA) under full programme coverage for the year 1985 has been made. This projection appears to be more disaggregative and informative as compared to earlier estimates at macro levels.

Bansil's (1971) projection of such credit requirements for the whole of India for the year 1973–74 may be mentioned in this respect. He has projected Rs. 987 crores as total credit needs for agriculture on the assumption that 50% of the value of the three major inputs for farm business would be made through borrowings and the estimated credit-needs for miscellaneous purposes during 1973-74 would be Rs. 274 crores.

Sharma and Prasad (1971) work on the assessment of production credit requirements is also a micro level study. The important findings of the study are (a)the per acre credit needs are Rs. 113 in the less advanced and Rs. 156 in the more advanced regions under the existing technology, (b) the variations in the average credit needs among different size-groups of farmers in two regions reveal that the trend is similar. Credit needs are the highest among medium-sized farms followed by small farms and the lowest in case of large farms in both regions.

Pany (1982) has studied (a) to assess the credit requirements of cultivators, (b) to examine the changes in the supply and direction of IC, (c) to investigate the pattern of utilisation and repayment of such credit, and (d) to examine the deficiencies of IC, if any. On the basis of his findings certain policy measures have been suggested by him to strengthen the IC structure to achieve the desired objective.

Sharma and Pandey (1988) commented that in the year 2000 the process of increasing modernisation of Haryana State will result in (1) a shift in C.P. in favour of green fodder crops, (2) sufficient surplus of all the food-grains (except maize and barley) and milk output, (3) greater demand for crucial inputs and (4) an increase in the capital and credit- particularly short-term credit requirement to a substantial extent.

Dibhavi's (1998) work concludes that the majority of small size farmer borrowers are still dependent on the non-IC agencies. As the farm size increase, the percentage of the farmer borrower also increases. Among the factors affecting share of IC in total investment on farms, fixed cost per hectare, per capita consumption expenditure, and the share of non-IC in total farm investment were formed to be relatively more important.

A study conducted by Desai (1988) remarks that the magnitude of the scaled down credit requirements for crop production only for the years 1990, 1995 and 2000 A.D. reveals that the credit deliver system has to be streamlined and strengthened to undertake the task ahead. There is no place for complacency.

## Supply of IC

The trend in the supply of IC is revealed from various studies. As per the survey conducted by the National Council of Applied Economic Research (NCAER) (1974) 30.3% of the total borrowings made by the cultivating households in the country was from three institutional agencies, viz. government, co-operatives and commercial banks in 1970-71. Patel (1975) has reviewed the growth of agricultural advances made by co-operatives and commercial banks in the country.

Gupta and others (1978) have assessed the extent of uneven distribution and concentration of co-operative credit in different states in India in the period 1970-71 and 1975-76. The co-operative credit when measured in terms of per capita or per hectare, appears to be very unevenly distributed among the states. The contributory factor for such distribution is the difference in technological progress between developed and under-developed regions.

Basu (1979) found that while commercial banks have successfully increased the share of agriculture in their total outstanding credit several-fold, it has failed to maintain minimum degree of uniformity in its regional distribution.

A study conducted by Kumar and Kahlon (1978) in Ludhiana district reveals that the average amount borrowed by large farmers from institutional sources was significantly higher than that of small and medium farmers.

The observation of the Working Group on Rural Credit (1978-83) is that access to IC for the weaker sections in the country as a whole continues to be limited.

## Utilisation of Credit

I.C. differs from non-IC mainly on the ground of supervision over the end-use of credit. This has been emphasized in the writings of Horace Bel Shaw (1959) and Gadgil (1975). The fact that misutilisation of credit is a common phenomenon in the field of IC for agriculture appears to be well established from the findings of several studies.

The study conducted by Programme Evaluation Organisation, Planning commission (1965) has revealed that 23% of the short-term and 35% medium-term credit were diverted to purposes other than those for which credit was advanced.

To indicate misutilisation of credit co-operative loans at the macro level, Vyas (1974) has used the parameter of the ratio of profit to loans advanced by primary land devolopment banks in some major states. While some of the states like Gujarat and A.P. have shown creditable results on this count.

The performance of some other states like Assam, Orissa, W.B. and Tripura have been grossly disappropriating.

Galgalikar and Gadve (1978) find in their case study of credit society in Akola district of Maharashtra, the diversion of funds for unproductive purposes to be the highest in case of marginal farmers.

## Repayment of Credit

Like misutilisation of credit, poor recovery performance of the credit institutions is another obstacle for recycling of more loanable funds for agriculture. As per the report of the Committee to Review Arragments for IC for Agriculture and Rural Development over the period 1970-71 to 1978-79 the recovery performance of Primary Agricultural Credit Societies (PACS) registered a deteriorating trend.

Another independent study made by Dadhichi (1971) on wilful defaults on co-operative credit in Rajasthan has brought to light some important revelations. Wilful defaulters are generally those having large holdings, of higher castes, of a higher level of education and are large borrowers and members or ex-members of the managing committees. Reasons for such a situation are relanding co-operative credit to make profit out of it, harassing present managing committees, uncertainty in avoiding of fresh credit and an element of fear about the future financial position of the society.

Dadhichi (1971)in another study has found strong association between repayment and irrigation.

In another micro study Singh and Sandhy (1980)have estimated the extent of overdues of different categories of farmers and the causes of such overdues.

## Problems of IC

Problems encountered by the IC agencies to achieve the desired objectives may be placed under two heads, viz. (a) structural and (b) operational. While structural problems are mainly associated with co-operatives, operational problems are common both in case of co-operative and commercial banks.

Chaudhury and Sharma's (1970) study of the crop loan systems in A. P. and Punjab indicates that denial of loan to owner–cultivators, faulty setting of credit limits, timings of loans, ill quality of loan and accompanying extension guidance, are the faults, among others in the operational aspects of the system.

A weak capital base is the reason for the poor performance of the long-term co-operative credit structure in many states of the country, is the opinion of the committee on Co-operative Land Development Banks (1975).

Both the expert group (ACSCB)and working group (1978)have observed the lack of co-ordination among different credit and non-credit agencies working at different levels. Many more studies have been made on I.C. The works of Bhattacharya (1994), Subbarao, Rao & Tripathi (1977) may be mentioned in this connection.

## Factors Influencing the Change in C.P./Farming Practices

Majid (1963) has studied the influence of size of holding on the selection of C.P. in U.P.for the period (1959-60 to 1961-62). He finds direct influence of the size of holdings on the pattern of crops, preponderance of food crop being observed in the case of small farmers. Earlier, Shah (1963) investigated into the same issue for Saurashtra and found that with perceptible change in C.P., cotton and groundnut directly competed with each other, the former being substitued by the latter. In this process, the big and small farmers participated almost equally. Meenakshi Malya (1963) shows that the proximity to an urban centre, better transport and efficient marketing facilities have more important influence on the CP and the influence of the size of holdings, debt per acre, number of cattle, etc., on the CP was not significant. A similar conclusion was arrived by Jasdanwalla (1966). The study conducted by Rao (1965) shows that urban fringes influence on efficiency is more noticeable among big and medium cultivators in an urban group, while the influence of market proximity in the non-urban group is reflected in the higher efficiency of medium and small groups of farmers.

The influence of irrigation on CP seems to vary from region to region. Ramlingam (1963) reports that the availability of irrigation water encouraged growing of cash crops on larger area. A contrary conclusion was arrived by Shah (1963). Sinha (1978) has examined the impact of lift irrigation schemes on CP and crop yield in some villages of Haryana. It has been concluded that changes in the CP and increase in crop yields do take place through lift irrigation systems,no drammatic results can be expected or achieved. To obtain best results, diffusion of knowledge of new technology, optimisation of water resources, etc. are as important and relevent as construction of lift irrigation canals. A study by Madalia (1985) has revealed that much of the success of the farms belonging to the progressive area in Mehsana district in Gujarat State has been attributed to factors like favourable soil conditions, better irrigation facility, etc.

Kharinta, Aggarwal and Kaur (1985) have shown that the profitability of grape orchard in comparison to other crop rotations and availability of credit facilities motivate the adoption of grape cultivation. Shrivastava (1985) has shown that the sample farms situated in zone 1 (nearer to industrial centre) of Ranchi district in Bihar have more diversified their farm economy as compared to that of other two zones (not nearer to industrial zones like zone 1).

Bhardwaj, Dixit and Gupta (1991) conclude in a case study in Haryana that the element of risk (yield and price) has a significant impact on CP and resource allocation of farmers under study. Gupta and Tewari (1985) have concluded in a study of Allahabad farms that larger farms and wealthier farms are relatively less diversified. Tenancy discourages diversification. Farms with higher irrigation intensity, and located nearer to market are relatively more diversified. Share cropping and family size do not affect the level of crop diversification. Desai (1977) has concluded study in Surat district that increasing availability of net irrigable land would shift the crop pattern in favour of more remunerative and also labour intensive crops. This shift would, in turn, increase the net income of an average farmer.

Sridharan and Radhakrishnan (1978) have shown in their study in Nilgiri district, Tamilnadu that though there is no shift in CP between 1966-67 and allocation between crops is influenced by physical,economic, biological, and sociological factors. Sridharan (1971)have concluded study in Coimbatore district that the factors influencing the CP can be broadly classified into physical, economic, biological, technological and institutional. The study made by Biswas and Khambeta (1980) attempts to throw light on (a) the type of CP that would increase or stabilise the crop- turn specially in climatically vulnerable areas; (b) give indication of probability of success and (c) also help to identify the time of minimum irrigation so as to get optimum yield. Shah (1978) has shown that the farmers intend to change traditional CP if and only if the irrigation, credit, etc. constraints are resolved. The study conducted by Thomas and Devi (1990) in Kerala concludes that the climatic condition in the state as well as farmer's high expectations towards future prices for the crops have resulted in the increasing trend in acerage of these crops despite the fact there was no significant shift in the CP. Kumar (1989) concludes that in the state of Bihar, as a whole, various environmental and techonological variables exercise a dominant effect on the area and yield of rice. Acharya and Agarwal (1974) have studied the factors affecting supply and growth rate of commercial crops in Rajasthan. These factors are relative prices, current year's rainfall, etc. Dikshit aned Singh (1974) have concluded that under the impact of technological changes introduced in agriculture in recent past, the CP is undergoing a significant changes in U.P.and this has created serious imbalances in the product-mix. Economic factors seem to be the major determinants of the shifts in the CP.

Bhanja (1971) has concluded in a study in Birbhum district that in order to popularize the new varieties of paddy the small and marginal farmers need sufficient I.C. Subrahmanyam and Patel (1973) have conducted a study in West Godavari district of A.P., which manifests: (i) Credit is an important factor for increasing the incomes of farmers through optimum utilisation of the available resources. (ii) Credit is important for the adoption of new technology and for bringing more area under HYVs. (iii) Credit policy should be biased in favour of the small farmers. (iv) The return from using the credit (short terms) is worth the risk of borrowing the amount (if properly used). Desai and Naik (1971) have examined the prospects of demand for short-term IC for HYVs. Athavale, Yadav and Mishra (1971) have made a study in two districts in Madhya Pradesh (M.P.), which concludes that the contribution of co-operative credit in the HYV programme has varied considerably in the selected two districts. Tewari and Sharma (1980) have concluded that small farmer who is really liquidity starved, can be benefitted a lot, if and only if a dependable IC programme is provided for the increased adoption of high yielding technology and thereby increased incomes. Desai ,Gupta and Singh (1988) have concluded that the degree of agricultural progress is associated with some factors related to IC. The studies made by Desai (1994), Gadgil(1994), Tripathi, Dev, and Dsharing (1994) have emphasized the need of IC for accelerating agricultural growth in India.

Mruthyanjya and Kumar (1989) examined the changes in input use, productivity, cost of production, profitability and employment in crops due to the crop production strategy followed in the post -green revolution period, and identified the change in CP in this respect. The study also sugested ways of controlling the imbalances in the CP.

Ray, Sau and Nayek (1994) have studied the impact of availability of institutional finance on changing CP in Midnapore district of W.B and concluded that the shift in CP has been truely influenced by the IC. But this study has not made an in-depth analysis,say, by considering the economics of CP and exact relationship between IC and change in CP at farm household level. Ray (1994) has studied the impact of IC on the change in CP in all India, W.B. Punjab and Haryana and some sample districts in W.B. ,viz., Burdwan, Midnapore, Murshidabad and Nadia during the period 1972-73 to 1988-

89. He concluded that there is a significant impact of IC on change in CP. But this study has not been extended to the farm household level and also not made an in depth analysis, say, considering the economics of CP in relation to credit in topographical area across the sample states.

Ray (2000) has highlighted the progress of commercial banking and its impact on the change in C.P.in eastern and north–eastern India on commercial or HYV crops for the sample states, districts and farm households. Ray (1998) has made a study on the growth of non-foodgrains (NF) in W.B. and concluded that (i) the percentage of the total N.F. area has been risen over years (which is an indication of change in CP) for most of the selected states or districts during 1972-73 to 1993-94, (ii) there exists a positive impact of consumption of fertiliser per hectare or percentage of gross irrigated area (GIA) to gross cropped area (GCA) or I.C. per hectare on the percentage of NF area and (iii) also positive impact of IC per hectare on both the consumption of fertiliser per hectare and the percentage of GIA to GCA in the studied regions, (iv) the correlation between the percentage of NF area, the consumption of fertiliser per acre and the percentage of GIA to GCA has found good in most of the studied areas.

Ray and Sau (1998) have studied the diverse changes in CP across states in India and concluded that (i) there exists pro-superior cereals and pro-foodgrains CP change, pro non-foodgrains type of CP change and pro-HYVs CP change in India. (ii) The factors responsible for the different CP across the sample 12 states are per capita income, agricultural exports and profitability ,infrastructural factors, like irrigation and finance and geographical factors like soil conditions. (iii) It is revealed that % of area under non-superior cereals and N.F is negatively correlated with per capita state domestic product, while higher profitability of commercial crops induces the small and marginal farmers to undertake more and more commercial cropping in the interest of increasing gross profit per unit of land. (iv) A spectacular recent increase in exports of oil cakes and raw cotton and rice has induced large increase in share of this agricultural crops in GCA. (v) Percentage of irrigated area in GCA is positively and significantly related to share of HYV area, rice and wheat area and rice-wheat and sugarcane area in GCA. The variation in CP index during 1970s and 1980s in state like Haryana, Punjab and W.B.is significantly explained by IC per hectare of GCA. The state level variation in share of NF and non-superior cereals area in 1985-86 and 1993-94 is significantly explained by % of drought prone area.

## Research Gap

From an in-depth and extensive study of the existing literature on change in CP and on IC it has been observed that no in-depth study has yet been undertaken to examine the impact of IC on change in CP/agricultural growth in India and its states. Against this backdrop, a research may be made for studying the impact of IC on change in CP/agricultural growth in India.

# CHAPTER–3

## DATA BASE AND METHODOLOGY

The importance of any of the empirical studies is generally examined and valued by its data base and methodology. The analysis on data base, in general, includes (1) the choice and selection of the field, (2) definition and recognition of the variables/parameters on which the data are to be collected, (3) methods and sources of data collection and finally (4) problems and limitation of the collected data. Methodology of a study, likewise, comprises proper choice and selection of the methods for the analysis and interpretation of the data set, consistent with the objectives of the study and in conformity with the nature of the data base of the study. The broad objective of the present study is to examine the impact of credit availability on change in CP. and the economics of CP across states of India, districts of W.B. and blocks of Midnapore. For this purpose we take resort to both the secondary and primary data.

### Data Base

In this study both the secondary and primary data have been used. Secondary data relate to area under crops (individual or all combined), credit-IC or non-IC (NIC), irrigated area, fertilizer consumption, per capita income (PCI), per capita state domestic product (PCSDP), indices of area, production and yield of crops ( individual or all combined), value of output, etc., at the national, regional, state, district and block levels. The primary data pertain to area under different crops, size of land holdings, credit (IC or NIC), own fund used for cultivation, value added, profitability of crops, irrigated area, composition of agricultural labours, etc. at the farm household level. This secondary data have been collected from Government of India and different State Government publications and also from different W. B. Government offices, office of Principal Agriculture Officer (Purba and Paschim Midnapore), United Bank of India (Head Office, Purba and Paschim Midnapore). On account of various deficiencies of secondary data, primary data have been collected from the farm households of purposively selected five villages located in alluvial, laterite and coastal or saline zones of Midnapore district of W.B.

**Study Area**

The study area covers the following areas:

(i) all India,

(ii) Purpossively selected 12 relatively agriculturally developed and under-developed states of four regions, namely, North-West Region, Eastern Region, Central Region, Southern Region in India. The developed states considered in the study area : Haryana, Punjab, Uttar Pradesh (U.P.), W. B., Tamil Nadu and A. P. The underdeveloped states are: Rajasthan, Gujarat, M.P, Orissa, Maharashtra and Himachal Pradesh (H.P.). The state of development has been assessed on the basis of yield of crops.

(iii) Different relatively (in respect of agricutural yield) more developed or less developed districts in W.B. Developed districts considered in our study are Burdwan, Birbhum, Nadia, Hooghly, Malda, Murshidabad, 24-Paraganas, and Midnapore. The under-developed districts considered in our study are Bankura, West Dinajpur, Jalpaiguri, Purulia, Darjeeling, Howrah and Coochbehar. The total number of districts is 15.

(iv) Different relatively more developed and less developed blocks of Midnapore district. The total number of blocks is 54. The state of development has been assessed on the basis of foodgrains productivity. Developed blocks are Jhargram, Binpur-I, Binpur-II, Jamboni, Salboni, Garbeta-I, Garbeta-III, Midnapore Sadar, Debra, Pingla, Dantan-I, Mohanpur, Kharagpur-II, Ghatal, Daspur-I, Tamluk, Panskura-I, Panskura-II, Moyna, Mahisadal, Nandigram-II, Khejuri-I, Khejuri-II, Ramnagar-I, Ramnagar-II, Egra-I, Chandrakona-I. Underdeveloped blocks are Nayagram, Sankrail, Gopiballavpur-I, Gopiballavpur-II, Keshpur, Garbeta-II, Keshiary, Dantan-II, Narayangarh, Sabang, Kharagpur-I, Chandrakona-II, Daspur-II, Sahid Matangini, Nandakumar, Nandigram-III, Nandigram-I, Sutahata, Haldia, Potashpur-I, Potashpur-II, Bhagawanpur-I, Bhagawanpur-II, Egra-II, Contai-I, Contai-II, Contai-III.

(v) For field level study data relating to our research problem have been collected from 160 cultivators who belong to six villages, viz. Chitra and Harasankar of Sahid Matangini andTamluk blocks respectively in the saline or coastal zone, Gobindanagar, Kaya and Maguri Jagannath Chak of Panskura-I in the alluvial zone, and Lapuria of Garbeta–I in the laterite zone of Midnapore district. Among the cultivators, we have taken 90 cultivators with credit and 70 without credit.

Of the total 90 sample households with credit, 20 households are from the villages in coastal area of the district, 50 from the villages in the alluvial zone and 20 from the villages in laterite zone. Of the total 70 sample households without credit, 20 households are from the villages in the saline or coastal area, and laterite zone each and 30 from the alluvial zone of the district. The break-up is illustrated in the table below.

**No. of Sample Households from Different Zones in the District**

| | *Saline* | *Alluvial* | *Laterite* | *Total* |
|---|---|---|---|---|
| With credit | 20 | 50 | 20 | 90 |
| Without credit | 20 | 30 | 20 | 70 |
| Total | 40 | 80 | 40 | 160 |

## Methodology

The methodology used in the study may be disscussed in the following manner:

### Method Used for Sample Design

In this study, the multi-stage purposive stratified random sampling has been applied. The stages are as follows:

(a) 1st stage: All India level.

(b) 2nd stage: Purposively selected different state level.

(c) 3rd stage: Selected district level.

(d) 4th stage: Selected block level

(e) 5th stage: Farm household level.

### Criteria Used for Selecting Study Area and Cultivator Households

In India, some relatively developed and under-developed states have been selected by using the agricultural productivity as an index of development. Following the same procedure (i.e. using the agricultural productivity index) all the district of W.B. and blocks of Midnapore have been classified into developed and underdeveloped groups. The credit availability has been used as a criteria to classify farmers into four categories, viz., cultivators with credit (both IC and NIC), cultivators without credit (either IC or NIC), cultivators with only IC and cultivators with only NIC. After the classification of farmers into the above categories, the selection of the farm households have been made at random. The district of Midnapore selected for a thorough study at the block level has been classified into three agro-climatic zones, viz. coastal or saline, alluvial and laterite by following the usual state government norms.

The criteria for selecting the sample study area, viz., the index of agricultural productivity and the credit availability, have been chosen with the objectives of our study in mind.

**Method of Collection of Primary Data**

The primary data have been collected both by questionaire and interview methods.

**Statistical Techniques Used for this Study**

Various statistical and econometric techniques have been employed for the analysis of data. They include, for example, growth rate analysis, regression analysis,coefficient of variation analysis, correlation analysis and percentage analysis.

To study the change in CP we have concentrated on some simplified and mechanical statistical procedures. Compound annual growth rates (henceforth CAGRs) have been estimated by fitting to the data a simple semi-logarithmic trend equation:

$$\log y = a + bt$$

where y is the variable for which the growth rate is to be calculated and t is the time variable.

In some cases CAGRs are separately computed for two sub-periods for all the data sets and for the entire period as well. Additionally, for verification of existence of statistically significant acceleration or deceleration in the dependent variables, such as production ,etc. we estimate a quadratic in time variable by fitting it to the data. The equation is:

$$\log y = a + bt + ct^2$$

The above exercises i.e., fitting trend equation and calculating CAGRs have been done for all India, the state , district and block levels separately.

The following multiple regression model has been used to examine the impact of change in CP on the agricultural growth:

$$P = a + b_1A + b_2Y + b_3CPI + U_1$$

where P = index number of agricultural production (all crops combined)

A = index number of agricultural area (all crops combined)

Y = index number of agricultural productivity (all crops combined)

CPI = index of CP

and other symbols have their usual meanings.

In this study the pro-HYV, pro-non food grains change in CP have been considered as an index of commercialization of agriculture. Thus the following simple regression model has been used to examine the impact of those types of change in CP leading to the commercialisation of agriculture on the total agricultural production:

$$P = a + bX + u_t$$

where P = index of agricultural production (all crops combined), X = % of area under HYV crops or % of area under non-food grains to GCA and other symbols have their usual meanings.

The other logic behind using this simple regression in place of multiple regression, as used above, is to examine the real impact of change in CP by eliminating the problem of multicollinearity between yield and CP. The multicollinearity problem has been eliminated by using the method of 'Dropping Variables'.

The study has been made for the whole period (1972-73 to 1996-97), the two sub-periods, viz. sub-period I (1972-73 to 1980-81) and the sub-period II (1981-82 to 1996-97) (1)to examine, in detail, the impact of C.P. change on agricultural growth in the post green revolution period, (2) to compare its effect upon the agricultural growth in the 70s and the period thereafter up to 1996-97.

The Chow Test (1960) has been done using the test statistic

$$F = \frac{(RRSS - URSS)/\ (K+1)}{URSS\ /\ (n_1 + n_2 - 2K-2)}$$

which has an F distribution with degree of freedom (K+1), ($n_1$+$n_2$-2K-2) to test statistically the difference in the effects of the explanatory variables between the two sub-periods. The Chow Test is not applicable in the case of high multicollinearity in the regressors and unequal error variances.

The simple/multiple regression analysis has been done to examine the impact of (a)credit and other related variables, viz. fertiliser consumption, irrigated area, PCI/PCSDP, etc. on change in C.P./agricultural growth, (b) size of land holdings on change in C.P., (c) the change in C.P. on profitability, (d) the change in C.P. on employment. The correlation matrix (either product moment or rank)has been constructed for the related dependent and independent variables.

To examine the impact of bank credit on the changes in CP in the pre- and post-bank nationalisation periods in respect of average level of production, we have used the following dummy variable model:

$$NFA = C + aX + bD + u$$

where NFA = % of area under non-foodgrains to gross cropped area (GCA) as an index of change in CP.

X = co-operative credit/hectare

D = dummy variable (representing the presence and / or absence of commercial banks'credit ) which takes the value O for 1950-51 to 1969-70 (pre-nationalisation period), and 1 for 1970-71 to 1992-93 (post-nationalisation period).

To examine the impact of credit on the change in CP across the sample farm households the following regression model has been used:

$$Yi = \beta_1 + \beta_2 X_{i2} + \in_i, \qquad i = 1,2,\ldots\ldots,n$$

where Yi = % of area under non-foodgrains in the GCA of the household;

$X_{i2}$ = Dummy variable (representing the presence and/ or absence of credit for cultivation) which takes the value 1 if Yi corresponds to the households with credit facilities and own fund (OF) and O if Yi corresponds to the households without credit facilities, i.e., only OF, $\beta_2$ measuring the difference in the percentage of area under non-foodgrains for the farm households with credit and the farm households (FHs) without credit. On the other hand, the impact of different types of credit ,viz. IC or NIC on the change in CP across the sample farm households has been estimated from the following model:

$$Yi = \alpha_1 + a_2 D_{i2} + \alpha_3 D_{i3} + \alpha_4 D_{i4} + \epsilon_i$$

where Yi = percentage of area under non-foodgrains in the GCA

$D_{i2}$ = 1 if Yi corresponds to FHs with IC and OF, 0 otherwise (i.e., if Yi corresponds to the FHs with OF only).

$D_{i3}$ = 1 if Yi corresponds to FHs with NIC + OF, 0 otherwise.

$D_{i4}$ = 1 if Yi corresponds to FHs with IC + NIC + OF , 0 otherwise.

**Rationale for Choosing the Time Period of the Study**

Since, the commercial banks in India were nationalised in 1969, so the first impact of nationalisation may be expected to start from 1969-70. In this study the data on co-operative credit and percentage of NF area have been used for the period (1950-51 to 1992-93) to examine the impact of commercial banks' nationalization on change in CP and hence on agricultural growth (by using the existence or non-existence of commercial banks' credit as dummy variables). Due to the reforms of banking system on the lines recommended by Narashimam committee (operated since 1992-93) the end time point of studying the impact of bank nationalisation has been selected as 1992-93. The impact of banking reforms which envisaged withdrawal of government financial support to the banks, deregulation by the government of the banking operations, etc. could be realized from 1993-94 onwards. Therefore, in order to measure the impact of nationalization of banks, the second sub-period has been to start from 1970-71 to 1992-93. It may be mentioned here that the main objective of bank nationalization was to provide sufficient credit to the weaker sections, namely marginal and small farmers, artisans, small traders and business, etc. through the banking system in the country, since in the pre-nationalisation period the private bank invested mainly in the industrial sector and in the speculative activities with the sole objective of maximizing profit, relegating objectives of serving the weaker sections in the society to the background. Naturally, the use of data on co-operative bank credit and % of NFarea for the period 1950-51 to1992-93 appears to be logical. We have chosen the period from 1972-73 to 1996-97 for examining the impact of CP on agricultural growth and the periods from 1972-73 to 1995-96 for examining the impact of I.C. and related factors on change in CP or agricultural growth in the sample states of India and districtsof W.B. and the period from 1992-93 to 2002-03 for examining the impact of I.C. and related factors on change in CP or agricultural growth in the blocks of Midnapore. For the sub-period analysis of the impact of CP on agricultural growth, we have selected sub-period-I as 1972-73 to 1980-81 and sub-period-II as 1981-82 to 1996-97. On the other hand, we have selected 1972-73 to 1982-83 as sub-period I and 1983-84 to 1995-96 as sub-period II for the sub-period analysis of the impact of IC and other factors on change in CP and agricultiral production.

**Concepts of Change in CP**

In this analysis the concept of the change in CP has been defined in the following ways:

(i) The trend of CPI has been considered to examine the change in CP.

(ii) Instead of considering all the crops individually at a time and measuring the change in the relative share of each of them in the gross cropped area (GCA), the change in the relative share of certain crops, taken together, say non-food grains, with respect to another group of crops, say foodgrains, will also indicate the change in CP. This change may be considered as a pro-non food grains change in CP.

(iii) There may occur a change in CP in favour of HYV crops. This type of increase in area under cultivation of HYV crops may be considered as a change in the CP and called pro HYV change in CP.

Since our primary focus is on the change in CP, there is a need to highlight the different dimensions of change in CP. Traditionally, the concept of C.P has been used as (i) the absolute area under different crops in an agricultural year; (ii) the percentage of area under different individual crops to GCA in an agricultural year. The first concept (i.e., the absolute area under different crops) may be conceived as absolute change in CP. But the second concept (i.e., the percentage of area under different crops) may be defined as the relative change in CP. Besides, these two traditional concepts of change in CP there has emerged another concept of change in CP in the post-green revolution era. This is characterised by also the changing composition of demand due to rapid industrialisation and urbanisation in the country.

In the post-green revolution era three types of change in CPmay be distinguished as follows: (i) pro-superior cereals, and pro-foodgrains CPchange, (ii) pro-non-foodgrains CPchange and (iii) pro-HYV CPchange. The first implies the rising trend of percentage of area under superior cereals and foodgrains. The second and third concepts imply the rising trend of percentage of area under non-foodgrains and HYV crops respectively.

The profit maximizing motive and/or food-security ensuring motive may be the factors behind the change in the CP. When there is an increase in area under cultivation of, say non-foodgrains crops, accompanied by a fall in the area under foodgrains, this change may be induced by the profit maximization motive of the cultivators, since normally the prices of many of the non-foodgrains are relatively higher than those of most of the foodgrains. If there is a remarkable shift towards the cultivation of the HYV crops whose prices have not gone up appreciably so as to bring the farmers higher revenue, that shift may be induced by the food security concern of the farmers, but food insecurity problem prompts them to go for cultivation of those crops whose production per hectare is quite high, even when the price is low. This concern for food insecurity arises from the fear that the prices of the foodgrains may rise so much so that it will be beyond their capacity to buy them when, they will require them, with the income earned from the sale of non-foodgrains. The food security problem is a problem of sustainable growth of foodgrains which requires formulating appropriate product plans. The farmers exactly do that on the basis of their experiences, i.e, what to produce and what not to produce. Absence of well-organised markets for agricultural products where the foodgrains will be available on time and at acceptable prices, may be a good reason for such a shift in the production of superior HYV crops away from the production of low yielding variety crops. If, on the other hand, organized markets are available within the reach of the farmers where foodgrains can be purchased from at acceptable prices and on time, the shift may be observed towards the cultivation of non-foodgrains, provided they fetch higher returns.

The cost of production, not the profit, may be another factor that may determine the CP in a particular region or for a particular group of farmers. The poor farmers and, as such, the regions inhabited mostly by the poor farmers can hardly meet the higher cost of production of, say, high income yielding non-foodgrains crops. These farmer's immediate concern is to buy the inputs and meet their costs, which they can hardly meet out of their own fund or even from the consolidated own and borrowed fund. To the contrary, the relatively better-off farmers normally do not face such problems and given the markets for foodgrains are available within their reach, these farmers normally go for the cultivation of higher profit ensuring non-foodgrains crops.

The Shift towards the cultivation of HYV foodgrains crops away from the non-foodgrain crops in order to successfully face the food insecurity problem may be accounted for by the average small holding size of the farmers. The marginal farmers can seldom make an optimum allocation of their small land holding among all available options – such as production of HYV foodgrain crops, traditional foodgrain crops (involving low-cost of production), and non-foodgrains. The constraint of small budget force them to choose only those options which satisfy their principal objective of ensuring food security and the budget constraint.

The relatively higher degree of profit maximizing or commercial motives of Indian farmers in the post green revolution period has led to the shift in the area towards the cultivation of HYV varieties of superior cereals and non-foodgrains. This type of CP change may be defined as pro-commercial (superior cereals and non-foodgrains) CP change, which may be used as an index of commercialization of agriculture.

Therefore, we may use alternatively the above definitions of CP change (namely, CPI, pro-superior cereals CP change; pro-HYV CP change, pro-commercial (superior cereals and non-foodgrains) CP change and pro-foodgrains CP change) on agricultural growth.

The present study emphasizes the relative change in CP including the pro-superior cereals and pro-foodgrains CP change, the pro-non-foodgrains CP change and pro-HYV CP change. The rationale behind such type of choice of the concepts of CP change is as follows. Since the present note tries to examine the impact of credit and related factors on change in CP, this impact can be best examined by studying the change in the relative share of the individual crops including HYV crops in the total cultivated area.

In the following chapters results derived from the application of the proposed methods are analysed.

# CHAPTER–4

## NATURE AND PATTERN OF AGRICULTURAL GROWTH IN INDIA: REGIONAL, STATE, DISTRICT AND BLOCK LEVEL ANALYSIS

Agricultural growth in India has shown considerable fluctuations over years. This chapter discusses the nature and pattern of agricultural growth across regions and sample states of India, districts of W. B and blocks of Midnapore.

Since independence, considerable progress has been made in the sphere of agricultural development in the country in terms of increase in crop production and productivity, technological developments and crop diversification. There were, however, periods of ups and downs in the growth of agriculture during the period of fifty years. The growth rate was appreciably higher during the initial years (upto the mid-sixties), but it had slowed down during the subsequent years, gathered momentum again during the eighties, and has decelerated considerably in the nineties. The deceleration in the growth during the nineties is more pronounced in the case of foodgrains and it raises concern from the point of view of food security. Moreover, it casts doubt regarding the future growth prospects of the economy, which depends largely upon the growth performance of agriculture (Mallik, 1997). This deceleration in the agricultural growth in the country can be accounted for by the slow down in the growth of irrigated area and a sharp deceleration in the rate of growth of fertiliser consumption (Rao, 1998). In this context of the growth performance of Indian agriculture, it is in the fitness of the things to study the nature and pattern of agritural growth and also the trend of growth of factors influencing agricultural growth. In the following sections, we make an in depth study on the nature and pattern of agricultural growth. There has been an impressive growth in the production of foodgrains in India; output recorded a fourfold rise from 50.8 million tonnes in 1950-51 to 198.2 millon tonnes in 1996-97. Most impressive has been the growth of non-foodgrains; production of major non-foodgrains like oilseeds, cotton and sugarcane recorded almost fivefold increase. The production indices of foodgrains, nonfoodgrains and all crops rose by 2.64%, 3.15% and 2.84% per annum, respectively, between the triennium ended (TE) 1951-52 and TE1995-96. Before green revolution the incremental output mainly emanated from area expansion, the relative contribution of area was as much as 46.9%, another 43.1% was the contribution of yield, the rest being the contribution of the interaction of the two. After the mid-sixties, the relative contribution of yield increased to 78.6%, and that of area decreased to 10.6%. In case of foodgrains, the relative contribution of area to incremental output was negative after the mid-eighties, it was –37.9% during the nineties. On the other hand in case of non-foodgrains ,the relative contribution of area to incremental output was 62.9% prior to the Green Revolution and only 54.3% during the nineties.

## Aggregate Performance of Indian Agriculture at Regional and State Levels

Examining the aggregate growth performance of the states in crop production it has been observed that at the all India level GDP and NDP generated in agricultural sector increased respectively at the rates of 2.16% and 2.15% during 1968-69 to 1981-82 and 3.28% and 3.34% during 1981-82 to 1990-91.

Let us highlight the CAGRs (%) of production (all crops combined) in India by sample states. It is observed that the CAGRs of agricultural production are significant in India and sample states excepting H.P. and Gujarat during 1972-73 to

1996-97 (Table 1). Again, the growth rate of area has been positive for most of the sample states during the same period. But the growth rate of yield has been positive for all sample states in that period. The massive use of HYV technology may be a major factor behind that positive and significant contribution of yield on agricultural production in the sample states of India.

**Table 1**
**Compound Annual Growth Rate (%) of Area, Production and Yield in States (1972-73 to 1996-97)**

| *States* | *Area* | *Production* | *Yield* |
|---|---|---|---|
| Haryana | 0.92* | 4.08* | 3.23* |
| H.P. | 0.3 | 0.08 | 1.8* |
| Punjab | 1.75* | 4.08* | 3.14* |
| U.P. | 0.2 | 2.02* | 2.65* |
| Orissa | 0.74* | 2.53* | 1.27* |
| W.B. | 0.11 | 3.56* | 2.28* |
| Gujarat | 0.80* | 0.4 | 2.41* |
| M.P. | 0.35 | 1.31* | 2* |
| Rajasthan | 1.00* | 2.02* | 2.2* |
| A.P. | -0.2 | 0.8* | 2.85* |
| Tamil Nadu | -0.18 | 0.8* | 2.55* |
| Maharashtra | 0.33 | 3.06* | 1.96* |

*Notes:* *indicates significant at 1% level.
*Source:* Area and Production of Principal Crops in India, Government of India.

Again, over the period 1962-65 to 1992-95, the highest growth rate in yield and output in India was recorded by wheat followed by cotton, rapeseed and mustard, rice and coarse cereals in that order. The minimum growth rate in yield and output was recorded by pulses, groundnut, and sugarcane. Table 2 gives details about the growth of area, yield and output of major crops in the sample states of India during 1972-73 to 1996-97. The table brings out that in most of the sample states the yield has made significant contribution to the production of food grains. But the growth rate of area has not been positive for all sample states. It became negative for the states like A.P., Gujarat, Tamil Nadu . The expansion of HYV technology may act as a major factor behind the positive and significant contribution of yield on the foodgrains production of sample states. In the case of some major non-foodgrains, viz., rapeseed and mustard, potato and sugarcane, both the area and yield have made significant contribution to production in most of the sample states.

**Table 2**
**State wise Compound Annual Growth rates (CAGRs) for Major Crops during 1972-73 to 1996-97**

*(Per cent)*

| | | *A.P.* | *Gujarat* | *Haryana* | *H.P* | *M.P.* | *Maharashtra* |
|---|---|---|---|---|---|---|---|
| Rice | A | 1* | 0.6 | 4.8* | -0.6 | 0.8 | 0.6 |
| | P | 3.5* | 3.2* | 8.3* | -0.3 | 2.3* | 2.1* |
| | Y | 2.4* | 2.5* | 3.3* | 0.3 | 1.6* | 1.6* |
| Wheat | A | -1.7 | -5.1* | 3.2* | 0.3 | 0.7 | -0.5 |
| | P | 4.3* | -6.1* | 6.1* | 3.5 | 4.1* | 3.4* |
| | Y | 6.2* | -1.1 | 2.9* | 2.7 | 3.3* | 3.9* |
| Total foodgrains | A | -0.8 | -0.5 | 0.2 | 0.3 | 0.2 | 0.2 |
| | P | 2.5* | 1.7** | 4* | 1.7 | 2.3* | 2.2* |
| | Y | 2.3* | 3.3* | 3.8* | 1.4 | 2.1* | 2* |
| Groundnut | A | 2.8* | 0.1 | -7.2* | | -1.2** | -0.7 |
| | P | 3.7* | 0.1 | -7.6* | | -0.7 | 1* |
| | Y | 0.9 | 0 | -0.6 | | 0.5 | 1.7* |
| Repseeds & Mustard | A | 10.1* | 12.5* | 6* | | 5.5* | |
| | P | 12.1* | 17.4* | 10* | | 10.2* | |
| | Y | 1.5* | 4.4* | 1.8* | | 4.5* | |
| Chillies | A | 1* | -1.7** | 3.1* | | 0.4 | -1** |
| | P | 5.6* | 1.4* | -1.1** | | -0.3 | -1.4** |
| | Y | 4.6* | 3.2* | 1.8* | | -0.6 | -0.4 |
| Patoto | A | | 6.8* | 4.3* | -0.8 | 2.9* | -0.3 |
| | P | | 7.6* | 4.8* | -3.3 | 3.4* | 0.1* |
| | Y | | 0.7 | 0.5 | 4 | 0.5 | 0.4 |
| Sugarcane | A | 1** | 5.7* | -0.2 | -2.4 | -0.7 | 3.7* |
| | P | 0.7 | 8.5* | 0.9 | -1.4 | 0.8 | 4.9* |
| | Y | -0.3 | 2.6* | -0.8 | 1.1 | 1.5** | 1.2** |

| | | *Orissa* | *Punjab* | *Rajasthan* | *U.P* | *T.N.* | *W.B.* |
|---|---|---|---|---|---|---|---|
| Rice | A | 0.1 | 8.3* | 0.5 | 0.9 | -1.2** | 0.7 |
| | P | 1.6** | 12.7* | 2.3* | 5.1 | 3.5* | 2.9* |
| | Y | 1.6** | 4* | 1.6* | 4.1 | 2.8* | 2.1* |
| Wheat | A | 4.1* | 2.2* | 1.6* | 2.3 | | 2.8* |
| | P | 5.7* | 4.9* | 5.5* | 5.3 | | 3.5* |
| | Y | 1.6** | 2.7* | 3.8* | 2.9 | | 1** |

*Contd...*

| | | *Orissa* | *Punjab* | *Rajasthan* | *U.P* | *T.N.* | *W.B.* |
|---|---|---|---|---|---|---|---|
| Total foodgrains | A | 1.1** | 1.9** | 0.1 | 0.9 | -0.9 | 0.4 |
| | P | 2.1* | 5.3* | 2.7* | 5.1 | 1.3* | 2.6* |
| | Y | 1** | 3.3* | 2.6* | 4.1 | 2.3* | 2.2* |
| Groundnut | A | 8.1* | -12.2* | -3.2* | | 0.6 | |
| | P | 8.8* | -11.9* | 3* | | 1.6* | |
| | Y | 0.6 | 0.2 | 4.8* | | 1* | |
| Rapseeds & Mustard | A | 4.2* | -0.8 | 11.9* | -2.5 | | 5.6* |
| | P | 5.3* | 2.2* | 16* | 0 | | 10.8* |
| | Y | 1** | 2.9* | 3.5* | 2.6 | | 4.9* |
| Chillies | A | 6* | -6.7* | 2* | 0.8 | -0.7 | |
| | P | 7.5* | -4.1* | 4.8* | 4.3 | -4.5* | |
| | Y | 1.4** | 2.6* | 2.8* | 3.4 | -3.8* | |
| Patoto | A | | 4.3* | | 3.5 | -3.8* | 4.3* |
| | P | | 5.9* | | 6.9 | 0.5 | 9.2* |
| | Y | | 1.5** | | 3.4 | 4.5* | 4.8* |
| Sugarcane | A | 1.3** | -1.5** | -2* | 2 | 2.3* | -3.5* |
| | P | 2.7* | 0.9 | 3.6* | 3.4 | 3.5* | -1.5** |
| | Y | 1.4* | 2.4* | 5.6* | 1.4 | 1.2** | 2* |

*Source:* Area & Production of Principal Crops in India, Government of India.

Let us analyse the regional pattern of agricultural development in India during 1970-73 to 1992-95. For the purpose of analysis, the whole period (1970-95) has been divided into two sub-periods, namely, 1970-73 to 1980-83 (first period) and 1980-83 to 1992-95 (second period). Levels and growth rates have been worked out by taking triennium averages for 1970-73, 1980-83 and 1992-95.

The new agricultural technology, which was introduced during the mid-1960s, was confined only to north-western region of India during the initial phase of green revolution. Later it spread extensively across the country over time and has led to significant growth in agricultural output. Despite considerable inter-state variations, most of the states in India shared the gains of new technology. Taking the whole period 1962-65 to 1992-95, total agricultural output in India at 1990-93 prices increased at a CAGR of 2.7%. There were important changes during the different sub-periods in the pattern of agricultural development. The period from 1970-73 to 1980-83 is characterized by the extension of new seed fertilizer technology in the case of wheat and rice. The states like Punjab and Haryana in the north-west region maintained their high rates of growth, although at a slightly lower level. U.P. also slightly accelerated its growth. However, the eastern states like W.B and Orissa were unable to adopt the new technology. There was a significant deceleration in the growth rates in W.B. because of the significant fall of area under HYV wheat. Orissa showed a higher growth rate during 1970-73 to 1980-83 than that recorded by it during 1962-65 to 1970-73. The production in the central region was characterised by wide fluctuations. The performance of southern region was rather mixed during 1970-73 to 1980-83. An analysis of incremental output during 1970-73 to 1980-83 confirms the above results.

The period from 1980-83 to 1992-95 marks a turning point in India's agricultural development with respect to the growth of agricultural output. The growth rate of crop output in all India level accelerated from 2.38% during 1970-73 to 1980-83 to 3.40% during 1980-83 to 1992-95. The noticeable feature of the 1980s was that agricultural growth penetrated to all the regions in India. In the north-western region the period 1980-83 to 1992-95 saw a slight slow down of growth in Punjab, but a significant acceleration in the growth rate in Haryana and U.P.

The most remarkable development was a notable acceleration of growth in the eastern region. Specially, remarkable was the performance of W.B. where the growth rate increased to an unprecedented level of 5.39% per annum. However, there was an deceleration in growth rates in Orissa. The acceleration of the growth in the densely populated but hitherto agriculturally stagnant states of western India was a development of major significance.

Despite the prevalence of mixed picture regarding growth performance of individual states in central region, the region recorded an acceleration in its growth rate during this period (1980-83 to 1992-95).

The southern region recorded even a higher growth rate than the north-western region. But most interesting development was an unprecedented rate of growth of 4.59%

recorded by Tamil Nadu during 1980-83 to 1992-95 compared with a negative growth rate registered by it during the earlier decade. An important characteristics was a sharp rise in the contribution made by both the southern and the eastern regions in incremental value of output and their recovery from the losses they had suffered during the 1970s. During 1980-83 to 1992-95 the contribution of the north-western region to total incremental output declined sharply from 38.24% during 1970-73 to 1980-83 to 28.41% during 1980-83 to 1992-95 and that of central region declined from 34.13% to 28.98%. Therefore, the 1980s restored to a large extent the imbalance that had been generated during the 1970s (Bhalla and Singh,1997).

Concentrating on cropwise value of output in India by regions and sample states it is observed that in all India, the value of output (at current prices) has risen over years for most of the individual crops during 1970–71 to 1992-93. On the other hand, from the percentage distribution of incremental value of output of different individual crops it is found that there exists (i) a rising trend for growth rate for total cereals, wheat, jowar, bajra, maize, urad, tea, coffee, other drugs and narcotis, black pepper, fruits and vegetables, banana, other fruits and vegetables, other crops and rubber, (ii) a declining trend or stagnancy for other crops.

In terms of value of output, in the eastern region, wheat gained substantially mainly at the cost of pulses. On the other hand, both in the central and the north-western region wheat gained at the cost of pulses, coarse cereals and oilseeds. While in the southern region the non-foodgrains gained at the cost of wheat, coarse cereals, and some other foodgrains and in all India wheat and non-foodgrains gained at the cost of coarse cereals, pulses, etc.

The main components of growth of value of output are growth of yield, the growth of area and changes in CP. With the yield-raising new technology gradually spreading to more areas, yield rather than area growth has become the predominant source of growth in Indian agriculture since the mid-1960s. During 1960-65 at the all India level while value of output grew by 2.71% per annum, yield (value of output per hectare of total crops sector) registered a growth rate of about 2.30% per annum, accounting for as much as 85% of growth of output during this period. The highest increase in yield has been registered by the north -western region followed by southern, central and eastern regions in that order. The CP changes involving shift from low value coarse cereals to high value crops like oilseeds was the main source of growth in the central region and also to some extent in the southern region. In the eastern region, it was primarily the rate of growth in rice productivity, which was the main vehicle for development. In the central region and also to some extent in the southern region, the main source of growth was CP changes involving shift from low value coarse cereals to oilseeds. In the southern region, rapid increase in both rice as well as oilseeds production played a major part in agricultural development in the region. However, in the north-western region except for Haryana, it was the growth of rice, wheat and cotton and to some extent sugarcane that made significant contribution to the agricultural growth in the region. In addition to yield growth, another source of agricultural output is growth of cropped area (Bhalla and Singh,1997).

From the CAGRs of area, production and yield of major HYV crops in the sample states of India (Table 2a) it is observed that in most of the sample states the yield has made significant contribution to the production of food grains. But the growth rate of area of different HYV crops have not been significant for a good number of sample states.

From the above analysis of trends in the out put of various crops and crop groups and in their yield and the CP in the states of India during 1962-65 to 1992-95 and also during different sub-periods it can be concluded that all the major crops excluding coarse cereals, gram, tur, jute and tobacco experienced high growth rates in their output in recent years. Again, for majority of them not only pace of production growth increased, productivity also increased quite significantly during 1980s. There was dramatic transition from low to very high growth accompanied by improvement in productivity growth in oilseeds sectors. The most remarkable feature of the

**Table 2a**

**Statewise Compound Annual Growth Rates for HYV Crops in India During 1972-73 to 1997-98 (%)**

| | | *Andhra Pradesh* | *Gujarat* | *Haryana* | *Himachal Pradesh* | *Madhya Pradesh* | *Maha-rashtra* | *Orissa* | *Punjab* | *Rajas-than* | *Tamil Nadu* | *Uttar Pradesh* | *West Bengal* | *All India* |
|---|---|---|---|---|---|---|---|---|---|---|---|---|---|---|
| Paddy | A | 1.0* | 0.6 | 4.8* | 0.3 | 0.8 | 0.6 | 0.1 | 8.3** | 0.5 | 1.2 | 0.9 | 0.7 | 1.1 |
| | P | 3.5** | 3.2* | 8.3* | 0.3* | 2.3* | 2.1 | 1.6 | 12.7** | 2.3 | 3.5* | 5.1 | 2.9* | 1.8 |
| | Y | 2.4* | 2.5* | 3.3* | 0.2 | 1.6 | 1.6 | 1.6* | 4.0* | 1.6 | 2.8* | 4.1* | 2.1* | 2.2* |
| Wheat | A | 0.7 | 1.1 | 3.2* | 0.3 | 0.7 | 0.5 | 4.1** | 2.2* | 1.6 | NA | 2.3 | 2.8* | 1.8 |
| | P | 4.3* | 2.1* | 6.1* | 3.5 | 4.1* | 3.4* | 5.7* | 4.9* | 5.5* | NA | 5.3* | 3.5* | 2.2 |
| | Y | 6.2* | 3.4* | 2.9* | 2.7* | 3.3* | 3.9* | 1.6 | 2.7* | 3.8* | NA | 2.9* | 1 | 2.0* |
| Maize | A | 0.8 | 2.6* | 1.6* | 0.6 | 0.8 | 1.3 | 0.4 | 1.2 | 0.8 | 0.8 | 1 | NA | 0.9 |
| | P | 2.3* | 3.8** | 2.2* | 1.2 | 1.8 | 2.6* | 0.6 | 1.8 | 0.6 | 1.2 | 1.8 | 1.2 | 1.8 |
| | Y | 4.5* | 5.9* | 4.8* | 2.3* | 2.7* | 3.8* | 1.2* | 2.0* | 1 | 1.7* | 2.4 | NA | 2.2* |
| Jowar | A | 0.6 | 1.2 | 1.1* | 1.2 | 0.8 | 1 | 0.2 | 0.8 | 1.2* | 0.7 | 0.04 | NA | 0.8 |
| | P | 1.2 | 2.3* | 1.6 | 2.3* | 1 | 1.8 | 0.6 | 1.6 | 1.8* | 2.2* | 0.6 | NA | 1.8* |
| | Y | 1.8* | 3.4** | 2.2* | 3.4* | 1.2* | 2.4* | 1.2* | 2.0* | 2.4* | 4.8** | 1.2* | NA | 2.0* |
| Bajra | A | 0.5 | 1.2 | 0.8 | 1.1 | 1 | 1.2 | NA | 0.3 | 7.8* | 2.2* | 0.02 | NA | 1 |
| | P | 1.3 | 2.4* | 1.3 | 2.4* | 1.2 | 2.4 | 0.01 | 1.1* | 8.8** | 4.9** | 0.4 | NA | 1.8* |
| | Y | 1.8* | 3.5* | 1.8* | 3.4* | 2.4* | 3.2* | NA | 0.8 | 8.9** | 6.8** | 0.8 | NA | 2.2* |

*Notes:* *, **, indicate significant at 1% & 5% level respectively.

*Source:* Govt Of India.

output growth during the period was that by and large all the states from the north-western, central and eastern region of India, performed well in respect of growth in foodgrains output, but did equally well with regard to their major non-foodgrains too, particularly oilseeds, cotton and sugarcane. Among them the performance of W.B., Punjab, Haryana and Rajasthan was much more impressive than that of others.

### Aggregate Performance of Agriculture in W.B. : District Level Analysis

In recent years W.B. has emerged as one of the largest rice producing states in India. CAGR of production which was around 1.1% between 1970-71 and 1980-81 has shot up to 4.8 % in the eighties. About 3/4ths of this growth can be attributed to growth in yield. While area expansion especially through growth in multiple cropping explains the rest. Rice is the most important crop comprising over 90% of total food-grains production in the states. This clearly suggest that the rural economy has had a shot in the arm in recent years.

Against this general background, we may analyse the scenario of agricultural growth in the districts of W.B. Using the index number of area, production and yield of the districts in W.B. we see that though the production and yield performances are encouraging in the districts, there exist wide variations in productivity or yield across the districts. Various factors like infrastructural, economic, climatic, etc. can explain this differential productivity across the districts of W.B. It is observed that the productivity differentials in terms of standard deviation increased over years for both rice and foodgrains ,but the coefficient of variation in respect of foodgrains declined while that in respect of rice increased. It is found that irrigation and other infrastructural factors played a significant role in explaining the productivity differentials in rice and foodgrains across the districts in the state. Among the infrastructural factors irrigation is the most significant one (Sau, 1998; Saha, 1997). The proportion of land under irrigation has appeared to be the most important determinant of the proportion of land to be allocated to HYV in the dry season.

The credit supply also becomes effective in expanding HYV paddy cultivation in the dry season only (Sasmal, 1998). It is also observed that the use of fertiliser is an important factor explaining higher yield response in the eighties than that in the period prior to 1980 in the districts of W. B. The prevalence of larger proportions of small and marginal farmers has acted as a driving force behind the larger proportion of HYV coverage and higher agricultural growth during 1980s in W.B. The small and marginal farmers are catching up their larger counterparts not only in terms of yield performance, but, more importantly, the newly emerging institutional structure is permitting them to increase their access to purchase inputs at a scale equivalent to the latter group. The analysis on the growth of non-foodgrains in the districts of W.B. shows that its growth is significantly influenced by the factors like institutional credit, fertiliser consumption, per capita income, irrigation, percentage share of vested land distributed ,etc. (Gharei, 1998; Roy, 1998).

It is observed that the agricultural growth has been significant during 1972-73 to 1996-97 in all the districts of W.B. (Table 3). The acceleration in agricultural growth is observed in almost all the districts. Though the growth of yield is significant in all the districts of W.B., that of area is significant in only eleven districts (Table 3). In W.B. as a whole, the acceleration in growth rates is observed in the case of rice, oilseeds, foodgrains, non-foodgrains, and all crops. There is no acceleration or deceleration in output growth in the case of potato and jute in W.B. There is deceleration in growth rates in case of wheat and pulses in W.B (Chattopadhyay and Das, 1997). In all the districts as well as in the state as a whole, the growth rate of area has not been significant ,but yield made significant contribution to the growth of production. This was due to the greater adoption of HYV technology during the eighties and later. The crop wise analysis across the districts of W.B. shows that the yield of rice has risen over years during 1972-73 to 1996-97 in all the districts excepts Darjeeling. On the other hand, the production and yield of total foodgrains, total non-foodgrains, oilseeds and potato have increased

**Table 3**
**Compound Annual Growth Ratcs of Area, Production and Yield (All Crops Combined) during 1972-73 to 1996-97 in the Districts of West Bengal**

*(Per cent)*

| | *Districts* | *Area* | *Production* | *Yield* |
|---|---|---|---|---|
| 1 | Burdwan | 0.60* | 4.08* | 3.05* |
| 2 | Birbhum | -0.2 | 3.05* | 3.05* |
| 3 | Bankura | 6.82* | 4.08* | 4.08* |
| 4 | Midnapore | 0.80* | 5.13* | 4.08* |
| 5 | 24-Paraganas | 0.50* | 3.05* | 3.04* |
| 6 | Nadia | 8.03* | 5.13* | 3.04* |
| 7 | Murshidabad | 0.20* | 3.05* | 3.05* |
| 8 | W-Dinajpur | 0.1 | 3.05* | 3.05* |
| 9 | Malda | 0.2 | 3.05* | 4.08* |
| 10 | Jalpaiguri | 0.1 | 2.02* | 1.00* |
| 11 | Darjeeling | 0.70* | 3.05* | 2.02* |
| 12 | Cochbehar | 0.70* | 3.05* | 2.02* |
| 13 | Purulia | 0.70* | 2.84* | 2.02* |
| 14 | Howrah | 1.00* | 5.13* | 3.05* |
| 15 | Hooghly | 1.00* | 5.13* | 3.05* |

*Notes:* *indicate significant at 1% and 5% levels respectively.
*Source:* Statisical Abstract & Economic Revew (various issues), Government of West Bengal.

overtime during the same period in all the districts of W.B. Table 4 shows the growth rate of area, production and yield of some major crops in the districts of West Bengal during the period 1972-73 to 1996-97. The area growth rate of total foodgrains has been positive and significant in the districts like Burdwan, Bankura, Midnapore, Howrah, Hooghly, 24-Paraganas, Nadia, West Dinajpur, Malda, Darjeeling and Purulia. The growth rate of total cereals was also significant in these districts and also in Murshidabad. The growth rates of production and yield for both the total foodgrains and cereals have been significant in all the districts except Jalpaiguri where the contribution of yield has been insignficant in case of foodgrains as a whole. So far as the total non-foodgrains is concerned, the area grcwth rate has been positive and significant in the districts like Birbhum, Bankura, Midnapore, 24-Paraganas, Nadia, Mursidabad, Purulia and Hooghly. But the growth rate of production of non-foodgrains as a whole, has been significant for all the districts.

**Table 4**

**Compound Annual Growth Rates (CAGRs) of Area, Production and Yield of Some Major Crops During 1972-73 to 1996-97 in the Districts of West Bengal**

*(Per cent)*

| | *Total Foodgrains* | | | *Total cereals* | | |
|---|---|---|---|---|---|---|
| *Districts* | *Area* | *Production* | *Yield* | *Area* | *Production* | *Yield* |
| Burdwan | 1.00** | 5.13* | 3.05* | 1.00* | 5.13* | 3.05* |
| Birbhum | 0.3 | 4.08* | 4.08* | 0.50** | 4.08* | 3.05* |
| Bankura | 1.00* | 6.18* | 4.08* | 1.00* | 6.18* | 4.08* |
| Midnapore | 0.39* | 5.13* | 4.08* | 1.00* | 5.13* | 4.08* |
| 24-Paraganas | 0.60* | 4.08* | 9.42* | 0.80* | 4.08* | 9.42* |
| Nadia | 0.90* | 5.13* | 4.08* | 2.02* | 7.25* | 3.05* |
| Murshidabad | 0.90* | 4.08* | 4.08* | 0.90* | 5.13* | 4.08* |
| W-Dinajpur | 9.42* | 4.08* | 4.08* | 0.002 | 4.08* | 4.08* |
| Malda | 0.60** | 4.08* | 4.08* | 1.00* | 5.13* | 4.08* |
| Jalpaiguri | 0.1 | 2.02** | 3.05* | 0.30** | 1.00* | 3.05* |
| Darjeeling | 0.70* | 3.05** | 6.18* | 0.70* | 3.05** | 6.18** |
| Cochbehar | 0.3 | 2.02* | 2.02* | 0.4 | 2.02* | 2.02* |
| Purulia | 2.02* | 4.08* | 3.05* | 1.00* | 3.05* | 2.02* |
| Howrah | 1.00* | 3.05* | 2.02* | 0.02* | 4.08* | 2.29** |
| Hooghly | 0.90* | 4.08* | 3.05* | 1.00* | 3.05* | 3.05* |

| | *Total Non-Foodgrains* | | | *Oilseeds* | | |
|---|---|---|---|---|---|---|
| *Districts* | *Area* | *Production* | *Yield* | *Area* | *Production* | *Yield* |
| Burdwan | 0.77 | 4.08* | 3.27* | 4.60* | 5.26* | 3.05* |
| Birbhum | 2.30* | 6.08* | 5.50* | 4.19** | 8.15* | 0.04 |
| Bankura | 3.51* | 9.20* | 5.75* | 2.94* | 6.08* | 1.00* |
| Midnapore | 7.46* | 12.50* | 5.40* | 10.74* | 15.84* | 0.2 |
| 24-Paraganas | 3.26* | 3.12* | 1.85* | 6.82* | 13.54* | 0.7 |
| Nadia | 1.99* | 5.62* | 2.62* | 2.55* | 12.86* | 2.02 |
| Murshidabad | 1.80* | 5.30* | 3.14* | 3.98* | 7.68* | 0.40* |
| W-Dinajpur | -0.56 | 3.2 | 3.52* | 1.31 | 7.14* | 0.90* |
| Malda | 0.79 | 1.10* | 4.84* | 6.08* | 5.02* | 1.00* |
| Jalpaiguri | -1.81 | 3.20* | -1.26 | 6.40* | 5.02* | 2.02* |
| Darjeeling | -1.67 | 3.02* | 5.92* | 3.56* | 8.44** | 0.09 |
| Cochbehar | 0.52 | 4.90* | 4.37* | 1.51* | 5.44* | 0.08 |
| Purulia | 3.90* | 8.35* | 10.64* | 2.02* | 10.19* | 0.4 |
| Howrah | 0.54 | 5.29* | -6.19 | -1.78 | -0.99 | 3.05* |
| Hooghly | 4.46* | 7.70* | 3.31* | 6.95* | 9.86* | 1.00* |

*Notes:* *,**, indicate significant at 1% and 5% levels respectively.

*Source:* Statistical Abstract & Economic Revew (various issues), Government of West Bengal.

The growth rate of yield of total non-foodgrains has been positive and significant in as many as 13 districts in the state. Among different non-foodgrains oilseed has a pride of place. It is observed that the growth rate of both area and production of oilseeds, during 1972-73 to 1996-97, has been positive and significant for all the districts except Howrah. However, the contribution of yield in the production of oilseeds has not been satisfactory in the districts of W.B. Table 4a shows the CAGRs of area, production and yield of HYV crops in the districts of W.B. It is found that both area and yield have made significant contribution to the growth of production of HYV foodgrains in most of the districts of W.B.

The above analysis shows that yield has made greater contribution than area to the production of total foodgrains and cereals in the districts like Burdwan, Bankura, Midnapore, Howrah, Hooghly, 24-Paraganas, Nadia, Malda and Purulia. This was due to massive introduction of HYV technology for multiple cropping in the districts of W.B. during the eighties. For other districts the yield-raising technology has also played a big role in the increase in the production of the crops. In the districts, the production of foodgrains ,in particular, rice, has increased because of the introduction of HYV technology, larger provision of credit to farmers, etc. On the other hand, in case of the growth of non-foodgrains as a whole, the contribution of yield has been greater than that of area. This has been possible due to the introduction of yield raising HYV technology. But, in the case of oilseeds contribution of area was more significant than that of yield in the districts of W.B.

## Agricultural Growth in the Blocks of Midnapore

Let us make a cropwise analysis of growth of area, production and yield in blocks of Midnapore. It is observed from Table 5 that the growth rates of production and yield of foodgrains have been significant for most of the blocks in the district. Side by side, the growth rates of production and yield of sample non-foodgrain crops have been significant for a good number of blocks in the district.

**Table 4a**

**Compound Annual Growth Rate of Various HYV Crops in the Districts in West Bengal During 1991-92 to 2001-02 (%)**

| *Districts* | *Total Paddy* | *Wheat* | *Total Food-grains* | *Total Cereals* | *Total Pulses* | *Total Oilseeds* | *Rape & Mustard* | *Jute* | *Potato* | *Sugar-cane* | *Maize* |
|---|---|---|---|---|---|---|---|---|---|---|---|
| Darjeelling | | | | | | | | | | | |
| A | (-) 1.36* | 1 | 2.23* | 2.18* | -1.03 | (-)2.12* | (-)7.9 | -2.05 | 1.02 | NA | (-) 9.2* |
| P | (-) 2.06** | -2.19 | 10.94* | 11.98* | 16.93* | (-)6.7* | 2.3 | 2.28 | 4.08* | NA | (-) 9.02* |
| Y | -50.67 | (-)61.3* | (-)80.64* | (-)87.1* | -1.92 | -1.4 | -12.82 | 2.6* | 8.14* | NA | (-) 26.5* |
| Jalpaiguri | | | | | | | | | | | |
| A | -1.25 | 1.49* | 1.56* | 1.46 | 3.9* | 2.06 | -1.03 | 2.64 | 2.95 | 2.01 | NA |
| P | 5.34 | 3.15* | 9.14* | 9.32** | 2.12 | -1.09 | -1.16 | 21.56* | 28.12* | 1.74* | NA |
| Y | 35.1 | 29.90* | 24.47** | 25.57** | (-)36.55** | -12.29 | -18.33 | 3.8* | 11.46* | 17.22 | NA |
| Hooghly | | | | | | | | | | | |
| A | 1.22 | -1.04 | 2.25 | 3.19 | 2.02 | -1.76 | -1.03 | 1.47 | 1.38 | 4.8* | NA |
| P | 6.17 | -2.1 | 6.11 | 6.01 | 2.02* | (-)1.31** | -1.27 | 15.97 | 46.09 | 5.11** | NA |
| Y | -34.09 | -13.7 | 18.98** | 19.30** | -1.07 | (-)20.07* | (-)25.27* | 2.8** | 13.53 | 39.23* | NA |
| Burdwan | | | | | | | | | | | |
| A | 9.73* | 2.06 | 10.79* | 10.25* | 5.93* | -1.11 | -1.62 | 2.52 | 1.20* | -1.01 | NA |
| P | 40.59* | 2.64* | 41.76* | 41.29* | 4.4* | 6.1 | 6.1 | 14.16* | 3.8 | 2.3 | NA |
| Y | 25.67 | 52.45* | 20.2** | 20.45** | 8.8* | 24.53* | 23.59* | 3.1** | 2.35 | (-)12.67* | NA |
| Birbhum | | | | | | | | | | | |
| A | 3.33 | 1.74* | 5.82* | 8.58* | -4.04 | -7 | 5.4 | 0.61 | 7.2* | (-)7.6* | NA |
| P | 31.62* | NA | 39.01 | 38.05* | -6.5 | 1.46* | 1.46* | 0.5 | 15.31* | 3.9 | NA |
| Y | 57.6 | 60.5* | 61.15* | 65.54* | -26.04 | 41.51* | 43.92* | 3.7* | 21.83 | 28.9* | NA |

*Contd...*

| Districts | Total Paddy | Wheat | Total Food-grains | Total Cereals | Total Pulses | Total Oilseeds | Rape & Mustard | Jute | Potato | Sugar-cane | Maize |
|---|---|---|---|---|---|---|---|---|---|---|---|
| Bankura | | | | | | | | | | | |
| A | -4.49 | -1 | -6.4 | -6.2 | (-)3.4* | (-)1.04* | (-)7.54* | -1.9 | 7.3* | -1.4 | 3.39 |
| P | 19.35* | 2.84 | 19.37* | 4.19 | (-)1.83* | 7.17 | -3.27 | -2.47 | 19.16* | 3.39 | 1.06 |
| Y | 46.05** | 27.01 | 49.27* | 46.6* | -22.7 | 5.09 | 12.95 | 3.7 | 42.25 | 41.31 | 71.62 |
| Malda | | | | | | | | | | | |
| A | (-)3.07* | 1.9* | (-)12.11* | (-)2.85* | (-)3.04* | NA | 1.53* | NA | 9.8* | 3.78** | (-)3.39* |
| P | 5.47 | 3.37* | 5.86 | 9.66* | -1.41 | NA | 3.03* | NA | 3.44* | 35.09* | 1.11 |
| Y | 13.6* | 20.77** | 55.76* | 55.68* | -24.02 | NA | 6.6* | NA | 9.40* | 2.70* | 3.07* |
| 24 Parganas(north) | | | | | | | | | | | |
| A | 2.99** | 5.89* | 3.45** | 3.58** | -2.83 | 7.9* | 8.22** | 3.58* | 4.40* | 1.31** | NA |
| P | 8.46 | 1.16 | 5.08 | 9.63 | 1.54 | 2.03 | 2.05 | 5.5* | 10.78* | 7.75** | NA |
| Y | -5.2 | -17.64 | 3.13 | 1.55 | 35.17* | (-)20.24* | (-)26.58* | 1.21 | 4.91* | 3.16** | NA |
| 24 parganas(south) | | | | | | | | | | | |
| A | 2.18 | 2.50* | 1.1 | 1.29 | (-)_6.6* | -1.1 | -1.27 | -7.76 | 3.44* | 1 | NA |
| P | 18.97* | 4.25* | 19.93* | 19.40* | (-)2.81* | -5.9 | -1.41 | -10.75 | 7.53* | 6.61 | NA |
| Y | 25.2 | -44.13 | 39.41* | 39* | 24.02* | -8.12 | -17.88 | -1.68 | 7.91* | 13.16** | NA |
| Howrah | | | | | | | | | | | |
| A | 1.23** | 5.1* | (-)1.34** | (-)1.19** | (-)1.76** | (-)2.33* | -4.2 | 3.68* | 6.79* | 0.2 | NA |
| P | -1.38 | 1.01* | -1.54 | -1.46 | -4.2 | -2.27 | -3.5 | 6.93* | 16.95* | 2.24 | NA |
| Y | -0.86 | 5.09 | 10.17 | 8.44 | 13.63 | -10.34 | -14.26 | 4.30** | 35.7 | -11.31 | NA |

*Contd...*

| Districts | Total Paddy | Wheat | Total Food-grains | Total Cereals | Total Pulses | Total Oilseeds | Rape & Mustard | Jute | Potato | Sugar-cane | Maize |
|---|---|---|---|---|---|---|---|---|---|---|---|
| Purulia | | | | | | | | | | | |
| A | -1.61 | 2.15* | 2.16 | -1.69 | -2.63 | (-)5.81* | (-)1.24* | NA | -3.7 | 2.5 | -1.8 |
| P | 17.78** | 4.92* | 18.59* | 18.90* | (-)3.33* | (-)3.2* | (-)5.7* | NA | 2.77 | 5.61* | 0.82 |
| Y | 11.31** | 5.45 | 66.04 | 69.5* | (-)28.29** | 3.19 | 27.8* | NA | 36.6* | 39.31* | 6.6 |
| Coochbehar | | | | | | | | | | | |
| A | (-)3.26* | 8.62* | (-)2.02** | (-)1.89** | 1.78 | 1.54* | 2.44* | 2.33* | 9.2* | 0.05 | -0.09 |
| P | 8.09** | 1.95* | 9.94* | 10.29* | -6.1 | 6.2 | 1.38** | 42.72* | 81.34* | 3.4** | 3.3** |
| Y | 98.55* | 19.23 | 39.06* | 40.83* | -46.89 | -2.16 | -1.79 | 3.7* | -17.41 | 28.48* | 30.7* |
| Mursidabad | | | | | | | | | | | |
| A | 1.46 | 5.38* | 5.61 | 4.05 | -1.14 | 7.33 | 5.4 | 5.48* | 2.35* | 3.62* | -2.6 |
| P | 12 | 15.15* | 28.00* | 26.98* | 1.51 | 1.80* | 1.58* | 9.00* | 7.78* | 27.19* | 1.4 |
| Y | 64.59 | 22.82 | 32.7* | 37.95* | 61.84* | 17.93* | 18.56* | 6.14* | 3.95* | 6.95 | 1.8* |
| Nadia | | | | | | | | | | | |
| A | 6.66* | 18.84* | 4.40** | 4.62** | -0.62 | 1.79 | 2.08* | 1.42 | 1.54** | 2.3 | 1.01** |
| P | 23.48* | 4.20* | 20.30* | 20.29* | 1.45* | 1.97* | 0.81 | 2.32 | 5.56 | 0.94 | -1.04 |
| Y | 15.18 | 2.7 | 26.02* | 24.99* | -7.64 | -8.27 | -12.36 | 2.2 | 7.16* | 10.5 | 1.82* |
| Paschim Medinipur | | | | | | | | | | | |
| A | 5.5 | 9.11 | 1.06 | 1.06 | 8.7 | 7.58 | (-)8.40* | (-)9.9* | 1.65* | 4.19 | 3.1 |
| P | 19.71* | 1.91* | 21.85* | 21.72* | 6.3 | 1.13** | -3.5 | 1.55 | 57.43* | 29.61** | 1.07 |
| Y | 76.56** | 14.71 | 32.78* | 49.34** | 80.21** | 8.96 | -1.7 | 6.8* | 39.6** | 9.6 | 7.27** |

Contd...

| Districts | Total Paddy | Wheat | Total Food-grains | Total Cereals | Total Pulses | Total Oilseeds | Rape & Mustard | Jute | Potato | Sugar-cane | Maize |
|---|---|---|---|---|---|---|---|---|---|---|---|
| Purba Medinipur | | | | | | | | | | | |
| A | 3.88* | 2.80* | 3.79* | 4.18* | 3.68* | 1.3 | 2.90** | 1.93* | 1.39 | 1.56* | NA |
| P | 26.16* | 8.95* | 26.14* | 26.24* | 4.19* | 4.09 | -1.02 | -1.86 | 4.35 | 1.21 | NA |
| Y | 75.33* | 6.71* | 35.39* | 34.41* | 47.71 | 17.44 | 18.96 | 2.75* | 5.68 | 6.10* | NA |
| All Over W.B. | | | | | | | | | | | |
| A | 10.56 | 14.55* | 17.84 | 21.94** | (-)4.09** | 4.67** | 1.25 | 13.39* | 11.28* | 9.86* | (-)2.37 |
| P | 28.91 | 38.56 | 32.04* | 14.14 | -0.9 | 5.5 | 2.7 | 2.72* | 2.73* | 7.77* | (-)5.86* |
| Y | 89.82* | 16.55* | 43.26* | 43.65* | 6.58* | 2.63 | 2.86 | 0.19* | 7.3 | 13.25* | -25.65 |

*Notes:* * & ** indicate significant at 1% & 5% level respectively

*Source:* Govt. of West Bengal.

**Table 5**

**Compound Annual Growth Rate of Various Crops in All Blocks in Midnapur District During 1992-93 to 2002-03 (%)**

| *Block* | *Aus Paddy* | *Aman Paddy* | *Boro Paddy* | *Jute* | *Wheat* | *Potato* | *Til* | *Mustard* |
|---|---|---|---|---|---|---|---|---|
| Jhargram | | | | | | | | |
| A | -1.06 | 0.35 | -1.12 | -4.58 | (-)18.92* | 8.86 | 12.68* | 5.23 |
| P | 2.02* | 6.27* | 3.04 | -4.58 | (-)19.36* | 7.53 | 6.42 | 4.25 |
| Y | 3.98* | 11.38* | 6.26 | 2.36 | 1.05 | -3.59 | -4.06 | 12.03* |
| Binpur-I | | | | | | | | |
| A | 0.98 | 0.03 | -1 | .3.34 | (-)13.07* | 8.03 | 6.54 | -4.08 |
| P | 4.12 | 5.45* | 1.98 | -2.98 | (-)12.95* | 8.12 | 3.36 | 3.57* |
| Y | 5.04* | 10.95* | 4.56 | 3.95 | 2.02 | -2.05 | 2.49 | 4.03* |
| Binpur-II | | | | | | | | |
| A | -1 | 0.29 | -1.08 | -3.58 | (-)12.85* | 7.8* | 6.68 | -3.42 |
| P | 2.98 | 5.68* | 2.42 | -3.98 | (-)11.95* | 7.7* | -3.26 | 2.25 |
| Y | 4.62* | 11.00* | 4.69 | 2.68 | 3.03 | -3 | 2.85 | 4.68* |
| Jamboni | | | | | | | | |
| A | -1.34 | 0.04 | 1 | -4.43 | (-)15.02* | 5.95* | 5.85* | -4.46 |
| P | 1.98 | 5.98* | 3.43 | -4.49 | (-)14.49 | 6.08 | -2.98 | 3.35* |
| Y | 3.56 | 11.42* | 5.06* | 2.05 | 1 | -2.29 | 4.8 | 4.08 |
| Nayagram | | | | | | | | |
| A | -1.26 | 0.34 | -1.08 | NA | (-)16.50* | 8.04 | 11.60* | 5.34 |
| P | 2.56 | 6.34** | 3.05 | NA | (-)16.52* | 8.12* | 6.43 | 5.12 |
| Y | 4.08 | 11.60* | 6.25 | NA | 2.58 | -2.26 | -2.09 | 11.42* |

*Contd...*

| Block | Aus Paddy | Aman Paddy | Boro Paddy | Jute | Wheat | Potato | Til | Mustard |
|---|---|---|---|---|---|---|---|---|
| Sankrail | | | | | | | | |
| A | -0.92 | 0.32 | -1.2 | -4.84 | (-) 18.43* | 8.64 | 12.6* | 5.56 |
| P | 3.34 | 6.3* | 3.67 | -4.8 | (-) 17.75* | 8.26* | 6.63* | 5.45* |
| Y | 5.08 | 11.49* | 6.58* | 2.85 | 2.85 | -3.58 | -4.05 | 12.09* |
| Gopiballabpur-I | | | | | | | | |
| A | 1.14 | 0.35 | -1 | NA | (-) 16.62* | 7.78 | 12.05* | 4.85 |
| P | 3.56 | 6.30** | 3.67 | NA | (-) 16.83* | 7.34* | 7.53 | 5.02** |
| Y | 5.49 | 10.34* | 6.64 | NA | 2.32 | -4.46 | 2.8 | 12.68* |
| Gopiballabpur-II | | | | | | | | |
| A | -1.08 | 0.36 | -1.08 | NA | (-) 16.60* | 7.72 | 12.09* | 4.86 |
| P | 2.98 | 5.85** | 2.76 | NA | (-) 16.75* | 7.64 | 6.98* | 4.92** |
| Y | 3.86* | 10.68* | 6.62* | NA | 3 | -4.49 | -2.9 | 11.56* |
| Debra | | | | | | | | |
| A | -1 | 0.3 | -3.04 | 5.12 | (-) 17.35* | 12.74* | (-)59.99* | 20.92** |
| P | 3.04* | 5.12** | 2.02 | 5.12 | (-) 16.18* | 13.88* | -12.51 | 27.12** |
| Y | 4.08* | 12.74* | 5.12* | -1 | 0.4 | (-) 22.14* | -3.04 | 5.12 |
| Pingla | | | | | | | | |
| A | 18.53* | 0.3 | 2.02 | -4.08 | (-) 39.09* | -8.32 | 18.53* | (-) 5.12* |
| P | 23.36* | 6.18* | 6.18* | 24.66** | (-) 34.98* | (-) 13.88* | 5.12 | -2.02 |
| Y | 5.12* | 10.51* | 5.12* | 11.62* | 3.04 | 6.18 | -8.32 | 3.04 |
| Keshiary | | | | | | | | |
| A | 16.76* | 0.39 | 2.53 | NA | (-)31.04* | -8.3 | 17.52* | -6.08 |
| P | 18.08* | 6.26* | 6.07 | NA | (-)30.56* | (-)10.80* | 5.42 | -3.56 |
| Y | 6.34* | 11.34* | 6.02 | NA | 2.05 | -7.45 | -8.3 | 2.78 |

*Contd...*

| Block | Aus Paddy | Aman Paddy | Boro Paddy | Jute | Wheat | Potato | Til | Mustard |
|---|---|---|---|---|---|---|---|---|
| Dantan-I | | | | | | | | |
| A | 16.50* | 0.38 | 2.86 | -4 | (-) 34.06* | -8.86 | 18.03* | -6.02 |
| P | 17.82* | 6.32 | 6.12 | 22.60* | (-) 33.85* | -10.56 | 9.06 | -3.52 |
| Y | 7.43* | 12.02* | 6.45 | 12.4 | 2.98 | -7.77 | -3.04 | 3.06 |
| Dantan-II | | | | | | | | |
| A | 6.66* | 0.73 | 1.46 | NA | 5.37 | 12.74* | 5.34 | -3.42 |
| P | 7.43* | 7.34 | 12 | NA | 15.14* | 13.88* | -4.46 | 3.57 |
| Y | 10.42* | 10.56* | 23.48* | NA | 22.82* | (-) 22.14* | 4.08 | -4.08 |
| Narayangarh | | | | | | | | |
| A | 19.71* | 0.3 | -2.02 | -4.08 | 18.84* | -8.32 | 18.53* | -5.14 |
| P | 76.56** | 6.18 | 6.18 | 24.6* | 4.2 | (-) 13.88* | 5.16 | -2.06 |
| Y | 7.42* | 11.34* | 6.12 | 11.62** | 9.11* | -6.18 | -8.3 | 3.07 |
| Mohanpur | | | | | | | | |
| A | 18.53* | 0.3 | -3.04 | NA | (-) 17.35* | 12.74* | (-)59.99* | 20.92* |
| P | 23.36* | 5.12 | 2.02 | NA | 16.18 | 13.88* | -12.25 | 27.12** |
| Y | 5.12* | 12.74* | 5.12 | NA | 0.4 | (-) 22.14* | -3.04 | 5.12 |
| Salboni | | | | | | | | |
| A | -1.04 | 0.08 | 1 | NA | (-) 23.36* | 19.72* | 30.92* | -5.12 |
| P | 3.04 | 4.08 | 4.08 | NA | (-) 22.14* | 33.31* | 22.14* | -2.02 |
| Y | 4.08 | 12.74 | 4.08 | NA | 1 | (-) 17.35* | -8.34 | 3.04 |
| Keshpur | | | | | | | | |
| A | 3.88* | 0.3 | 2.86 | -4.07 | 2.8 | 2.35 | 1.57 | 2.08 |
| P | 26.16** | 6.18 | 6.12 | 23.68* | 8.95* | 7.78* | 5.12 | 18.56* |
| Y | 75.33** | 10.51* | 6.45 | 10.62* | 67.16** | 39.5* | -8.32 | (-) 12.36* |

*Contd...*

| Block | Aus Paddy | Aman Paddy | Boro Paddy | Jute | Wheat | Potato | Til | Mustard |
|---|---|---|---|---|---|---|---|---|
| Garbeta-I | | | | | | | | |
| A | 10.56* | 1.46 | 3.88 | 5.48 | 9.11 | 1.54 | -8.4 | 5.4 |
| P | 28.91** | 12.00* | 26.15* | 90.07* | 1.91 | 5.56 | -3.5 | 1.57 |
| Y | 89.82** | 64.59* | 74.95* | 5.14 | 14.71* | 71.69** | -1.73 | 18.56* |
| Garbeta-II | | | | | | | | |
| A | 6.66 | 5.5 | 1 | 1.42 | 2.8 | 1.65 | 1.79 | 2.08 |
| P | 12.00* | 19.71* | 4.08 | 23.17* | 8.95 | 57.42** | 1.96 | 8.12* |
| Y | 23.48* | 76.56* | 4.08 | 2.22 | 67.16* | 39.63* | -8.27 | (-)12.36* |
| Garbeta-III | | | | | | | | |
| A | 6.58 | 5.48 | 1.05 | 1.45 | 2.79 | 1.59 | NA | 2.01 |
| P | 11.95* | 19.76* | 4.06 | 22.98* | 8.96 | 26.89* | NA | 7.84 |
| Y | 23.40* | 25.85* | 4.04 | 2.08 | 22.34* | 38.53* | NA | (-)12.28* |
| Midnapur | | | | | | | | |
| A | 10.56 | 3.88 | 5.5 | NA | 9.12* | 1.39 | NA | 2.9 |
| P | 28.91* | 26.16* | 19.72* | NA | 1.9 | 4.35 | NA | 1.02 |
| Y | 29.82* | 25.32* | 75.54** | NA | 13.98* | 26.68* | NA | 16.18* |
| Sabang | | | | | | | | |
| A | 12.74* | 0.08 | 1 | 1 | -23.36 | 19.72* | 30.98* | 6.18* |
| P | 19.72* | 4.08 | 4.08 | -4.08 | -22.14 | 32.31* | 22.14* | 8.32* |
| Y | 5.12* | 12.74* | 4.08 | 5.12 | 1 | -17.35 | -8.32 | 16.18* |
| Kharagpur-I | | | | | | | | |
| A | 5.64 | 4.4 | 1.06 | NA | 2.8 | 11.28* | 29.08* | 1.24 |
| P | 23.00* | 20.3* | 21.85* | NA | 8.95 | 27.34* | 21.04* | 2.71 |
| Y | 32.7* | 20.02*. | 32.70* | NA | 27.16* | 23.09* | -7.98 | 2.86 |

Contd...

| Block | Aus Paddy | Aman Paddy | Boro Paddy | Jute | Wheat | Potato | Til | Mustard |
|---|---|---|---|---|---|---|---|---|
| KHaragpur-II | | | | | | | | |
| A | (-)3.07* | 5.48 | 1.05 | 3.54 | 9.17* | 9.8* | 1.53 | 6.17* |
| P | 5.47* | 20* | 4.06 | 55.11* | 3.37 | 3.44 | 3.03 | 8.3 |
| Y | 13.6* | 25.85* | 4.05 | 1.21 | 20.77* | 9.4* | 65.86** | 16.17* |
| Chandrokona-I | | | | | | | | |
| A | 19.70* | 0.07 | 1.03 | 1 | (-)23.50* | 19.76* | 30.96* | 6.02 |
| P | 5.13* | 4.07 | 4.07 | -4.07 | (-)22.16* | 32.30* | 22.14 | 8.03 |
| Y | 12.72* | 12.75 | 4.07 | 5.15 | 1 | (-)17.36* | -8.34 | 16.81* |
| Chandrokona-II | | | | | | | | |
| A | 2.99** | -3.07 | 1.56 | -0.77 | 5.89* | 4.4 | 1.79 | 8.2* |
| P | 8.46 | 5.47 | 21.80* | (-)10.75* | 1.16 | 10.78* | 1.96 | 2.05 |
| Y | -5.2 | 13.68* | 33.30* | -1.68 | -17.63 | 49.1* | -7.45 | 26.58* |
| Ghatal | | | | | | | | |
| A | (-)2.13* | 2.18 | 1.04 | 3.6 | 2.5 | 3.44 | 1.68 | 1.27 |
| P | (-)2.27* | 18.97* | 4.08 | 6.92 | 4.2 | 7.26* | 1.92 | 1.41 |
| Y | (-)10.33* | 25.2* | 4.09 | 4.3 | 44.12* | 7.91 | 6.1 | 17.88* |
| Daspur-I | | | | | | | | |
| A | 2.18 | (-)12.07* | 3.45 | 2.33 | 5.1 | 6.18* | 1.68 | -2.33 |
| P | 18.97* | 5.85 | 9.59 | 42.72* | 10.1* | 16.95* | 1.92 | -2.27 |
| Y | 25.2* | 55.76* | 3.12 | 3.7 | 50.94* | 35.7* | -8.43 | (-)10.33 |
| Daspur-II | | | | | | | | |
| A | 2.16 | (-)11.98* | 3.43 | 2.31 | 5.16 | 6.17 | 1.62 | -2.36 |
| P | 17.99* | 5.8 | 9.62 | 42.70* | 10.14* | 16.80* | 1.83 | -2.39 |
| Y | 24.25* | 50.02** | 3.15 | 3.75 | 50.82* | 35.72* | -8.42 | -9.99 |

Contd...

| *Block* | *Aus Paddy* | *Aman Paddy* | *Boro Paddy* | *Jute* | *Wheat* | *Potato* | *Til* | *Mustard* |
|---|---|---|---|---|---|---|---|---|
| Tamluk-I (Tamluk) | | | | | | | | |
| A | (-)1.23** | 0.08 | 1.06 | 1.06 | 2.15 | 3.66 | NA | -4.2 |
| P | -1.38 | 4.08* | 4.05* | -4.05 | 4.92 | 9.2* | NA | -3.5 |
| Y | -0.86 | 12.74** | 4.05 | 5.13* | 0.54 | 6.18* | NA | (-)14.8* |
| Tamluk-II (Sahid Matangini) | | | | | | | | |
| A | -1.61 | 0.06 | 1.08 | 1.08 | 1.16 | 3.68 | NA | NA |
| P | 17.78** | 4.07* | 4.07* | -4.07 | 4.99* | 9.23* | NA | NA |
| Y | 11.31* | 11.94** | 4.07 | 5.16 | 0.55 | 6.16 | NA | NA |
| Panskura-I | | | | | | | | |
| A | (-)3.26* | 1.52 | 2.7 | 2.33 | 8.6* | 9.2* | 1.63 | 2.42 |
| P | 8.09* | 3.78 | 5.53 | 42.72** | 1.95 | 31.33* | 1.92 | 1.36 |
| Y | 28.55** | 9.5* | 6.23 | 3.78 | 19.23* | (-)17.41* | -6.42 | -1.79 |
| Panskura-II | | | | | | | | |
| A | (-)3.3* | 1.56 | 2.78 | 2.32 | 8.72* | 9.02* | 1.62 | 2.4 |
| P | 7.99* | 3.72 | 5.03 | 42.71* | 1.98 | 30 | 1.9 | 1.35 |
| Y | 27.95** | 10.02* | 6.46 | 3.77 | 19.20* | (-)16.52* | -6.44 | -1.72 |
| Moyna | | | | | | | | |
| A | 0.22 | 0.25 | 0.19 | -0.05 | 0.01 | 4.08 | 1.24 | (-)7.9* |
| P | 6.17 | 6.11 | 6 | 0.28 | -0.19 | 0.02 | 2.71 | 0.3 |
| Y | (-)34.09* | 18.98* | 19.30* | 2.6 | (-)61.3* | 8.14* | 2.86 | 2.6 |

*Contd...*

| Block | Aus Paddy | Aman Paddy | Boro Paddy | Jute | Wheat | Potato | Til | Mustard |
|---|---|---|---|---|---|---|---|---|
| Mahisadal-I (Nandakumar) | | | | | | | | |
| A | 0.25 | 1.56 | 1.46 | 6.5 | 1.49 | 9.5* | NA | -0.03 |
| P | 5.34 | 9.14* | 9.32* | 21.56* | 3.15 | 28.12* | NA | -0.16 |
| Y | 35.10* | 24.47* | 25.57* | 3.8 | 29.9* | 11.46* | NA | (-)18.33* |
| Mahisadal-II (Mahisadal) | | | | | | | | |
| A | 0.23 | 1.48 | 1.47 | 6.56* | 1.52 | 9.26* | 0.07 | -0.04 |
| P | 4.98 | 9.16* | 9.85* | 22.55* | 4.2 | 27.10* | -0.17 | -0.15 |
| Y | 34.09** | 23.98* | 25.67* | 3.89 | 30.25* | 10.82* | (-)12.30* | (-)17.85* |
| Nandigram-I | | | | | | | | |
| A | (-)1.36* | 2.23 | -0.03 | -0.05 | 0.01 | 0.02 | NA | (-)7.9* |
| P | (-)2.06** | 10.94* | 16.93* | 0.28 | -0.19 | 4.08 | NA | 0.3 |
| Y | (-)50.67* | (-)80.64* | -1.92 | 2.6 | -6.13 | 8.14* | NA | (-)12.82* |
| Nandigram-II | | | | | | | | |
| A | -0.25 | 1.56 | 3.9 | 6.5 | 1.49 | 0.95 | NA | -0.03 |
| P | 5.34 | 9.14* | 0.12 | 21.56* | 3.15 | 28.12* | NA | -0.16 |
| Y | 35.1* | 24.47** | (-)36.55* | 3.8 | 29.9* | 11.46* | NA | (-)18.33* |
| Nandigram-III (Chandipur) | | | | | | | | |
| A | (-)1.32* | 2.26 | -0.02 | -0.07 | 0.05 | 0.06 | NA | -7.9 |
| P | -2.05 | 9.98* | 15.92* | 0.3 | -0.17 | 4.07* | NA | 0.4 |
| Y | (-)48.60* | (-)78.65* | -1.96 | 2.7 | (-)58.68* | 8.16 | NA | (-)11.98* |

Contd...

| Block | Aus Paddy | Aman Paddy | Boro Paddy | Jute | Wheat | Potato | Til | Mustard |
|---|---|---|---|---|---|---|---|---|
| Sutahata-I (Sutahata) | | | | | | | | |
| A | 0.22 | 0.25 | 0.02 | 0.47 | -0.04 | 1.38 | -0.76 | -0.03 |
| P | 6.17 | 6.11* | 0.02 | 15.97* | -0.1 | 46.46* | -1.31 | -0.27 |
| Y | (-)34.09* | 18.98* | -1.07 | 2.82 | (-)13.72* | 13.56* | (-)20.07* | (-)25.27* |
| Sutahata-II (Haldia) | | | | | | | | |
| A | 0.23 | 0.23 | 0.03 | 0.48 | -0.05 | 2 | NA | -0.06 |
| P | 6.18 | 6.13* | 0.03 | 15.87* | -0.13 | 46.34* | NA | -0.26 |
| Y | (-)32.05* | 17.99* | -1.08 | 2.72 | (-)13.72* | 13.56* | NA | (-)25.26* |
| Potashpur-I | | | | | | | | |
| A | 9.73* | 10.79* | 5.93 | 5.2 | 2.06 | 1.2 | -1.11 | -0.62 |
| P | 40.95* | 41.75* | 4.44 | 14.16* | 41.29** | 38.92* | 0.61 | -0.61 |
| Y | 25.67* | 20.2* | 8.57 | 3.1 | 20.44* | 23.5* | 24.52* | 23.59* |
| Potashpur-II | | | | | | | | |
| A | 9.23* | 9.85* | 5.89* | 5.28* | 2.08 | 1.29 | -1.15 | -0.65 |
| P | 39.02* | 40.05* | 4.46 | 14.18* | 41.32* | 34.99* | 0.56 | -0.6 |
| Y | 24.98* | 20.26* | 8.36 | 3.18 | 20.46* | 23.56* | 25.02* | 23.68* |
| Khejuri-I | | | | | | | | |
| A | 3.33 | 5.8 | -8.03 | 0.06 | 1.74 | 7.93* | NA | -0.75 |
| P | 31.62* | 39.07* | 38.07* | 0.05 | 38.56 | 15.30* | NA | -0.32 |
| Y | 57.60** | 61.15* | 65.53* | 3.37 | 6.54* | 21.82* | NA | 12.95 |

Contd...

| Block | Aus Paddy | Aman Paddy | Boro Paddy | Jute | Wheat | Potato | Til | Mustard |
|---|---|---|---|---|---|---|---|---|
| Khejuri-II | | | | | | | | |
| A | 3.35 | 5.86 | -8.05 | 0.08 | 1.79 | 7.88* | NA | -0.78 |
| P | 32.06* | 40.01* | 39.06* | 0.07 | 39.57* | 15.26* | NA | -0.36 |
| Y | 58.07* | 60.16* | 66.55* | 3.39 | 61.55* | 22.83* | NA | 12.96* |
| Bhagawanpur-I | | | | | | | | |
| A | -0.49 | -0.63 | -6.27 | -0.01 | -0.01 | 0.75 | NA | -0.75 |
| P | 19.35* | 19.37* | 4.18 | -0.24 | 0.28 | 19.16* | NA | -0.32 |
| Y | 46.05** | 49.37* | 46.6* | 3.7 | 27.00* | (-)42.25* | NA | 18.95* |
| Bhagawanpur-II | | | | | | | | |
| A | -0.48 | -0.65 | -6.29 | -0.05 | 0.02 | 0.78 | -2.05 | -0.78 |
| P | 18.40* | 20.33* | 4.12 | -0.26 | 0.25 | 20.21* | 0.75 | -0.35 |
| Y | 47.02* | 48.35* | 46.60* | 3.79 | 27.56* | (-)45.30* | 0.52 | 19.48* |
| Ramnagar-I | | | | | | | | |
| A | 1.46 | 5.64 | 4.04 | 5.48 | 5.37 | 2.35 | 0.73 | 0.54 |
| P | 11 | 28.00* | 26.98* | 90.00* | 15.14* | 7.78* | 1.8 | 1.57 |
| Y | 64.59* | 32.76* | 37.95* | 6.14 | 22.82* | 39.50* | 17.92* | 18.56* |
| Ramnagar-II | | | | | | | | |
| A | 2.02 | 6.02 | 4.02 | 5.69* | 5.41* | 2.76 | 0.78 | 1.06 |
| P | 10.5 | 27.05* | 25.35* | 89.23* | 15.18* | 7.88* | 2.01 | 2.01 |
| Y | 63.52* | 33.02* | 36.58* | 6.18 | 21.03* | 40.00* | 18.06* | 18.30* |
| Egra-I | | | | | | | | |
| A | 0.05 | 1.06 | 1.04 | -0.09 | 9.12* | 1.72 | 0.75 | -0.84 |
| P | 19.71* | 21.85* | 21.72* | 0.15 | 1.92 | 57.42* | 1.12 | -0.35 |
| Y | 76.56** | 32.78* | 49.34* | 0.68 | 14.73* | 39.63* | 8.98* | -0.17 |

*Contd...*

| *Block* | *Aus Paddy* | *Aman Paddy* | *Boro Paddy* | *Jute* | *Wheat* | *Potato* | *Til* | *Mustard* |
|---|---|---|---|---|---|---|---|---|
| Egra-II | | | | | | | | |
| A | 0.07 | 1.05 | 1.08 | -0.12 | 10.03* | 1.93 | 0.72 | -0.88 |
| P | 20.72* | 22.02* | 22.03* | 0.15 | 2.06 | 56.08* | 2.02 | -0.36 |
| Y | 75.32** | 33.03* | 50.36* | 0.69 | 1.05 | 39.06* | 8.56 | -0.18 |
| Contai-I | | | | | | | | |
| A | -1.02 | 0.35 | -3.08 | 5.16 | (-)17.36* | 12.78* | (-)60.03* | 21.07* |
| P | 3.04 | 5.13 | 2.02 | 5.16 | (-)16.12* | 13.28* | (-)12.28* | 26.05* |
| Y | 4.05 | 12.76* | 5.12 | -1.02 | 0.46 | (-)21.02* | -3.02 | 5.16 |
| Contai-II | | | | | | | | |
| A | -1.12 | 0.32 | -3.05 | 5.18 | (-)17.35* | 13.02* | (-)61.05* | 22.05* |
| P | 3.14 | 5.16 | 2.06 | 5.18 | (-)16.13* | 13.26* | (-)12.22* | 26.09** |
| Y | 4.05* | 12.72* | 5.13 | -1.01 | 0.48 | (-)22.05 | -3.03 | 5.19 |
| Contai-III | | | | | | | | |
| A | -1.01 | 1.32 | -4.09 | 5.06 | (-)17.32* | 13.12* | (-)60.00* | 21.09* |
| P | 3.04 | 5.14 | 2.12 | 5.06 | (-)16.00* | 13.02* | (-)12.02* | 26.08* |
| Y | 4.15* | 12.73* | 5.23 | -1.12 | 1.2 | (-)21.08* | -3.13 | 5.02 |

*Notes:* *, **, indicate significant at 1% & 5% level respectively.
*Source:* District Statistical Handbook, Midnapur(Various Issues).

### A Summing-Up

In India as a whole the growth rate was appreciably high during the sixties, which slowed down during the seventies, to revive appreciably during the eighties and declined again in the early nineties. The said growth rate has been significant for most of the sample states and for India as a whole. The noticeable feature of the nineties was that the agricultural growth penetrated to all the regions of India. In all the districts of W.B. also, agricultural growth has been statistically significant during the eighties and early nineties. There has also been acceleration in agricultural growth in the districts of Midnapore, Bankura, Purulia, Hooghly, 24-Paraganas and West Dinajpur. The significant agricultural growth has been observed in the blocks of Midnapore.

# CHAPTER–5

# IMPORTANCE OF CROPPING PATTERN IN AGRICULTURAL GROWTH IN INDIA: STATE AND DISTRICT LEVEL ANALYSIS

In the previous chapter we have observed that there has occurred an impressive growth in agricultural production in India , sample states, and districts of W.B. In this perspective, the question that may arise is: what has been the importance of CP in agricultural growth in India, sample states, and districts? In this chapter an attempt has been made to examine the impact of change in CP on agricultural growth. Before a detailed discussion on change in CP and factors influencing it in India, there is a need to examine the impact of change in CP on the agricultural growth. The objectives of the study are (i) to examine the relative role of different components including CP in the growth of agricultural production in India by states and W.B. by districts, (ii) to examine the role of commercialization of agriculture through change in CP on the agricultural growth of India by states and W.B. by districts.

## Change in CP and Agricultural Growth: State Level Analysis

In this section we shall examine first the relative role of change in CP in agricultural growth and then the impact of change in CP in favour of commercialisation of agriculture on the agricultural growth in all India and sample states.

## Relative Role of CP Change in Agricultural Growth

Let us examine the relative influence of CP change on agricultural growth as reflected in the change in agricultural production in India and sample states. The estimates of regression analysis are given in Table 1. The change in CP has no statistically significant impact on agricultural growth in India over the longer period of 1972-73 to 1996-97. But, interestingly, the CP made statistically significant contribution to the agricultural growth in the sub-period I, while its contribution in the sub-period II was insignificant. To examine the stability of coefficients of two sub-periods in regression equations, the Chow test has been applied and the estimated F = 104.14. Therefore, the hypothesis that there is no difference in the impact of CP on agricultural growth in the country between the two sub-periods is not accepted at 5 percent probability level. This means that while the CP had significant influence on agricultural production in the seventies, it had no significant influence in the eighties as well as in the early 1990s for a country like India.

The change in CP has no statistically significant impact on agricultural growth in Punjab. Here the Chow F is estimated to be 9.20 (so the null hypothesis (Ho) is rejected at 1% level of significance) . It is observed that the change in CP has greater influence on agricultural production in sub-period I as compared to sub-period II. The yield component has also statistically significant effect on production in whole period and sub-periods. The phenomenon that there was no significant influence of CP on agricultural production in either of the sub-periods or in the whole period in Punjab can be explained by the fact that Punjab produces mainly wheat. It occupies major portion of the total cultivable land, thus leaving little scope for any other crops to encroach upon the wheat land. Wheat being a HYV crop and at the same time highly remunerative, the farmers hardly think about opting for alternative crops. But in case of Haryana there took place change in CP in favour of non-foodgrains during the sub-periods and that had a significant effect on agricultural production. The Chow test estimate of F = 4.52 shows that there

**Table 1**

**Estimates of Multiple Regression Equations concerning Agricultural Production and Cropping Pattern in the Sample States in India**

| | *Whole Period* | | | | | *Sub-Period I* | | | | | *Sub-Period II* | | | | |
|---|---|---|---|---|---|---|---|---|---|---|---|---|---|---|---|
| | *Explanatory Variables* | | | | | *Explanatory Variables* | | | | | *Explanatory Variables* | | | | |
| | ***Constant*** | *A* | *Y* | *CPI* | $R^2$ | ***Constant*** | *A* | *Y* | *CPI* | $R^2$ | ***Constant*** | *A* | *Y* | *CPI* | $R^2$ |
| Andhra Pradesh (A.P.) | -65.04 | -0.02 | 0.7** | 1.2 | 0.45 | -169.75 | -0.076 | 0.61*** | 2.15 | 0.25 | 46.54 | 0.53 | 1.39** | -1.37 | 0.91 |
| | (-0.48) | (-0.12) | .(3.42) | .(0.83) | | (-0.66) | (-0.38) | .(1.37) | .(0.99) | | .(0.23) | .(1.06) | .(8.32) | (-0.9) | |
| Haryana | 17.02 | 0.07* | 0.96* | -0.12 | 0.99 | -169.37 | 1.07** | 0.69** | 0.87*** | 0.98 | -12.12 | 0.012 | 0.99** | 0.39* | 0.99 |
| | (0.53) | .(1.91) | .(22.91) | (-0.36) | | (-2.27) | .(3.13) | .(5.76) | .(1.55) | | (-2.21) | .(0.77) | .(41.83) | .(2.02) | |
| Rajastan | -365.5 | 3.08* | 0.89* | 1.13 | 0.49 | -68.01 | 0.58 | 2.28** | -1.47 | 0.79 | -328.64 | 3.51* | 0.07 | -1.67 | 0.37 |
| | (-2) | .(2.06) | .(1.83) | .(1.15) | | (-0.2) | .(0.16) | .(3.57) | (-1.48) | | .(0.97) | .(1.88) | .(0.05) | .(0.79) | |
| Madhya Pradesh (M.P.) | -189.46 | 1.97** | 0.92** | -0.02 | 0.96 | -109.22 | 1** | 1.09** | 0.002 | 0.99 | -189.5 | -2.35 | 0.48 | 0.101 | 0.83 |
| | (-4.36) | .(3.89) | .(9.53) | (-0.29) | | (-7.89) | .(6.11) | .(31.56) | .(0.097) | | (-1.9) | (-2.42) | .(1) | .(10.45) | |
| Orissa | 378.46 | 1.16 | 0.75 | -4.46 | 0.5 | -536.41 | 1.35** | 1.04** | 3.95* | 0.99 | -2019.41 | -7.87 | 0.61 | 11.33* | 0.55 |
| | .(0.62) | .(0.74) | .(1.35) | (-0.92) | | (-2.06) | .(2.62) | .(6.67) | .(1.81) | | (-2.88) | (-2.7) | .(0.22) | (-2.35) | |
| Tamil Nadu | -12.41 | 1.06** | 0.36* | -0.097 | 0.78 | 165.31*** | 1.42** | 8.7** | *-2.6* | 0.88 | -34.87 | *-0.64*** | 0.2 | 1.74** | 0.87 |
| | (-0.58) | .(4.3) | .(2.46) | (-0.45) | | .(1.37) | .(5.58) | .(3.32) | (-1.84) | | (-0.65) | (-1.7) | .(0.97) | .(2.71) | |
| Punjab | -88.79 | 0.95* | 0.63** | 0.35 | 0.98 | -250.62 | 1.97** | 0.21*** | 1.35 | 0.97 | -14.73 | -0.45 | 1.27** | -0.27 | 0.99 |
| | (-1.38) | .(1.76) | .(6.23) | .(0.5) | | (-2.24) | .(3.43) | .(1.58) | .(0.93) | | (-0.46) | (-0.83) | .(12.17) | (-0.45) | |
| Gujarat | 20.75 | -0.05 | 0.82** | 0.04 | 0.84 | 3.16 | 0.28 | 0.70** | 0.02 | 0.85 | -14.69 | 0.37 | 0.77** | 0.005 | 0.97 |
| | .(1.40) | (-0.17) | .(6.65) | .(0.21) | | .(0.1) | .(0.47) | .(3.05) | .(0.07) | | (-1.46) | .(1.13) | .(2.83) | .(0.04) | |

| | Whole Period | | | | | Sub-Period I | | | | | Sub-Period II | | | | |
|---|---|---|---|---|---|---|---|---|---|---|---|---|---|---|---|
| | Explanatory Variables | | | | | Explanatory Variables | | | | | Explanatory Variables | | | | |
| | **Constant** | *A* | *Y* | *CPI* | $R^2$ | **Constant** | *A* | *Y* | *CPI* | $R^2$ | **Constant** | *A* | *Y* | *CPI* | $R^2$ |
| Uttar | -45.94 | 0.03 | 1.14** | 0.23 | 0.81 | -190.82 | 0.22 | 1.52*** | 1.23 | 0.65 | 45.35** | 0.70** | 0.11* | 0.005 | 0.96 |
| Pradesh (U.P.) | (-1.52) | .(0.26) | .(7.1) | .(0.73) | | (-2.27) | .(0.14) | .(1.98) | .(1.55) | | .(5.46) | .(3.67) | .(10.83) | .(0.23) | |
| Himachal | -264.24 | -2.46 | 0.66* | 0.27 | 0.29 | 225.67 | -3.84 | 1.21** | 1.49 | 0.6 | 33.61 | -0.42 | -0.46 | 1.73* | 0.6 |
| Pradesh (H.P.) | (-1.18) | (-1.22) | .(2.14) | .(0.13) | | .(0.54) | (-0.79) | .(2.52) | .(0.32) | | .(0.24) | (-0.38) | (-2.86) | .(2.32) | |
| West Bengal | -151.39 | 1.22** | 1.09** | 0.16 | 0.87 | -111.97 | 1.04** | 1.03** | 0.05 | 0.91 | -155.63 | 1.44** | 1.11*** | 0.35 | 0.95 |
| (W.B.) | (-5.75) | .(6.45) | .(25.64) | .(0.82) | | (-21.37) | .(41.88) | .(90.36) | .(1.21) | | (-21.62)* | .(18.16) | .(76.1) | .(0.9) | |
| Maharashtra | -476.79 | 5.9** | 0.49 | 0.042 | 0.8 | -41.85 | -0.77 | 1.70** | 0.72* | 0.95 | -1137.73 | 11.3* | 1.09* | -0.62 | 0.87 |
| | (-429) | .(4.26) | .(1.13) | .(0.069) | | (-0.38) | (-0.47) | .(3.8) | .(1.8) | | (-5.33)* | .(4.8) | .(2.11) | (-0.69) | |
| All India | -214.45 | 2.56* | 0.60* | 0.064 | 0.86 | -231.76 | 1** | 1.11** | 1.101* | 0.99 | 94.75 | -0.15 | 1.22** | 0.58 | 0.69 |
| | (-2.005) | .(3.61) | .(8.55) | .(0.082) | | (-14.14) | .(11.56) | .(27.28) | .(9.69) | | .(1.7) | (-0.23) | .(2.67) | .(1.08) | |

*Notes:* (i) A = index number of area (all crops combined),
(ii) Y = index number of yield or productivity (all crops combined),
(iii) CPI = Cropping pattern index,
(iv) Dependentvariable = index of production (all crops combined),
(v) Figures in brackets indicate t-ratios.
(vi) *,**,*** indicate coefficients significant at 1%, 5% & 10% level respectively.

*Source:* (i) Area & Production of Principal crops in India , Govt. of India
(ii) Fertiliser statistics, Govt. of India.

are statistically significant differences between the regression coefficients for two sub-periods at 5% probability level.

The change in CP has a statistically significant impact on agricultural production in sub-period II as compared to both the sub-period I and whole period in Tamil Nadu. Here F = 4.80 (estimated by Chow test) which indicates the differential influences of area, yield and CP exist at 5% level of significance. The yield and area components have relatively more significant impact on production in the whole and the first sub-period.

The change in CP has no significant impact on the agricultural production in M.P. The Chow test (F = 0.80) further reveals that there are no statistically significant differences in the influences of area, yield and CP on agricultural production between the two sub-periods. However, for the whole period and the first sub-period the impacts of area and yield were quite high and statistically significant at 5% level of significance. In the second sub-period, only area is seen to have significant effect on agricultural production. Neither yield nor CP is seen to have any significant effect during this period.

The regression results in case of Orissa show the statistically significant influence of change in CP on agricultural production in the sub-periods only. The Chow F = 17.60. This means that there exist statistically significant differences in the influences of area, yield and CP on agricultural production between the two sub-periods. The yield component has good effect on the production.

In A.P., the effect of yield is found to be statistically significant in the whole period as well as in both the sub-periods at 5% probability level. The effect of CP is not significant in any period, even the effect of area is also not significant. This may be explained by the fact that in A.P. during the period under study, there has not been any significant change in the CP, which is the explanatory variable in the model. The Chow test applied to test the statistical significance of the differences in the effect of area, yield and CP between two sub-periods shows that there is no significant difference (Chow F = 1.22).

In Maharastra, while area played a significant role during the whole period and the sub-period II , yield played a significant role in both the sub-periods, as well as during the entire period. The change in CP has statistically significant impact on agricultural growth in sub-period II only. The Chow test (F = 7.67) shows that there has been significant difference in the effects of area, yield and CP during the sub-period I and sub-period II. The change in CP has played a relatively greater role in sub-period I (as compared to sub-period II) in explaining agricultural growth in this state.

The change in CP has a statistically significant impact on agricultural production in the sub period II in H.P. The Chow test indicates that there are stability of coefficients (F = 2.82) between the two sub-periods. The yield component has relatively better effect on production.

The change in CP has no statistically significant effect on agricultural production in W.B. But the yield and area components have significant impacts. The Chow test ( F= 9.28) implies the significant difference in the effects of area, yield and CP taken together between two sub-periods.

The estimated values of the regression coefficients indicate that there is no statistically significant impact of CP on agricultural production in Rajasthan. Area is found to have significant effect upon agricultural production during the whole period, and in the second sub-period. Yield has significant effect upon agricultural production during whole period as well as in the sub-period I. The Chow test (F = 1.53) shows no statistically significant difference in the effects of the variables on agricultural production between two sub-periods.

In Gujarat yield played a significant role in the increase in agricultural production. No other factors such as area and CP had any significant impact on agricultural production. The estimated F = 0.40 (Chow test) indicates that there is no significant difference in the effects of area, yield and CP taken together on agricultural production between two sub-periods.

In U.P. yield played a significant role in both whole period and sub-period I in increasing agricultural production. CP did not have any impact on production in either of the periods. In the sub-period II the area was the only factor that significantly contributed to the growth of agricultural production. The Chow test implies that the relationship is stable (F = 0.005) in the two sub-periods, which means that there has been no significant difference in the effects of area, yield and CP on agricultural production between the two sub-periods.

From the above analysis it is clear that the relative contribution of CP change on agricultural growth is statistically significant for both the developed states, viz, Haryana, Tamilnadu and underdeveloped states, viz, Orissa, H.P. in different sub-periods. The differential influences of CP change on agricultural growth between two sub-periods have also been observed for India & most of the sample states.

## Impact of Change in CP in Favour of Commercialisation of Agriculture on Agricultural Growth

It is seen from the regression results given in Table 2 that the commercialisation of agriculture through change in CP as measured by the percentage of area under HYV in GCA (HYVA) had significant influence upon the agricultural production during 1972-73 to 1996-97 in India and the states like A. P., Rajasthan, Haryana, Orissa, U.P., Maharashtra. On the other hand, the impact of commercialization of agriculture through change in CP as measured by the change in the percent of area under non-food grains to GCA(NFA) on agricultural production (Table 3) shows that for as many as eight states ,namely, Haryana, Rajasthan, M.P., Tamil Nadu, Gujarat, H.P., W.B. and Maharashtra the relationships are not statistically significant and for two states , viz., A.P. and U.P. positive and significant; for both Punjab and Orissa it was found to be negative and significant . Thus, for Punjab the result indicates that as NFA decreases, total agricultural production (P) increases. In other words, as FA ( i.e., percentage of area under food grains in the total area) increases, total agricultural production increases since there is an inverse relationship between NFA and FA–as one increases, the other decreases. This is what has exactly happened in the case of Punjab as it is evident from the regression result ,i.e.,

$$P = -409.19 + 8.33^{*}FA, \quad R^2 = 0.75 \text{ (*indicates significant at 5\% level)}$$
$$(7.94)$$

That is, the agricultural growth in Punjab has taken place through change in CP in favour of foodgrains which are treated as commercial crop.

**Table 2**
**Estimates of Simple Regression Equations Concerning Agricultural Production & HYVA for India by States (1972-73 to 1996-97)**

| *State* | *Constant* | *HYVA* | $R^2$ | *F* |
|---|---|---|---|---|
| A.P. | 107.23* | 1.37** | 0.44 | 17.29 |
| | (10.25) | (3.97) | | |
| Haryana | 36.10* | 2.96** | 0.71 | 53.86 |
| | (1.87) | (6.08) | | |
| M.P. | 124.76** | 0.52 | 0.072 | 1.71 |
| | (13.95) | (1.25) | | |
| Rajasthan | 153.74** | 0.76** | 0.19 | 5.16 |
| | (20.26) | (2.16) | | |
| Orissa | 93.94** | 1.23** | 0.39 | 14.07 |
| | (12.52) | (3.56) | | |
| Tamil Nadu | 93.71** | 1.10* | 0.16 | 4.19 |
| | (4.17) | (1.93) | | |
| Punjab | 150.82** | 0.105* | 0.16 | 4.19 |
| | (9.91) | (1.94) | | |
| Gujarat | 100.65** | 0.22 | 0.03 | 0.68 |
| | (15.85) | (0.78) | | |
| U.P. | 73.63** | 12.24** | 0.73 | 59.48 |
| | (10.37) | (7.29) | | |
| H.P. | 75.49** | 0.66** | 0.57 | 29.16 |
| | (7.12) | (3.45) | | |
| W.B. | 146.96** | -0.34 | 0.02 | 0.45 |
| | (8.53) | (-0.56) | | |
| Maharashtra | 80.72** | 2.73* | 0.65 | 40.86 |
| | (6.39) | (6.13) | | |
| All India | 17.79* | 3.39* | 0.75 | 66.00 |
| | (1.43) | (7.86) | | |

*Notes:* (i) Dependent variable = Index of Agricultural Production (all crops combined).
(ii) HYVA =% area under HYV crops to gross cropped area.
(iii) Figure in brackets indicate t-ratios.
(iv) *,** indicate coefficients significant at 1%, 5% respectively.

*Source:* Same as Table 1.

**Table 3**
**Estimates of Simple Regression Equations Concerning Agricultural Production and NFA for India by States (1972-73 to 1996-97)**

| *State* | *Constant* | *NFA* | $R^2$ | *F* |
|---|---|---|---|---|
| A.P. | 120.69** | 0.81* | 0.14 | 3.58 |
| | (7.83) | (1.77) | | |
| Haryana | 159.80** | 0.18 | 0.002 | 0.04 |
| | (7.00) | (0.18) | | |
| Rajasthan | 113.65* | 1.88 | 0.04 | 0.92 |
| | (1.95) | (0.91) | | |
| M.P. | 118.72** | 0.75 | 0.04 | 0.92 |
| | (6.53) | (0.90) | | |
| Orissa | 119.02** | -0.024** | 0.13 | 3.29 |
| | (24.81) | (-1.72) | | |
| Tamil Nadu | 136.28 | 2.58 | 0.02 | 0.45 |
| | (44.97)** | (0.71) | | |
| Punjab | 384.81 | -9.02 | 0.67 | 44.67 |
| | (10.86)** | (-6.37)** | | |
| Gujarat | 100.52 | 0.09 | 0.002 | 0.04 |
| | (4.24) | (0.21) | | |
| U.P. | 79.98 | 2.36 | 0.21 | 5.85 |
| | (4.24)** | (2.33)* | | |
| H.P. | 114.43 | -0.017 | 0.00003 | 0.00 |
| | (16.62)** | (-0.03) | | |
| W.B. | 89.19 | 2.61 | 0.075 | 1.78 |
| | (2.26)* | (1.27) | | |
| Maharashtra | 109.85 | -1.28 | 0.027 | 0.61 |
| | (3.67)** | (-0.74) | | |
| All India | -223.81 | 12.40 | 0.26 | 8.94 |
| | (-1.77)** | (2.64)** | | |

*Notes:* (i) Dependent variable = Index of Agricultural Production (all crops combined).
(ii) Figure in brackets indicate t-ratios.
(iii) NFA = Percentage of Area under non-food grains to gross cropped area.
(iv) * ,** indicate coefficients significant at 1%, 5% respectively.

*Source:* Same as Table 1.

In case of Orissa even considering FA as an explanatory variable instead of NFA it is found that there does not exist any significant relations between FA and P. Absence of any definite trend in either NFA or FA and P may be the reason behind the non-existence of any significant relation between the change in CP and agricultural production.

### Change in CP and Agricultural Growth: District Level Analysis

In this section we shall analyse first the relative role of change ,in CP in agricultural growth and then the impact of change in CP in favour of commercialisation of agriculture on the agricultural growth in all W.B. & its districts.

### Relative Role of CP Change in Agricultural Growth

Let us examine the relative influence of CP change on agricultural growth in the districts of W.B. The regression results in Table 4 show that the CP change has no statistically significant influence on agricultural production in Birbhum. The indices of yield do not have statistically significant effect on index of agricultural production. Area has ,on the other hand, had very significant effect on agricultural production in all the periods. The Chow test calculates F = 14.73 . This means that there is significant difference in the effects of the explanatory variables–area, yield and CP–between the two sub periods.

In the case of Coochbehar the relative influence of CP change is statistically significant during the whole period, but that becomes statistically insignificant in the sub period analysis. On the other hand F = 3.49 (calculated by Chow test) implies the stability of coefficients between the two sub-periods.

The regression results for both Midnapore and Burdwan show that though the area and yield have statistically significant effect on the agricultural production, the CP has no such significant effect. The Chow test indicates the rejection of null hypothesis ($H_0$) at 5% probability level for Burdwan (F = 21.01) and at 1% level for Midnapore (F=21.01).

The regression results reveal the relatively less significant role of change in CP in agricultural production in Murshidabad. The Chow test (F=5.67) implies the instability of relationship among the coefficients in two sub-periods.

In case of other districts the CP has made significant influence on agricultural production only for Nadia and West Dinajpur during whole period, for Malda, Bankura, and Purulia during sub-period I. No such significant effect during

**Table 4**

**Estimates of Multiple Regression Equations Concerning Agricultural Production and Cropping Pattern in the Districts of West Bengal**

| | *Whole Period* | | | | | *Sub-Period I* | | | | | *Sub-Period II* | | | | |
|---|---|---|---|---|---|---|---|---|---|---|---|---|---|---|---|
| | *Explanatory Variables* | | | | | *Explanatory Variables* | | | | | *Explanatory Variables* | | | | |
| *Districts* | ***Constant*** | *A* | *Y* | *CPI* | $R^2$ | ***Constant*** | *A* | *Y* | *CPI* | $R^2$ | ***Constant*** | *A* | *Y* | *CPI* | $R^2$ |
| Burdwan | -152.08 | 1.24** | 1.16 | 0.05 | 0.89 | -126.41 | 1.2** | 1.04** | 0.01 | 0.9 | -125.17 | 0.99* | 1.21** | 0.042 | 0.87 |
| | (-13.2) | .(43.63) | .(0.95) | .(0.95) | | (-15.61) | .(26.85) | .(51.33) | .(0.25) | | (-4.63) | .(4.93) | .(18.27) | .(0.37) | |
| Murshidabad | -125.02 | 1.2** | 1.03** | 0.013 | 0.91 | -116.43 | 1.15** | 0.99** | 0.01 | 0.95 | -172.5 | 1.68** | 1.04** | -0.016 | 0.93 |
| | (-20.06) | .(21.08) | .(10.24) | .(0.55) | | (-10.92) | .(18.33) | .(30.67) | .(0.18) | | (-16.77) | .(15.86) | .(128.34) | (-0.59) | |
| Malda | -155.35 | 1.34** | 0.95** | 0.2 | 0.9 | -132.62 | 1.08** | 0.82** | 0.36** | 0.97 | -228.25 | 1.07 | 1.23** | 0.53 | 0.77 |
| | (-3.06) | .(3.64) | .(11.23) | .(0.7) | | (-5.25) | .(7.01) | .(10.99) | .(2.69) | | (-1.89) | .(0.93) | .(4.1) | .(0.83) | |
| Nadia | -278.07 | 2.1** | 0.7 | 0.92* | 0.83 | -122.04 | 1.17** | 1.04** | 0.004 | 0.97 | 29.83 | 1.75** | -0.015 | -1.33 | 0.4 |
| | (-3.38) | .(2.83) | .(0.76) | .(1.85) | | (-17.12) | .(20.68) | .(62.54) | .(0.16) | | 0.73 | .(1.63) | (-0.04) | (-0.85) | |
| Jalpaiguri | -114.16 | 1.02** | 1.09** | 0.09 | 0.89 | -108.07 | 0.98** | 1.1** | 0.006 | 0.91 | -130.1 | 1.17** | 1.11** | -0.005 | 0.92 |
| | (-47.1) | .(49.88) | .(119.86) | .(1.13) | | (-55.02) | .(96.89) | .(189.63) | .(0.45) | | (-22.97) | .(29.9) | .(75.51) | (-0.37) | |
| Bankura | -114.96 | 1.06** | 1.2** | 0.07 | 0.98 | -88.97 | 0.81** | 1.05** | 0.04*** | 0.99 | -109.42 | 1.9** | 0.81** | (-0.51) | 0.94 |
| | (-4.3) | .(4.59) | .(14.37) | .(0.45) | | (-19.89) | .(17.93) | .(39.23) | .(1.63) | | (-1.34) | .(2.24) | .(2.86) | (-0.55) | |
| Howrah | -139.96 | 1.19** | 1.21** | -0.02 | 0.93 | -100.6 | 0.88** | 1.12** | 0.02 | 0.89 | -169.38 | 1.35** | 1.3** | (-0.05) | 0.91 |
| | (-16.87) | .(19.56) | .(43.04) | (-0.18) | | .(14.62) | .(21.89) | .(39.94) | .(0.04) | | (-29.1) | .(38.33) | .(53.73) | -1.23 | |
| Hooghly | -190.84 | 1.69** | 1.13** | 0.004 | 0.83 | -132.39 | 1.17** | 1.08** | 0.05 | 0.89 | -216.33 | 1.86** | 1.19** | (-0.04) | 0.91 |
| | (-9.45) | .(14.45) | .(33.47) | .(0.03) | | (-19.05) | .(24.79) | .(46.79) | .(1.2) | | (-6.61) | .(12) | .(14.03) | -0.19 | |
| 24 - Parganas | -5.19 | -0.47 | 1.19 | 0.34 | 0.9 | -110.62 | 1.06** | 1.04** | 0.006 | 0.92 | 121.36 | -2.27 | 1.52** | .(0.51) | 0.78 |
| | (-0.05) | (-0.69) | .(0.000001) | .(0.39) | | (-30.02) | .(28.33) | .(162.11) | .(0.4) | | .(0.52) | (-0.89) | .(3.51) | 0.71 | |
| Birbhum | -103.58 | 1.12** | 0.94** | -0.007 | 0.92 | -87.71 | 0.91** | 0.97** | 0.0008 | 0.92 | -135.38 | 1.48** | 0.93** | (-0.02) | 0.89 |
| | (-10.930) | .(15.56) | .(61.09) | (-0.14) | | (-8.92) | .(9.1) | .(17.12) | .(0.015) | | (-23.01) | .(33.09) | .(77.24) | -0.65 | |

Contd...

| Districts | Whole Period | | | | | Sub-Period I | | | | | Sub-Period II | | | | |
|---|---|---|---|---|---|---|---|---|---|---|---|---|---|---|---|
| | Explanatory Variables | | | | | Explanatory Variables | | | | | Explanatory Variables | | | | |
| | **Constant** | A | Y | CPI | $R^2$ | **Constant** | A | Y | CPI | $R^2$ | **Constant** | A | Y | CPI | $R^2$ |
| Purulia | -103.06 | 1.08** | 0.94** | 0.04 | 0.92 | -101.99 | 0.87** | 0.9** | 0.23** | 0.93 | -147.87 | 1.43** | 1.04** | (-0.02) | 0.91 |
| | (-11.37) | .(13.82) | .(27.58) | .(0.51) | | (-9.06) | .(14.3) | .(32.99) | .(2.01) | | (-11.11) | .(13.9) | .(21.93) | -0.49 | |
| Coochbehar | -103.16 | 0.85 | 1.23** | 0.04** | 0.84 | -82.82 | 0.68** | 1.13** | 0.103 | 0.85 | -120.66 | 1.01** | 1.2** | (-0.003) | 0.89 |
| | (-11.57) | .(0.73) | .(32.11) | .(12.69) | | (-5.6) | .(6.51) | .(11.13) | .(0.56) | | (-19.23) | .(24.18) | .(4.48) | -0.04 | |
| West Dinajpur | -82.08 | 0.3 | 0.99** | 0.57** | 0.99 | -91.28 | 0.57** | 1.11** | 0.26 | 0.94 | -24.6 | -0.65 | 0.82** | .(1.24) | 0.95 |
| | (-2.05) | .(1.11) | .(22.65) | .(2.02) | | (-2.62) | .(1.72) | .(7.16) | .(0.1) | | (-0.11) | (-0.27) | .(1.84) | 0.96 | |
| Darjeeling | -143.72 | 1.26** | 1.13** | 0.02 | 0.84 | -121.35 | 1.18** | 1.05** | -0.03 | 0.86 | -157.91 | 1.33** | 1.17** | .(0.02) | 0.87 |
| | (-40.62) | .(41.44) | .(78.04) | .(1.3) | | (-23.85) | .(37.93) | .(47.88) | (-0.97) | | (-25.47) | .(32.65) | .(55.46) | .(0.75) | |
| Midnapore | -178.34 | 1.67** | 1.03** | 0.06 | 0.89 | -100.73 | 0.97** | 1.02** | 0.02 | 0.86 | -191.14 | 1.76** | 0.08** | 0.008 | 0.93 |
| | (-13.27) | .(14.2) | .(34.48) | .(0.76) | | (-19.17) | .(20.89) | .(86.82) | .(0.88) | | (-18.65) | .(21.04) | .(36.82) | .(0.11) | |

*Notes:* (i) A = index number of area (all crops combined),
(ii) Y = index number of yield or productivity (all crops combined),
(iii) CPI = Cropping pattern index,
(iv) Dependentvariable = index of production (all crops combined),
(v) Figures in brackets indicate t-ratio.
(vi) *, ** & *** indicate coefficients significant at 1%, 5% & 10% respectively.

*Source:* (i) Economic Review , Govt. of W.B.
(ii) Statistical Abstract, Govt. of W.B.

sub-period II is found for other districts. But both the area and yield have made significant impact on agricultural production for most of the other districts in the whole period and both the sub-periods. The Chow test reveals that F=0.59,4.82,26.26, 3.32, 18.20, 0.60, 0.71, 4.32, 14.10, 0.59, respectively for Malda, Hooghly, Howrah, Nadia, Purulia, 24 Parganas, Bankura, Darjeeling, Jalpaiguri, West Dinajpur. That is, there is a significant difference in the effects of area, yield and CP on agricultural production in the districts like, Hooghly, Howrah, Nadia, Purulia, Darjeeling, Jalpaiguri between two sub- periods.

**Impact of Change in CP in Favour of Commercialisation of Agriculture on the Agricultural Growth**

It has already been mentioned that the commercialisation of agriculture through change in the CP may be expressed as either HYVA or NFA.From Table 5, it is seen that there has been positive and significant impact of HYVA on the agricultural production in six districts, viz., Murshidabad, Malda, Nadia, Howrah, Birbhum and Midnapore ,but no significant for other districts during 1972-73 to 1996-97. In case of three districts , viz., Burdwan, Hooghly and 24 Parganas that impact be significantly negative. The change in CP in favour of non-HYV crops may be the reason behind this result. As far as the impact of NFA on the agricultural Production in the districts of W.B. during 1972-73 to 1996–97 is concerned, it is found from the Table 6 that the NFA has positive and statistically significant impact on agricultural production in seven districts, viz., Burdwan, Murshidabad, Nadia, Bankura, Birbhum, Coochbehar, and Midnapore ,but no such significant effect in the districts like Malda, Jalpaiguri, 24 -Parganas, Purulia. On the other hand, the impact be negative(either significant or insignificant) in case of Howrah, Hooghly, Darjeeling and West Dinajpur.

In the districts like Burdwan, Hooghly and 24-parganas the negative relationship between HYVA and agricultural production is found to be statistically significant. Even considering the percent of area under non-HYV as an

**Table 5**
**Estimates of Simple Regression Equations Concerning HYVA and Agricultural Production in W.B. by Districts (1972-73 to 1996-97)**

| *District* | *Constant* | *HYVA* | $R^2$ | *F* |
|---|---|---|---|---|
| Burdwan | -14.66 | 3.46** | 0.66 | 31.17 |
| | (-0.51) | (-5.76) | | |
| Murshidabad | 22.25 | 3.29** | 0.39 | 10.73 |
| | (0.68) | (3.27) | | |
| Malda | 43.36** | 3.37** | 0.84 | 87.73 |
| | (4.17) | (9.37) | | |
| Nadia | 68.18* | 3.81* | 0.61 | 26.86 |
| | (3.72) | (5.15) | | |
| Jalpaiguri | 221.94 | -29.00 | 0.01 | 0.20 |
| | (0.12) | (-0.45) | | |
| Bankura | 134.91** | -0.39 | 0.01 | 0.29 |
| | (6.69) | (-0.54) | | |
| Howrah | 24.001 | 3.26** | 0.71 | 39.18 |
| | (1.402) | (6.26) | | |
| Hooghly | 235.08** | -1.502* | -0.24 | 6.21 |
| | (8.61) | (-2.49) | | |
| 24-Parganas | 158.90** | -0.68* | 0.27 | 18.20 |
| | (8.71) | (-1.97) | | |
| Birbhum | 2.39 | 2.87** | 0.44 | 13.28 |
| | (0.076) | (3.64) | | |
| Purulia | 114.58** | 0.19 | 0.34 | 0.38 |
| | (8.55) | (1.29) | | |
| Coochbehar | 14.15** | -0.21 | 0.35 | 12.20 |
| | (19.35) | (-1.25) | | |
| West Dinajpur | 122.33** | -0.04 | 0.003 | 0.008 |
| | (10.35) | (-0.07) | | |
| Darjeeling | 94.29* | 2.17 | 0.05 | 0.83 |
| | (2.31) | (0.91) | | |
| Midnapore | 40.82* | 3.11** | 0.68 | 35.67 |
| | (2.73) | (5.97) | | |

*Notes:* (i) Dependent variable = Index of Agricultural Production (all crops combined).
(ii) HYVA = % Area under HYV crops to gross cropped area.
(iii) Figures in brackets indicate t-ratios.
(ii) *,** indicate coefficients significant at 1%, 5% respectively.

*Source:* (i) Economic Review, Govt. of W.B.
(ii) Statistical Abstract, Govt. of W.B.
(iii) Directorate of Agriculture, Govt. of W.B.

Table 6
Estimates of Simple Regression Equations Concerning NFA and Agricultural Production in W.B. by Districts (1972-73 to 1996-97)

| District | Constant | NFA | $R^2$ | F |
|---|---|---|---|---|
| Burdwan | 39.41 | 0.35 | 0.39 | 14.06 |
| | (1.17) | (3.55)** | | |
| Murshidabad | 38.16 | 4.36 | 0.37 | 12.95 |
| | (1.30) | (3.45)** | | |
| Malda | 112.78 | 1.99 | 0.04 | 0.92 |
| | (3.25)** | (0.90) | | |
| Nadia | 34.11 | 4.39 | 0.26 | 7.73 |
| | (0.66) | (2.67)** | | |
| Jalpaiguri | 118.17 | 1.58 | 0.003 | 0.07 |
| | (0.68) | (0.27) | | |
| Bankura | 10.21 | 11.42 | 0.46 | 18.74 |
| | (0.36) | (4.15)** | | |
| Howrah | 163.43 | -2.304 | 0.04 | 0.92 |
| | (5.22)** | (-0.89) | | |
| Hooghly | 176.51 | -0.13 | 0.069 | 1.63 |
| | (7.83)** | (-0.22) | | |
| 24-Parganas | 107.31 | 2.38 | 0.069 | 1.63 |
| | (3.57)** | (1.21) | | |
| Birbhum | 107.83 | 0.43 | 0.54 | 25.83 |
| | (18.07)** | (4.86)** | | |
| Purulia | 117.39 | 0.11 | 0.0002 | 0.004 |
| | (10.48)** | (0.066) | | |
| Coochbehar | 12.84 | 4.23 | 0.75 | 66.00 |
| | (1.35)*** | (9.01)** | | |
| West Dinajpur | 137.52 | -0.78 | 0.025 | 0.56 |
| | (-5.96)** | (-0.72) | | |
| Darjeeling | 206.44 | -2.17 | 0.27 | 8.14 |
| | (8.12)** | (-2.71)** | | |
| Midnapore | 47.85 | 11.95 | 0.35 | 11.85 |
| | (1.64)*** | (3.28)** | | |

Notes: (i) Dependent variable = Index of Agricultural Production (all crops combined).
(ii) NFA = Percentage of Area under non-foodgrains to gross cropped area.
(iii) Figures in brackets indicate t-ratios.
(iv) *, ** indicate coefficients significant at 1%, 5% respectively.

*Source:* (i) Economic Review, Govt. of W.B.
(ii) Statistical Abstract, Govt. of W.B.
(iii) Directorate of Agriculture, Govt. of W.B.

explanatory variable instead of HYVA it is found that there does not exist any significant relations between the percent of area under non-HYV crops and the agricultural production for those districts. Similarly, considering the FA (instead of NFA) as an index of CP change in case of Howrah, Hooghly, Darjeeling and West Dinajpur no significant relation between FA and agricultural production is observed. Absence of any definite trend in the explanatory variables like NFA, FA, percentage of area under non-HYV crops, HYVA and the agricultural production may be the reason behind the non-existence of any significant relation between the change in CP and agricultural production in those districts. But it is true that the commercialization of agriculture through change in CP has made a significant impact on agricultural growth in many districts in the state.

## A Summing-Up

From the above discussion it is seen that the relative contribution of CP change on agricultural growth is statistically significant for both the developed states, viz., Haryana, Tamilnadu and underdeveloped states, viz, Orissa, H.P. in different sub-periods. The differential influences of CP change on agricultural growth between two sub-periods have also been observed for India & most of the sample states. The same is true for the districts of W.B. But the change in CP as defined by CPI does not have any significant effect upon agricultural production in many states and districts either in the whole period or in the two sub-periods. However, the change in CP as measured by HYVA or NFA which are considered as an index of commercialisation of agriculture has significant effect upon agricultural production in larger number of states and districts. But it is also seen that the impact of change in CP on agricultural growth has been insignificant for a large number of states and districts. In the context of declining land-man ratio, resource chrunch , changing composition of demand, etc. change in CP can be considered as an important factor for accelerating agricultural growth in India.

# Chapter–6

## NATURE OF CHANGES IN CROPPING PATTERN IN INDIA BY STATES

In the previous chapter, we have observed the significant role of change in CP in agricultural growth in the country. So, there is a need to examine the nature of change in C.P. in India and sample states. This chapter examines this aspect in detail.

The nature of changes in CP in India and the sample states may be discussed by analyzing the following two aspects of it:

(i) Nature of change in CP as a whole.

(ii) Nature of HYV CP change, i.e., nature of area shift among HYV crops. The main features of CP change in India by States are discussed below.

### CP Change in India and the Sample States

First, the percentage share of total food grains to GCA in all India has shown a declining trend during 1972-73 to 2000-01. However, within food grains category, the areas under cultivation of rice, wheat, tur (arhar) as percentage of the GCA have shown a moderately increasing trend over time, while those under jowar, bazra, ragi and maize have moved in the opposite direction. The percentage shares of sugarcane, potato and oilseeds as percentage of the GCA have shown rising trends during this period and the area under cultivation of lineseed, cotton tobacco, opium have shown a declining trend

(Table 1). The percentage share of total non-food crops has, however, increased over years during 1972-73 to 2000-01 (Table 1) at a moderate rate.

To discuss the change in CP at the state level we have selected twelve states, viz, A.P., Gujarat, Haryana, H.P., M.P., Maharashtra, Orissa, Punjab, Rajasthan, Tamil Nadu, U.P. and W.B. Let us now analyse the change in CP in these selected states.

In A.P. the percentage of area under cultivation of ragi, wheat, jowar, bajra, has shown a declining trend during 1972-73 to 2000-01. On the other hand, the percentage area under cultivation of rice, maize, oilseeds and sugarcane has increased over years. As a whole, the percentage share of total food-grains in the GCA has decreased over years during 1972-73 to 2000-01.

In Gujarat, it is observed that the percentage of area under cultivation of rice, maize, oilseed, potato, wheat, sugarcane has risen overtime during 1972-73 to 2000-01 and that of bajra, ragi and jowar decreased over years during the same period. In sum, the percentage share of total food grains in the GCA has declined over years during 1972-73 to 2000-01.

It is seen in H.P. that the percentage of area under cultivation of rice, potato, sugarcane, jowar has shown a declining trend during 1972-73 to 2000-01 ,while the trend of that percentage area has been in the opposite directions in the case of maize, wheat, oilseeds and ragi. As a whole, the percentage share of total food-grains in the GCA has shown a rising trend during 1972-73 to 2000-01.

In M.P. the percentage of area under cultivation of ragi, rice, wheat, jowar, bajra, maize and oilseeds has almost decreased over years during 1972-73 to 2000-01, while that under potato moved in the reverse direction during the same period. Again, the fluctuation has been observed for the percentage of area under cultivation of sugarcane and wheat over the period. As a whole, the percentage share of total food grains in the GCA., though fluctuated over time, has declined during 1972-73 to 2000-01.

In the case of Haryana, it is observed that the percentage area under cultivation of rice, wheat, oilseeds, potato and

**Table 1**
**Percentage Share of Different Crops in Gross Cropped Area in India by Sample States**

| *State/ Year* | *Rice* | *Wheat* | *Jowar* | *Ragi* | *Bajra* | *Maize* |
|---|---|---|---|---|---|---|
| A.P. | | | | | | |
| 1972-73 | 23.77 | 0.16 | 23.13 | 1.94 | 4.11 | 2.23 |
| 1977-78 | 29.21 | 0.17 | 18.72 | 2.61 | 4.54 | 2.43 |
| 1982-83 | 21.07 | 0.15 | 16.58 | 1.90 | 3.86 | 2.61 |
| 1987-88 | 21.87 | 0.017 | 1.35 | 0.14 | 0.28 | 0.25 |
| 1993-94 | 24.14 | 0.08 | 7.75 | 1.06 | 1.22 | 2.35 |
| 1996-97 | 36.56 | 0.08 | 7.85 | 1.06 | 1.14 | 3.46 |
| 2000-01 | 31.64 | 0.07 | 5.04 | 0.68 | 0.99 | 3.93 |
| Haryana | | | | | | |
| 1972-73 | 5.61 | 24.48 | 3.64 | - | 17.42 | 2.16 |
| 1977-78 | 6.82 | 25.02 | 3.20 | - | 16.31 | 1.16 |
| 1982-83 | 9.21 | 32.45 | 2.19 | - | 17.42 | 1.06 |
| 1987-88 | 10.00 | 37.48 | 2.94 | - | 10.18 | 0.85 |
| 1993-94 | 14.49 | 38.45 | 1.73 | - | 9.80 | 0.56 |
| 1996-97 | 15.05 | 36.57 | 2.35 | - | 10.37 | 0.47 |
| 2000-01 | 15 | 60 | 3 | | 3 | 0.42 |
| Gujarat | | | | | | |
| 1972-73 | 3.94 | 3.62 | 9.89 | 0.49 | 16.72 | 2.44 |
| 1977-78 | 4.65 | 6.59 | 9.94 | 0.49 | 13.18 | 2.03 |
| 1982-83 | 4.34 | 5.91 | 8.99 | 0.44 | 12.74 | 2.80 |
| 1987-88 | 4.13 | 2.30 | 6.27 | 0.42 | 9.39 | 3.28 |
| 1993-94 | 7.12 | 5.79 | 4.48 | 0.30 | 14.42 | 4.41 |
| 1996-97 | 6.21 | 5.62 | 3.36 | 0.19 | 10.37 | 3.86 |
| 2000-01 | 4.47 | 5.19 | 1.25 | 0.14 | 7.58 | 3.06 |
| Punjab | | | | | | |
| 1972-73 | 8.01 | 40.53 | 0.11 | - | 2.17 | 9.47 |
| 1977-78 | 13.06 | 41.64 | 0.03 | - | 1.64 | 7.00 |
| 1982-83 | 19.07 | 44.06 | 0.02 | - | 0.61 | 4.41 |
| 1987-88 | 24.69 | 44.95 | 0.004 | - | 0.20 | 3.37 |
| 1993-94 | 32.52 | 49.76 | 0.002 | - | 0.24 | 2.92 |
| 1996-97 | 31.53 | 47.15 | 0.03 | - | 0.05 | 2.42 |
| 2000-01 | 16 | 63 | | | 0.13 | 4.4 |
| Rajasthan | | | | | | |
| 1972-73 | 0.84 | 8.69 | 6.14 | - | 32.52 | 4.98 |
| 1977-78 | 1.08 | 10.83 | 4.28 | - | 24.62 | 4.43 |
| 1982-83 | 0.64 | 11.25 | 5.17 | - | 24.14 | 4.83 |
| 1987-88 | 0.73 | 11.72 | 7.47 | - | 27.22 | 6.12 |
| 1993-94 | 0.86 | 12.3 | 4.03 | - | 26.60 | 5.60 |
| 1996-97 | 0.81 | 13.63 | 3.43 | - | 25.93 | 5.79 |
| 2000-01 | 0.8 | 11.17 | 3.26 | | 22.37 | 5.68 |

*Contd...*

| *State/ Year* | *Rice* | *Wheat* | *Jowar* | *Ragi* | *Bajra* | *Maize* |
|---|---|---|---|---|---|---|
| Uttar Pradesh | | | | | | |
| 1972-73 | 19.07 | 26.76 | 3.18 | 1.13 | 4.74 | 6.47 |
| 1977-78 | 20.84 | 28.95 | 2.89 | 0.75 | 4.26 | 5.11 |
| 1982-83 | 20.48 | 33.58 | 2.28 | 0.66 | 3.86 | 4.39 |
| 1987-88 | 19.51 | 35.52 | 2.45 | 0.71 | 3.04 | 4.86 |
| 1993-94 | 22.16 | 37.59 | 1.92 | 0.65 | 2.62 | 4.36 |
| 1996-97 | 22.40 | 36.93 | 1.74 | 0.58 | 3.21 | 4.34 |
| 2000-01 | 24.08 | 37.37 | 1.34 | 0.32 | 3.71 | 0.59 |
| T. Nadu | | | | | | |
| 1972-73 | 37.02 | 0.01 | 8.83 | 3.62 | 5.56 | 0.20 |
| 1977-78 | 35.81 | 0.02 | 9.67 | 3.74 | 5.63 | 0.23 |
| 1982-83 | 31.20 | 0.003 | 16.79 | 3.42 | 4.68 | 0.33 |
| 1987-88 | 20.81 | 0.001 | 11.41 | 2.96 | 4.39 | 0.31 |
| 1993-94 | 30.33 | 0.001 | 7.81 | 2.28 | 3.09 | 0.55 |
| 1996-97 | 30.76 | 0.001 | 49.6 | 2.76 | 2.81 | 0.19 |
| 2000-01 | 32.11 | 0.001 | 19.89 | | 3.03 | 3.03 |
| H.P. | | | | | | |
| 1972-73 | 10.18 | 33.86 | - | 1.42 | - | 27.03 |
| 1977-78 | 9.44 | 34.02 | - | 1.27 | - | 29.29 |
| 1982-83 | 9.20 | 38.35 | 0.13 | 0.90 | - | 29.97 |
| 1987-88 | 9.99 | 41.54 | 0.02 | 0.78 | - | 32.51 |
| 1993-94 | 9.13 | 44.08 | 0.02 | 0.54 | - | 36.51 |
| 1996-97 | 9.59 | 41.95 | - | 1.51 | 0.01 | 36.08 |
| 2000-01 | 10 | 42 | | 1.68 | | 37 |
| M.P. | | | | | | |
| 1972-73 | 22.07 | 15.80 | 10.73 | 0.10 | 1.08 | 3.61 |
| 1977-78 | 22.00 | 16.52 | 9.04 | 0.10 | 0.77 | 3.25 |
| 1982-83 | 21.88 | 16.18 | 9.16 | 0.09 | 0.77 | 3.57 |
| 1987-88 | 21.78 | 16.40 | 8.87 | 0.08 | 0.71 | 3.86 |
| 1993-94 | 18.71 | 17.10 | 6.16 | 0.06 | 0.56 | 3.36 |
| 1996-97 | 16.51 | 15.18 | 1.19 | 0.01 | 0.16 | 1.04 |
| 2000-01 | 20.60 | 20.00 | 2.61 | 0.02 | 0.55 | 3.43 |
| Maharashtra | | | | | | |
| 1972-73 | 7.56 | 4.11 | 29.41 | 1.15 | 7.53 | 0.12 |
| 1977-78 | 7.56 | 6.12 | 33.68 | 1.13 | 8.87 | 0.33 |
| 1982-83 | 7.16 | 5.05 | 32.80 | 1.11 | 6.86 | 0.38 |
| 1987-88 | 5.01 | 2.74 | 24.89 | 6.08 | 6.78 | 0.36 |
| 1993-94 | 7.49 | 3.60 | 29.47 | 0.79 | 8.56 | 1.16 |
| 1996-97 | 6.91 | 3.74 | 26.64 | 0.70 | 9.11 | 1.51 |
| 2000-01 | 6.93 | 3.45 | 23.36 | 0.71 | 8.25 | 1.20 |
| Orissa | | | | | | |
| 1972-73 | 64.52 | 0.74 | 0.30 | 2.42 | 0.05 | 1.12 |
| 1977-78 | 55.53 | 0.85 | 0.45 | 3.50 | 0.09 | 1.63 |

*Contd...*

| State/ Year | Rice | Wheat | Jowar | Ragi | Bajra | Maize |
|---|---|---|---|---|---|---|
| 1982-83 | 46.13 | 0.76 | 0.45 | 3.58 | 0.10 | 1.99 |
| 1987-88 | 21.54 | 0.26 | 0.18 | 1.36 | 0.04 | 0.91 |
| 1993-94 | 50.00 | 0.35 | 0.28 | 1.35 | 0.09 | 0.97 |
| 2000-01 | 51.28 | 0.40 | 0.15 | 0.97 | 0.08 | 0.74 |
| West Bengal | | | | | | |
| 1972-73 | 71.48 | 5.19 | - | - | - | 0.66 |
| 1977-78 | 68.86 | 6.15 | - | - | - | 0.50 |
| 1982-83 | 67.02 | 3.67 | - | - | - | 0.70 |
| 1987-88 | 65.10 | 4.45 | - | - | - | 0.65 |
| 1993-94 | 73.76 | 3.85 | - | - | - | 0.66 |
| 1996-97 | 74.74 | 4.52 | 0.01 | 0.16 | 0.00 | 0.44 |
| 2000-01 | 72.34 | 4.88 | | 0.14 | | 0.40 |
| All India | | | | | | |
| 1972-73 | 22.75 | 11.52 | 10.30 | 1.47 | 17.58 | 8.58 |
| 1977-78 | 23.39 | 12.35 | 9.35 | 1.56 | 6.39 | 3.32 |
| 1982-83 | 22.12 | 13.57 | 9.52 | 1.36 | 6.36 | 3.29 |
| 1987-88 | 23.50 | 13.86 | 9.59 | 1.40 | 5.33 | 3.40 |
| 1993-94 | 24.26 | 14.38 | 7.43 | 1.10 | 5.49 | 3.46 |
| 1996-97 | 24.18 | 14.49 | 6.46 | 1.02 | 5.58 | 3.49 |
| 2000-01 | 23.11 | 12.57 | 5.01 | 1.06 | 7.00 | 4.00 |

| State/ Year | Food grains | Oil seeds | Potato | Sugar cane | Non-food grains | Non-superior cereals |
|---|---|---|---|---|---|---|
| A.P. | | | | | | |
| 1972-73 | 71.53 | 15.37 | - | 1.09 | 28.47 | 76.07 |
| 1977-78 | 73.83 | 11.58 | - | 1.35 | 26.17 | 70.62 |
| 1982-83 | 68.70 | 13.89 | - | 1.32 | 31.30 | 78.78 |
| 1987-88 | 62.48 | 18.92 | - | 0.13 | 37.52 | 98.12 |
| 1993-94 | 51.78 | 24.41 | - | 1.28 | 48.22 | 73.79 |
| 1996-97 | 65.78 | 27.58 | - | 1.83 | 34.82 | 73.70 |
| 2000-01 | 41.20 | - | - | 1.76 | 58.80 | 71.20 |
| Haryana | | | | | | |
| 1972-73 | 84.61 | 4.24 | 0.10 | 2.62 | 15.39 | 69.91 |
| 1977-78 | 77.30 | 3.36 | 0.18 | 3.60 | 22.70 | 68.16 |
| 1982-83 | 71.67 | 3.24 | 0.17 | 2.77 | 28.33 | 58.34 |
| 1987-88 | 86.57 | 5.94 | 0.21 | 3.07 | 13.43 | 52.52 |
| 1993-94 | 74.96 | 12.02 | 0.23 | 2.14 | 25.04 | 47.06 |
| 1996-97 | 72.98 | 12.27 | 0.23 | 2.95 | 27.02 | 47.05 |
| 2000-01 | 80.00 | 12.31 | | 3.76 | 20.00 | 41.00 |

*Contd...*

| *State/ Year* | *Food grains* | *Oil seeds* | *Potato* | *Sugar cane* | *Non-food grains* | *Non-superior cereals* |
|---|---|---|---|---|---|---|
| Gujarat | | | | | | |
| 1972-73 | 42.07 | 18.86 | 0.03 | 0.37 | 57.93 | 92.44 |
| 1977-78 | 42.98 | 22.01 | 0.07 | 0.78 | 57.02 | 88.76 |
| 1982-83 | 43.73 | 23.32 | 0.10 | 0.87 | 56.27 | 89.75 |
| 1987-88 | 32.27 | 16.70 | 0.06 | 0.97 | 67.73 | 93.57 |
| 1993-94 | 38.68 | 29.16 | 0.22 | 1.53 | 61.32 | 87.09 |
| 1996-97 | 38.68 | 27.50 | 0.22 | 1.69 | 61.32 | 87.32 |
| 2000-01 | 24.15 | 21.98 | | | 75.85 | 81.00 |
| Punjab | | | | | | |
| 1972-73 | 67.54 | 5.86 | 0.37 | 1.73 | 32.46 | 51.44 |
| 1977-78 | 69.62 | 4.64 | 0.56 | 1.81 | 30.38 | 45.36 |
| 1982-83 | 72.32 | 2.55 | 0.40 | 1.50 | 27.68 | 36.87 |
| 1987-88 | 76.12 | 3.50 | 0.40 | 1.49 | 23.88 | 30.36 |
| 1993-94 | 87.41 | 2.66 | 0.48 | 1.16 | 12.59 | 17.72 |
| 1996-97 | 83.12 | 3.38 | 0.48 | 3.38 | 16.88 | 17.82 |
| 2000-01 | 83.53 | | | 2.68 | 16.47 | 17.00 |
| Rajasthan | | | | | | |
| 1972-73 | 75.75 | 6.58 | – | 0.21 | 24.25 | 84.47 |
| 1977-78 | 71.60 | 6.18 | – | 0.36 | 28.40 | 88.09 |
| 1982-83 | 76.13 | 7.15 | – | 0.21 | 23.87 | 88.11 |
| 1987-88 | 74.55 | 10.17 | – | 0.20 | 25.45 | 87.55 |
| 1993-94 | 70.90 | 21.93 | – | 0.13 | 29.10 | 86.84 |
| 1996-97 | 70.96 | 21.38 | – | 0.14 | 29.03 | 86.80 |
| 2000-01 | 54.97 | | | | 45.03 | 87.00 |
| Uttar Pradesh | | | | | | |
| 1972-73 | 84.28 | 15.63 | 0.79 | 5.70 | 15.72 | 54.17 |
| 1977-78 | 86.72 | 16.51 | 0.86 | 7.01 | 13.28 | 50.21 |
| 1982-83 | 80.51 | 11.40 | 1.09 | 7.21 | 19.49 | 45.94 |
| 1987-88 | 78.64 | 8.09 | 1.30 | 7.69 | 21.36 | 44.97 |
| 1993-94 | 84.23 | 7.55 | 1.42 | 7.38 | 15.77 | 40.25 |
| 1996-97 | 82.76 | 7.10 | 1.42 | 8.47 | 17.24 | 40.20 |
| 2000-01 | 79.13 | | | 7.62 | 20.87 | 41.00 |
| T. Nadu | | | | | | |
| 1972-73 | 78.01 | 17.46 | 0.15 | 1.87 | 21.99 | 62.37 |
| 1977-78 | 68.16 | 17.44 | 0.14 | 2.15 | 31.84 | 64.17 |
| 1982-83 | 64.92 | 18.04 | 0.17 | 2.89 | 35.08 | 68.80 |
| 1987-88 | 62.49 | 14.95 | 0.04 | 3.06 | 37.51 | 79.19 |
| 1993-94 | 59.45 | 18.88 | 0.09 | 3.21 | 40.55 | 69.66 |
| 1996-97 | 58.03 | 18.78 | 0.09 | 3.67 | 41.97 | 70.60 |
| 2000-01 | 59.81 | | | 3.48 | 40.19 | 71.00 |

*Contd...*

| *State/ Year* | *Food grains* | *Oil seeds* | *Potato* | *Sugar cane* | *Non-food grains* | *Non-superior cereals* |
|---|---|---|---|---|---|---|
| H.P. | | | | | | |
| 1972-73 | 87.00 | 2.10 | 1.51 | 0.40 | 13.00 | 55.96 |
| 1977-78 | 87.40 | 2.20 | 1.45 | 0.43 | 12.60 | 56.54 |
| 1982-83 | 89.25 | 2.40 | 1.36 | 0.32 | 10.75 | 52.45 |
| 1987-88 | 90.23 | 1.90 | 0.15 | 0.31 | 9.77 | 48.47 |
| 1993-94 | 97.01 | 2.41 | 1.23 | 0.29 | 2.99 | 46.79 |
| 1996-97 | 96.86 | 2.32 | - | 0.39 | 3.14 | 45.70 |
| 2000-01 | 97.00 | | | | 3.00 | 47.00 |
| M.P. | | | | | | |
| 1972-73 | 82.13 | 9.86 | 0.08 | 0.25 | 17.87 | 62.13 |
| 1977-78 | 81.92 | 9.70 | 0.10 | 0.26 | 18.08 | 61.48 |
| 1982-83 | 82.10 | 10.00 | 0.12 | 0.21 | 17.90 | 61.94 |
| 1987-88 | 82.50 | 12.50 | 0.13 | 0.30 | 17.50 | 61.82 |
| 1993-94 | 69.49 | 19.29 | 0.16 | 0.19 | 30.51 | 64.19 |
| 1996-97 | 61.78 | 7.23 | 0.17 | 0.07 | 38.22 | 65.20 |
| 2000-01 | 60.00 | | | 0.23 | 40.00 | 62.14 |
| Maharashtra | | | | | | |
| 1972-73 | 57.78 | 9.13 | 0.05 | 0.86 | 42.22 | 88.33 |
| 1977-78 | 71.51 | 9.91 | 0.05 | 1.24 | 28.49 | 86.32 |
| 1982-83 | 70.11 | 10.12 | 0.05 | 1.61 | 29.89 | 87.79 |
| 1987-88 | 78.92 | 9.85 | 0.04 | 1.09 | 21.08 | 92.25 |
| 1993-94 | 64.50 | 13.60 | 0.06 | 1.65 | 35.50 | 88.41 |
| 1996-97 | 64.57 | 12.68 | 0.06 | 2.41 | 35.43 | 89.51 |
| 2000-01 | 61.38 | | | 2.75 | 38.62 | 90.00 |
| Orissa | | | | | | |
| 1972-73 | 84.93 | 9.13 | - | 0.40 | 15.07 | 34.74 |
| 1977-78 | 81.66 | 9.91 | - | 0.54 | 18.34 | 43.62 |
| 1982-83 | 78.92 | 10.12 | - | 0.51 | 21.08 | 53.11 |
| 1987-88 | 68.18 | 9.85 | - | 0.24 | 31.82 | 78.20 |
| 1993-94 | 80.23 | 13.60 | - | 0.22 | 19.77 | 49.68 |
| 2000-01 | 60.67 | | | 0.23 | 39.33 | 51.00 |
| West Bengal | | | | | | |
| 1972-73 | 86.13 | 2.54 | 1.07 | 0.46 | 13.87 | 23.33 |
| 1977-78 | 83.44 | 1.36 | 1.60 | 0.40 | 16.56 | 24.99 |
| 1982-83 | 77.76 | 3.55 | 1.69 | 0.43 | 22.24 | 29.31 |
| 1987-88 | 74.98 | 2.75 | 2.14 | 0.13 | 25.02 | 30.45 |
| 1993-94 | 81.99 | 3.10 | 2.90 | 0.13 | 18.01 | 22.39 |
| 1996-97 | 83.01 | 6.76 | 0.31 | 0.32 | 16.99 | 17.76 |
| 2000-01 | 82.00 | | | 0.22 | 18.00 | 18.00 |

*Contd...*

| *State/ Year* | *Food grains* | *Oil seeds* | *Potato* | *Sugar cane* | *Non-food grains* | *Non-superior cereals* |
|---|---|---|---|---|---|---|
| All India | | | | | | |
| 1972-73 | 74.15 | 9.34 | 0.32 | 2.00 | 25.85 | 64.26 |
| 1977-78 | 72.40 | 9.22 | 0.42 | 2.08 | 27.60 | 64.31 |
| 1982-83 | 72.78 | 8.91 | 0.51 | 2.02 | 27.22 | 62.64 |
| 1987-88 | 72.85 | 15.47 | 0.50 | 1.95 | 27.15 | 61.36 |
| 1993-94 | 80.78 | 15.00 | 0.59 | 1.95 | 19.23 | 20.49 |
| 1996-97 | 69.57 | 14.98 | 0.60 | 2.32 | 30.43 | 31.25 |
| 2000-01 | 61.23 | | 0.6 | 2.15 | 38.77 | 32 |

*Source:* (i) Area and production of principal crops in India, G.O.I.
(ii) Indian Agricultural Statistics, G.O.I.

sugarcane has increased over years during 1972-73 to 2000-01 and that of jowar, bajra and maize moved in the opposite direction. As a whole, the percentage share of total food grains to GCA, though fluctuated over time, has shown a declining trend during 1972-73 to 2000-01.

In Maharashtra the percentage of area under cultivation of wheat, rice, jowar and ragi, has almost declined over years during 1972-73 to 2000-01, while that under maize and bajra, showed an increasing trend during the same period. In sum, the percentage share of total food-grains to GCA though fluctuated over time, has shown a rising trend. In Orissa it is observed that there is a declining trend for the percentage of area under cultivation of wheat, oilseeds, maize, ragi and jowar, but in the case of sugarcane, rice, bajra, the trend was an increasing one during the period from 1972-73 to 2000-01. As a whole, the percentage share of total food grains in GCA. has fluctuated over time,, but increased during1972-73 to 2000-01.

So far as Rajasthan is concerned, the percentage of area under cultivation of wheat, maize, oilseeds has increased over years during 1972-73 to 2000-01 ,but that of rice, jowar, bajra, and sugarcane decreased over time. As a whole, the percentage share of total food grains in the GCA. has shown a decreasing trend during 1972-73 to 2000-01.

In Punjab, the percentage of area under cultivation of rice, wheat, maize, potato, sugarcane has risen during 1972-73 to

2000-01, while that under jowar, bajra and oilseeds has declined during the same period. On the whole, the percentage share of total food grains in the G.C.A. has almost increased over years.

In the case of U.P., the percentage of area under cultivation of jowar, bajra, maize, oilseeds, ragi has shown a declining trend during 1972-73 to 2000-01, but that of rice, wheat, potato and sugarcane exhibited opposite trend. In sum, the percentage share of total food grains to GCA., though fluctuated over time, has shown a declining trend in recent time.

In the case of Tamil Nadu, there were a rising trend in the percentage area under cultivation of jowar, oilseeds, sugarcane, but a declining trend is observed in the case of rice, potato, maize, bajra, and wheat during 1972-73 to 2000-01. On the whole, the percentage share of total food grains in the GCA., though fluctuated over years, has shown declining trend during 1972-73 to 2000-01.

In the case of W .B., the percentage of area under cultivation of rice, oilseeds and potato has shown rising trend during 1972-73 to 2000-01, while that of wheat, maize and sugarcane has shown an opposite trend. As a whole, it is observed that the percentage share of total food grains in the GCA. has shown a declining trend during 1972-73 to 2000-01. That is, the change in CP has taken place in favour of non-food grains (Table 1)

To sum up the above findings, the percentage share of total food grains in the GCA has shown an increasing trend for Punjab, H.P., Maharashtra and Orissa, but a declining trend for Haryana, W.B., U.P., Gujarat, A.P, M.P., Rajasthan and Tamil Nadu. In other words, while the change in CP has been in favour of cultivation of food grains for Punjab, H.P, Maharashtra, and Orissa, it has been in favour of cultivation of non-food grains for other selected states, namely, Haryana, W.B., U.P., Gujarat, A.P.,M.P, Rajasthan and Tamil Nadu during 1972-73 to 2000-01. As a whole, the pro-superior cereals and pro-food grains C.P change is observed to have occurred in four out of sample 12 states, namely, Punjab, H.P ,Maharastra and Orissa, while the pro-non-food grains CP change has been observed in 8 sample states viz., M.P.,

Haryana, W.B., U.P, Gujarat, A.P., Rajasthan and Tamil Nadu. In India, as whole, the pro-superior cereals CP change has occurred. The pro-HYV CP change has occurred in almost all the states including all India.

## Change in CP-A Regional Analysis

It is observed from a study by Bhalla and Singh (1997) that in India, area allocation among various crops has shown a measure of structural rigidity that reflects the traditional character of Indian agriculture wherein foodgrains remain the most predominant crop. Although the introduction of new technology did make some dent in this pattern, the impact was not significant even up to the beginning of the 1980s. It is only during the period from 1980-83 to 1992-95 that many important changes took place in the CP in India.

During the period of 18 years from 1960-65 to 1980-83 for India, as a whole, GCA increased from 157.7 million hectares in 1962-65 to 174.5 million hectares in 1980-83 and area under 43 crops increased from 151.31 million hectares to 165.70 million hectares during the same period. Foodgrains, which accounted for 77.85 per cent of cropped area under 43 crops in 1962-65, still claimed 76.63 per cent of area during 1980-83. Not only that, the share of foodgrains in the total value of crop output (at 1990-93 prices) which was 60.02 per cent during 1962-65 was nearly constant at 60.27 per cent during 1980-83. There was hardly any diversification away from food grains (Table 1a). In the green revolution belt located in the northwestern region, because of higher yields and a remunerative price regime both wheat and rice registered significant increases in their share.

Consequently, in the northwestern region, although the share of total food grains in cropped area remained nearly constant at 79 per cent, their share in total value of output increased substantially from 62.53 per cent during 1962-65 to 69.30 per cent during 1980-83.During 1962-65 to 1980-83, the changes in the CP in other regions were not that marked. There was a substantial decline in the share of coarse cereals in the southern region and a small decline in the share of coarse cereals

**Table 1a**
**Cropping Pattern Changes–Regionwise**

| *Region/ Period* | *Rice* | *Wheat* | *Coarse Cereals* | *Pulses* | *Food-grains* |
|---|---|---|---|---|---|
| North-Western Region | | | | | |
| 1962-65 | 15.30 | 19.98 | 23.20 | 20.99 | 79.47 |
| 1980-83 | 18.95 | 33.83 | 16.13 | 10.76 | 79.68 |
| 1992-95 | 22.49 | 38.47 | 12.33 | 9.20 | 82.48 |
| Eastern-Region | | | | | |
| 1962-65 | 62.71 | 2.90 | 7.37 | 15.63 | 88.62 |
| 1980-83 | 58.54 | 7.51 | 7.58 | 12.55 | 86.17 |
| 1992-95 | 61.83 | 8.41 | 4.78 | 9.64 | 84.66 |
| Central-Region | | | | | |
| 1962-65 | 10.55 | 9.66 | 38.14 | 16.68 | 75.03 |
| 1980-83 | 10.65 | 10.66 | 35.94 | 17.80 | 75.05 |
| 1992-95 | 10.83 | 10.16 | 29.70 | 17.54 | 68.23 |
| Southern-Region | | | | | |
| 1962-65 | 24.99 | 1.01 | 36.81 | 9.66 | 72.47 |
| 1980-83 | 24.88 | 1.10 | 29.92 | 11.39 | 67.30 |
| 1992-95 | 23.56 | 0.70 | 20.82 | 12.19 | 57.27 |
| All-India | | | | | |
| 1962-65 | 23.78 | 8.92 | 29.20 | 15.93 | 77.85 |
| 1980-83 | 23.96 | 13.68 | 25.08 | 13.91 | 76.63 |
| 1992-95 | 24.91 | 14.32 | 19.81 | 13.17 | 72.21 |

| *Region/ Period* | *Non-food grains* | *Oil-seeds* | *Fibre Crops* | *Cotton* | *Surgar cane* | *Remaining Crops* |
|---|---|---|---|---|---|---|
| North-Western Region | | | | | | |
| 1962-65 | 20.53 | 12.28 | 2.49 | 2.18 | 4.61 | 1.15 |
| 1980-83 | 20.32 | 10.68 | 2.89 | 2.75 | 4.72 | 2.04 |
| 1992-95 | 17.52 | 7.02 | 3.16 | 3.12 | 5.47 | 1.86 |
| Eastern-Region | | | | | | |
| 1962-65 | 11.38 | 3.30 | 4.16 | 0.09 | 1.01 | 2.91 |
| 1980-83 | 13.83 | 5.74 | 3.37 | 0.03 | 0.86 | 3.86 |
| 1992-95 | 15.34 | 6.43 | 3.04 | 0.03 | 0.74 | 5.13 |
| Central-Region | | | | | | |
| 1962-65 | 24.97 | 12.06 | 9.52 | 9.27 | 0.44 | 2.95 |
| 1980-83 | 24.95 | 12.02 | 8.09 | 7.90 | 0.69 | 4.15 |
| 1992-95 | 31.77 | 20.10 | 6.91 | 6.82 | 0.87 | 3.90 |
| Southern-Region | | | | | | |
| 1962-65 | 27.53 | 12.21 | 6.17 | 5.76 | 0.87 | 8.28 |
| 1980-83 | 32.70 | 13.94 | 5.66 | 5.24 | 1.68 | 11.42 |
| 1992-95 | 42.73 | 23.04 | 5.17 | 4.89 | 2.15 | 12.38 |
| All-India | | | | | | |
| 1962-65 | 22.15 | 10.59 | 6.31 | 5.36 | 1.56 | 3.69 |
| 1980-83 | 23.37 | 10.92 | 5.55 | 4.78 | 1.85 | 5.05 |
| 1992-95 | 27.79 | 15.31 | 5.04 | 4.42 | 2.13 | 5.32 |

*Source:* Government of India, Area and Production of Principal Crops in India (various issues), Ministry of Agriculture, New Delhi.

in the central region. While both the southern and central regions showed a small increase in their share of area under pulses, the area under oilseeds marginally increased in the southern region and remained almost constant in the central region.

The period 1980-83 to 1992-95 was characterized by significant changes in the CP, the like of which had not been seen in India during the last 30 years. The most important feature of the changes was that at the all India level the proportion of area under food-grains, which had registered a sharp decline by 4.42 per cent from 76.63 per cent of total area in 1980-83 to 72.21 per cent during 1992-95. On the other hand, the area under both rice and wheat actually registered a notable increase. The main shift took place from coarse cereals to oilseed, whose share in cropped area increased from 10.92 per cent in 1980-83 to 15.31 per cent in 1992-95. Among food grains, unlike the earlier period, there was a slow down in the increase of area under wheat and there was an increase in the area under rice. Further, there was a significant decline in the area under coarse cereals. As seen above, there was a big increase in the share of area under oilseeds. While there was a slight increase in the share of area under sugarcane, the area under fibre crops actually tended to decline marginally. Simultaneously, there was a small increase in the area under the remaining crops.

A significant decline in the proportion of area under coarse cereals took place in the central region comprising Gujarat, M.P., Rajasthan and to some extent Maharashtra. The southern region also registered a significant decline in area under food grains, mainly coarse cereals.

A related significant development was that in the central and southern region there was a major shift in CP in favour of oilseeds at the cost of coarse cereals .It is, however, important to note that during the period 1980-83 to 1992-95, despite a notable decline in area under cultivation at a rate of 1.48 per cent per annum, the output of coarse cereals registered a modest growth rate of 0.91 per cent, at the all India level because of a significant rate of growth of productivity at 2.08 per cent per annum. Notably, the CP changes in the northwestern region were in contrast to those in the central and southern regions. In the north-western region, area under foodgrains increased primarily as a result of increase in the share of area under rice and, to some extent, wheat, although there was a significant decline in the area under coarse cereals and pulses. On the other hand, unlike the central and southern region, the northwestern region as a whole registered a significant decline in area under oilseeds. In the eastern region, the share of area under total food grains remained almost constant during 1980-83 to 1992-95. However, within food grains while area under rice increased significantly, that under coarse cereals and pulses declined sharply. But the eastern region, unlike central and southern regions, did not register any increase in area under oilseeds.

## Change in CP in Terms of HYV and Non-HYV Crops: A State wise Analysis

The change in CP in different states and regions described above is an outcome of the implementation of different agricultural policies aiming at higher agricultural growth through the expansion of area under HYVs. In this context let us examine the direction of this change in CP between HYV and non-HYV crops.

It is observed that the percentage of area under HYV crops in the GCA has increased over years at different rates, ranging from moderate to high, for all the selected states and all India during 1972-73 to 1997-98. Therefore, it can be concluded that change in CP particularly towards cultivation of HYV crops has led to rapid agricultural growth in all the selected states and in India. (Table 2).

Let us now examine the change in HYV CP for the sample states. Among the HYV crops in India, the percentage of area under cultivation of wheat and bajra has almost declined over years ,while that under others, viz., rice, jowar, maize has increased moderately during 1972-73 to 1997-98 (Table 3).

In M.P., the percentage area under HYV cultivation of maize has risen over time with some fluctuations during 1972-73 to 1997-98, while that under paddy, bajra, jowar and wheat decreased over years with few exceptions.

**Table 2**
**Percentage of HYV Cropped Area in the G.C.A.in India by States**

| *States* | *1972-1973* | *1977-1978* | *1982-1983* | *1987-1988* | *1993-1994* | *1994-1995* | *1997-1998* |
|---|---|---|---|---|---|---|---|
| Andhra Pradesh | 12.83 | 23.50 | 31.89 | 32.92 | 30.34 | 37.60 | 36.85 |
| Gujarat | 12.21 | 17.84 | 20.04 | 15.43 | 21.75 | 23.53 | 18.93 |
| Hrayana | 25.52 | 32.13 | 48.09 | 51.60 | 51.63 | 52.60 | 75.84 |
| Himachal Pradesh | 34.82 | 40.64 | 46.65 | 47.87 | 60.46 | 65.72 | 59.27 |
| Madhya Pradesh | 5.30 | 12.96 | 17.33 | 22.18 | 24.64 | 28.41 | 31.4 |
| Maharashtra | 8.56 | 20.25 | 23.35 | 40.29 | 38.13 | 37.47 | 41.34 |
| Orissa | 7.37 | 9.48 | 21.90 | 10.22 | 46.50 | 47.39 | 42.84 |
| Punjab | 38.59 | 52.49 | 64.15 | 68.46 | 82.56 | 80.72 | 70.46 |
| Rajasthan | 5.59 | 8.31 | 13.91 | 15.64 | 17.99 | 15.61 | 18.65 |
| Tamil Nadu | 37.11 | 35.99 | 21.19 | 41.54 | 40.74 | 41.20 | 29.28 |
| Uttar Pradesh | 18.15 | 38.12 | 36.55 | 43.15 | 57.66 | 59.93 | 58.61 |
| West Bengal | 14.60 | 23.83 | 28.46 | 36.37 | 45.56 | 55.34 | 53.82 |
| India | 13.77 | 22.6 | 27.32 | 33.17 | 38.65 | 40.95 | 45.82 |

*Source:* Fertiliser Statistics, Government of India.

In the case of Rajasthan, the percentage of area under HYV cultivation of wheat, paddy, maize and jowar has almost declined over time during 1972-73 to 1997-98, but that under bajra increased during the same period.

In U.P. it is observed that the percentage of area under HYV cultivation of paddy, jowar, maize and bajra has increased over years during 1972-73 to 1997-98 with few exceptions ,while that under wheat declined over time.

In Orissa, it is seen that the percentage of area under HYV cultivation of wheat has almost declined during 1972-73 to 1997-98, but that under paddy shows an increasing trend. On the other hand, the percentage of area under HYV cultivation of maize, jowar and bajra has shown fluctuations during the same period.

In case of Gujarat, the percentage of area under HYV cultivation of paddy, maize and jowar has shown an increasing trend during 1972-73 to 1997-98. On the other hand, the percentage of area under HYV wheat has fluctuated during the same period without showing any secular trend. But the percentage of area under bajra has shown a declining trend during 1972-73 to 1997-98.

Table 3
**Percentage of Area Under Different HYV Crops in the Total HYV Cropped Area in India by States**

| | | *1972-1973* | *1977-1978* | *1982-1983* | *1987-1988* | *1993-1994* | *1994-1995* | *1997-1998* |
|---|---|---|---|---|---|---|---|---|
| Andhra Pradesh | Paddy | 75.94 | 81.90 | 75.49 | 73.69 | 78.29 | 72.45 | 78.71 |
| | Wheat | 4.43 | 1.08 | 0.66 | 0.34 | 0.17 | 0.19 | 0.14 |
| | Maize | 3.16 | 2.85 | 2.45 | 5.47 | 6.20 | 6.48 | 6.85 |
| | Jowar | 8.22 | 8.41 | 12.27 | 13.10 | 11.87 | 18.18 | 12.74 |
| | Bajra | 8.22 | 5.73 | 9.10 | 7.40 | 3.46 | 2.70 | 1.53 |
| Gujarat | Paddy | 6.86 | 6.83 | 14.49 | 19.27 | 22.89 | 20.42 | 25.09 |
| | Wheat | 18.59 | 28.63 | 20.69 | 12.61 | 15.56 | 21.02 | 19.25 |
| | Maize | 0.55 | 3.14 | 5.24 | 9.44 | 10.44 | 9.45 | 9.91 |
| | Jowar | 0.23 | 1.73 | 5.56 | 13.46 | 8.89 | 6.25 | 8.4 |
| | Bajra | 73.74 | 56.72 | 54.01 | 45.20 | 42.22 | 42.85 | 37.34 |
| Harayana | Paddy | 6.94 | 14.43 | 16.86 | 16.39 | 15.43 | 17.65 | 18.24 |
| | Wheat | 75.52 | 70.10 | 62.14 | 70.04 | 72.98 | 69.40 | 68.65 |
| | Maize | 0.98 | 1.46 | 0.78 | 0.63 | 0.52 | 0.46 | 0.42 |
| | Bajra | 16.54 | 14.31 | 20.20 | 12.32 | 11.07 | 12.48 | 12.68 |
| Himachal Pradesh | Paddy | 21.73 | 17.10 | 19.23 | 15.04 | 17.08 | 14.29 | 12.97 |
| | Wheat | 56.52 | 65.78 | 63.53 | 63.65 | 61.23 | 63.24 | 60.06 |
| | Maize | 21.73 | 17.10 | 17.22 | 21.29 | 21.69 | 22.64 | 26.95 |
| Madhya Pradesh | Paddy | 45.14 | 47.66 | 38.61 | 36.19 | 46.50 | 42.67 | 44.49 |
| | Wheat | 39.60 | 30.05 | 36.82 | 38.04 | 36.03 | 40.16 | 38.45 |
| | Maize | 2.81 | 3.19 | 4.28 | 4.99 | 6.07 | 6.62 | 6.51 |
| | Jowar | 11.17 | 9.86 | 19.29 | 9.69 | 10.36 | 9.93 | 10.1 |
| | Bajra | 1.27 | 1.21 | 0.98 | 1.07 | 1.06 | 1.22 | 0.43 |

*Contd...*

| | | 1972-1973 | 1977-1978 | 1982-1983 | 1987-1988 | 1993-1994 | 1994-1995 | 1997-1998 |
|---|---|---|---|---|---|---|---|---|
| Maharashtra | Paddy | 14.92 | 22.85 | 19.48 | 16.66 | 16.02 | 17.20 | 15.06 |
| | Wheat | 21.32 | 24.90 | 16.75 | 14.54 | 8.59 | 8.82 | 8.26 |
| | Maize | 0.20 | 1.26 | 0.86 | 1.24 | 2.49 | 2.78 | 2.92 |
| | Jowar | 23.17 | 45.34 | 47.20 | 50.13 | 52.63 | 50.47 | 53.49 |
| | Bajra | 20.08 | 5.62 | 15.69 | 17.41 | 20.28 | 20.72 | 20.37 |
| Orissa | Paddy | 78.20 | 86.03 | 93.31 | 92.83 | 96.16 | 96.22 | 83.08 |
| | Wheat | 7.82 | 8.77 | 3.45 | 2.57 | 6.12 | 0.72 | 0.36 |
| | Maize | 1.36 | 4.78 | 2.96 | 4.26 | 3.03 | 1.55 | 2.36 |
| | Jowar | 0.29 | 0.39 | 0.21 | 0.32 | 0.20 | 0.06 | 0.04 |
| | Bajra | 0.00 | 0.00 | 0.05 | 0.05 | 0.01 | 0.13 | |
| Punjab | Paddy | 16.57 | 22.89 | 28.77 | 32.07 | 36.47 | 37.93 | 37.64 |
| | Wheat | 82.30 | 73.10 | 68.44 | 65.46 | 60.26 | 59.36 | 59.58 |
| | Maize | 0.83 | 2.71 | 2.61 | 2.31 | 3.08 | 2.59 | 2.71 |
| | Bajra | 0.34 | 1.28 | 0.72 | 0.14 | 0.18 | 0.11 | 0.05 |
| Rajasthan | Paddy | 2.77 | 3.19 | 1.17 | 1.56 | 1.60 | 1.46 | 1.11 |
| | Wheat | 66.66 | 62.40 | 45.74 | 47.60 | 50.37 | 54.72 | 45.19 |
| | Maize | 2.22 | 2.20 | 1.17 | 1.12 | 0.78 | 0.80 | 0.62 |
| | Jowar | 0.55 | 0.71 | 1.25 | 1.85 | 0.34 | 0.91 | 0.33 |
| | Bajra | 27.77 | 31.48 | 50.68 | 47.85 | 46.91 | 42.10 | 52.72 |
| Tamil Nadu | Paddy | 78.57 | 87.16 | 73.17 | 71.66 | 67.48 | 68.21 | 70.76 |
| | Maize | 1.01 | 1.35 | 1.82 | 0.71 | 1.57 | 2.06 | 2.3 |
| | Jowar | 0.77 | 3.96 | 12.19 | 21.20 | 19.03 | 18.08 | 17.35 |
| | Bajra | 19.63 | 7.50 | 12.80 | 9.45 | 7.40 | 6.85 | 6.2 |

Contd...

| | | 1972-1973 | 1977-1978 | 1982-1983 | 1987-1988 | 1993-1994 | 1994-1995 | 1997-1998 |
|---|---|---|---|---|---|---|---|---|
| Uttar Pradesh | Paddy | 22.23 | 30.33 | 27.03 | 24.85 | 30.79 | 33.13 | 33.51 |
| | Wheat | 76.90 | 67.41 | 70.95 | 73.19 | 63.28 | 61.62 | 59.33 |
| | Maize | 0.48 | 1.12 | 0.47 | 1.22 | 2.79 | 3.05 | 4.03 |
| | Jowar | 0.04 | 0.44 | 0.02 | 0.00 | 0.77 | 0.00 | |
| | Bajra | 0.72 | 0.67 | 1.50 | 0.72 | 2.13 | 2.19 | 3.11 |
| West Bengal | Paddy | 66.64 | 74.21 | 85.55 | 87.77 | 92.64 | 92.75 | 92.7 |
| | Wheat | 33.32 | 25.78 | 13.16 | 12.22 | 7.36 | 7.25 | 7.29 |
| | Maize | 0.00 | 0.00 | 1.28 | 0.00 | 0.00 | 0.00 | |
| | Jowar | 0.00 | 0.00 | 0.00 | 0.00 | 0.00 | 0.00 | |
| | Bajra | 0.00 | 0.00 | 0.00 | 0.00 | 0.00 | 0.00 | |
| All India | Paddy | 36.60 | 41.40 | 39.67 | 41.12 | 42.11 | 43.47 | 43.69 |
| | Wheat | 45.60 | 40.58 | 37.58 | 36.39 | 32.45 | 32.62 | 31.04 |
| | Maize | 2.68 | 3.18 | 3.64 | 3.97 | 4.42 | 4.78 | 4.92 |
| | Jowar | 3.89 | 8.06 | 9.20 | 11.20 | 9.91 | 7.00 | 10.85 |
| | Bajra | 11.20 | 6.75 | 9.91 | 7.30 | 7.64 | 10.26 | 7.97 |

*Source:* Fertiliser Statistics, Government of India.

In case of Maharashtra, a rising trend is observed for the percentage of area under HYV cultivation of paddy, maize and jowar during 1972-73 to 1997-98.Again, the percentage of area under HYV cultivation of wheat and bajra has fluctuated over years during the same period.

In W.B., it is found that the percentage of area under HYV cultivation of paddy has increased with some fluctuations during 1972-73 to 1997-98 ,while that under wheat exhibits a declining trend during the same period.

In Haryana, though the percentage of area under HYV cultivation of wheat has shown a declining trend during 1972-73 to 1997-98, that of paddy has shown a rising tendency over time during the same period. On the other hand, the area under HYV maize and bajra has fluctuated over years.

In H. P., the percentage of area under HYV cultivation of Paddy has shown a declining trend during 1972-73 to 1997-98, but that under wheat and maize increased over years.

In case of Punjab, the percentage of area under HYV cultivation of paddy and maize has risen over time with some fluctuations during 1972-73 to 1997-98 ,while that under wheat declined over years. On the other hand, in case of HYV bajra the percentage of area has fluctuated over years.

In case of A. P., the percentage of area under HYV paddy, maize and jowar has risen over time with some fluctuations during 1972-73 to 1997-98 ,but that declined for wheat. Again, the percentage of area under HYV bajra has fluctuated over years.

In Tamil Nadu, the percentage of area under HYV jowar has almost risen during 1972-73 to 1997-98 , but that under HYV bajra declined. On the other hand, there have been fluctuations over years in the area under HYV paddy and maize in the state.

In summing-up, the above findings it may be stated that among the selected HYV crops, paddy and wheat have been the dominating crops in almost all the states as well as in India.

## Change in CP Across Different Landholding Size Classes in W. B.

In W.B. the CP has moved in favour of non-food crops for each size group of holding during 1985-86 to 1990-91

(Table 4). Examining the percentage of area under the different food crops to total area under all food crops during 1985-86 to 1990-91 the following observations (Table 5) may be made. (No analysis could be made for the period succeding 1990-91 since census data were not available beyond 1990-91).

**Table 4**
**Percentage of Area under Food and Nonfood Crops in Major Size Groups of Holding During 1985-86 and 1990-91 in W.B.**

| | *Food crops* | | *Non-food cropes* | |
|---|---|---|---|---|
| *Size group (ha)* | *1985-86* | *1990-91* | *1985-86* | *1990-91* |
| Below 1.0 | 85.147 | 83.190 | 14.852 | 16.880 |
| 1.0-2.0 | 86.450 | 83.396 | 13.549 | 16.603 |
| 2.0-4.0 | 87.424 | 84.573 | 12.575 | 15.426 |
| 4.0-10.0 | 99.870 | 84.407 | 0.129 | 15.592 |
| 10.0 & above | 55.788 | 8.555 | 44.211 | 91.444 |
| All groups | 85.560 | 82.520 | 14.439 | 17.479 |

*Source:* Agricultural Census (various Issues), Government of West Bengal.

**Table 5**
**Percentage Share of Area Under Food Crops in the Total Area Under All Food Crops in Major Groups of Holding in W.B.**

| | *Percentage of Area* | | | | | |
|---|---|---|---|---|---|---|
| | *1985-86* | | | *1990-91* | | |
| *Size group (ha)* | *Rice* | *Wheat* | *Pulses* | *Rice* | *Wheat* | *Pulses* |
| Below 1.0 | 76.65 | 4.48 | 5.78 | 80.24 | 4.61 | 3.31 |
| 1.0-2.0 | 80.26 | 4.36 | 5.30 | 81.16 | 4.47 | 3.21 |
| 2.0-4.0 | 82.25 | 4.07 | 4.43 | 83.32 | 4.14 | 3.22 |
| 4.0-10.0 | 82.59 | 4.14 | 4.67 | 85.54 | 3.36 | 2.45 |
| 10.0 & above | 97.41 | 0.60 | 0.05 | 78.17 | 3.14 | 2.19 |
| All groups | 80.23 | 4.25 | 5.10 | 81.51 | 4.46 | 3.20 |

*Source:* Agricultural Census (various Issues), Government of West Bengal.

First, the percentage of area under rice has risen over years for marginal, small and medium farmers ,but declined for large farmers.

Secondly, the percentage of area under wheat has risen for marginal, small and large farmers, ,but declined for medium farmers.

Thirdly, the percentage of area under pulses has declined for all the size classes of landholdings except for the large one.

The analysis of change of percentage of area under different non-food crops (to total area under all non food crops) across the different size classes of holding during 1985-86 to 1990-91 highlights the following (Table 6).

**Table 6**
**Percentage Share of Area Under Food Crops in the Total Area under All Non-food Crops in Major Size Groups of Holding in W.B. during 1985-86 and 1990-91**

| | *Percentage of Area* | | | | | |
|---|---|---|---|---|---|---|
| | *1985-86* | | | *1990-91* | | |
| *Size group (ha)* | *Jute* | *Oil Seeds* | *Tea* | *Jute* | *Oil Seeds* | *Tea* |
| Below 1.0 | 51.59 | 29.02 | 0.00 | 44.06 | 32.97 | 0.00 |
| 1.0-2.0 | 55.91 | 31.52 | 0.00 | 42.18 | 33.88 | 0.00 |
| 2.0-4.0 | 56.66 | 33.31 | 0.00 | 41.30 | 35.97 | 0.00 |
| 4.0-10.0 | 52.93 | 37.58 | 0.00 | 36.26 | 35.87 | 0.03 |
| 10.0 & above | 0.32 | 0.54 | 98.93 | 0.67 | 0.95 | 96.22 |
| All groups | 49.64 | 2.89 | 0.08 | 39.35 | 31.55 | 7.19 |

*Source:* Agricultural Census (various Issues), Government of West Bengal.

First, the percentage of area under jute cultivation has declined over years for all the size classes except the large one.

Secondly, the percentage of area under oilseeds has increased over time for all the size classes of farmers except for the medium one.

## Change in CP Across Different Land Holding Size Classes in India

In India the CP has moved in favour of non- foodgrains cultivation for each size group of holdings during 1985-86 to 1990-91 (Table 7).

**Table 7**
**Percentage Distribution of Area under Food and Non-Food Crops in Major Size Groups of Holdings in India**

| | 1985-86 | | 1990-91 | |
|---|---|---|---|---|
| *Size Group (in Ha.)* | *Food Crops* | *Non-Food Crops* | *Food Crops* | *Non-Food Crops* |
| Below1.0 | 88.6 | 11.4 | 86 | 14 |
| 1.0-2.0 | 85.2 | 14.8 | 81.1 | 18.9 |
| 2.0-4.0 | 82 | 18 | 78.1 | 21.9 |
| 4.0-10.0 | 77.5 | 22.5 | 73.3 | 26.7 |
| 10.0& above | 73.2 | 26.8 | 68.2 | 31.8 |
| All Groups | 80.3 | 19.7 | 77.3 | 22.7 |

*Source:* Indian Agriculture in Brief, 27th Edition, Jan. 2000.

## A Summing-Up

The pro-superior cereals and foodgrains CP change has occurred in 8 sample states, viz., Punjab, H. P., A.P., Gujarat, Haryana, U.P., and W.B. But the pro-non-foodgrains CP change has been observed in other 4 sample states and India. The pro-HYV CP change has occurred in all the sample states as well as in India. In W.B. and India, the CP has moved in favour of non-food crops for each size group of land holdings during 1985-86 to 1990-91.

# CHAPTER–7

## NATURE OF CHANGES IN CROPPING PATTERN IN WEST BENGAL: DISTRICT AND BLOCK LEVEL ANALYSIS

In this chapter , we shall discuss the nature of changes in CP in the districts of W.B. and in the blocks of Midnapore.The nature of changes in CP in the districts of W.B. is discussed in the following paragraphs.

### Change in CP in the Districts of W.B.

The nature of CP change reveals the following facts (see Table 1). First, in case of West Dinajpur, the percentage area under cultivation of rice, total cereals, total oilseeds, potato has slowly increased during 1972-73 to 2001-02 ,while that under wheat, total pulses, sugarcane and jute declined over years. On the other side, the percentage area under cultivation of wheat and til has fluctuated over years during the same period. As a whole, the percentage share of total foodgrains in the GCA though fluctuated over years has shown a decreasing trend during 1972-73 to 2001-02.

In case of Malda, the percentage of area under cultivation of rice, wheat, total non-foodgrains, total cereals, til, jute, potato, oilseeds has shown a rising trend during 1972-73 to 2001-02 ,while the opposite trend is found for the cultivation of total pulses and sugarcane. As a whole, the percentage share of total foodgrains to GCA has declined over years.

In Jalpaiguri, the percentage of area under cultivation of total cereals, wheat, total oilseeds, til, jute and potato has slowly increased over years during 1972-73 to 2001-02 ,while that under rice, total cereals, wheat, total pulses and sugarcane decreased roughly. On the whole, the percentage share of total foodgrains in the GCA has shown a decreasing trend during 1972-73 to 2001-02.

In Birbhum, it is found that the percentage area under cultivation of rice, sugarcane, total cereals, total oilseeds and potato has shown an increasing trend during 1972-73 to 2001-02 ,but the opposite trend is seen for the cultivation of wheat, total pulses and jute. In case of cultivation of til the percentage area has fluctuated over time during the same period. On the whole, it is observed that the percentage share of total foodgrains in the GCA has declined over time during 1972-73 to 2001-02.

In case of Bankura, the percentage of area under rice, wheat, total oilseeds, total cereals, til and potato has shown a rising trend during 1972-73 to 2001-02, but that under total pulses, jute and sugarcane has declined during the same period. As a whole, the percentage share of total foodgrains in the GCA has declined over time during 1972-73 to 2001-02.

In case of Howrah, the percentage of area under cultivation of rice, total cereals, til and potato has almost increased over years during 1972-73 to 2001-02 ,while that under oilseeds, jute, sugarcane, wheat and pulses, has declined during the same period. On the whole, the percentage share of total foodgrains has increased over time during 1972-73 to 1996-97, but declined over years during 1996-97 to 2001-02.

In Hooghly, it is observed that the percentage area under cultivation of jute, til, total oilseeds and potato has almost increased over years during 1972-73 to 2001-02 ,while that of rice, wheat, total cereals, total pulses and sugarcane has decreased during the same period. The percentage share of total foodgrains to GCA has declined, in general, over years during 1972-73 to 2001-02 in the district.

In 24 Paraganas, the percentage of area under rice, cereals,til, total oilseeds and potato has slowly risen during

**Table 1**
**Percentage Share of Different Crops in Gross Cropped Area in West Bengal by Districts**

| *District/ Year* | *Rice* | *Wheat* | *Total cereals* | *Total Pulses* | *Total food grains* | *Til* |
|---|---|---|---|---|---|---|
| Burdwan | | | | | | |
| 1972-73 | 83.55 | 4.84 | 88.48 | 2.74 | 91.22 | 0.15 |
| 1977-78 | 83.01 | 4.26 | 87.37 | 3.97 | 91.34 | 1.1 |
| 1982-83 | 80.59 | 1.72 | 82.41 | 1.72 | 84.13 | 4.5 |
| 1987-88 | 80.6 | 7.8 | 82.4 | 0.67 | 83.08 | 2.79 |
| 1993-94 | 85.73 | 0.31 | 86.04 | 0.24 | 86.29 | 1.1 |
| 1996-97 | 85.39 | 0.6 | 85.74 | 10.74 | 86.84 | 1.79 |
| 2001-02 | 83.69 | 0.2 | 84.33 | 0.68 | 84.95 | 0.65 |
| Birbhum | | | | | | |
| 1972-73 | 79.14 | 8.02 | 87.57 | 6.84 | 94.41 | 0.22 |
| 1977-78 | 75.02 | 12.06 | 87.49 | 7.46 | 94.95 | 0.76 |
| 1982-83 | 80.91 | 4.33 | 85.63 | 5.3 | 90.94 | 2.05 |
| 1987-88 | 73.49 | 6.36 | 80.06 | 4.75 | 84.81 | 1.36 |
| 1993-94 | 84.44 | 2.91 | 87.85 | 2.25 | 90.1 | 0.41 |
| 1996-97 | 83.69 | 4.56 | 88.41 | 3 | 91.42 | 0.4 |
| 2001-02 | 83.92 | 5.93 | 89.85 | 0.15 | 90 | 0.23 |
| Hooghly | | | | | | |
| 1972-73 | 75.91 | 6 | 81.95 | 2.38 | 84.33 | 0.06 |
| 1977-78 | 68.6 | 5.25 | 73.85 | 2.28 | 76.12 | 2.39 |
| 1982-83 | 66.56 | 2.05 | 68.61 | 2.18 | 70.79 | 8.18 |
| 1987-88 | 65.79 | 2.54 | 68.33 | 0.41 | 68.74 | 8.86 |
| 1993-94 | 66.63 | 0.26 | 66.88 | 0.09 | 66.98 | 7.15 |
| 1996-97 | 56.78 | 0.31 | 57.09 | 0.15 | 57.24 | 9.01 |
| 2001-02 | 74.56 | 4.27 | 78.83 | 0.37 | 80.93 | 4.24 |
| 24-Paraganas | | | | | | |
| 1972-73 | 79.17 | 2.34 | 81.56 | 5.22 | 86.78 | |
| 1977-78 | 82.34 | 1.64 | 83.99 | 5.49 | 89.48 | 0.32 |
| 1982-83 | 84.69 | 1.39 | 86.16 | 3.33 | 89.49 | 0.62 |
| 1987-88 | 79.48 | 1.81 | 81.29 | 3.16 | 86.46 | 1.77 |
| 1993-94 | 83.21 | 0.89 | 84.11 | 2.65 | 86.77 | 1.01 |
| 1996-97 | 83.36 | 0.09 | 84.15 | 2.09 | 86.24 | 1.32 |
| 2001-02 | 73.88 | 0.09 | 73.97 | 2.05 | 78.67 | 0.68 |
| Bankura | | | | | | |
| 1972-73 | 88.62 | 3.87 | 92.68 | 2.25 | 94.92 | 0.42 |
| 1977-78 | 89.86 | 5.57 | 90.91 | 2.3 | 93.21 | 3.01 |
| 1982-83 | 88.05 | 2.21 | 90.84 | 1.65 | 92.49 | 4.17 |
| 1987-88 | 83.83 | 2.68 | 87.06 | 1.26 | 88.32 | 4.93 |
| 1993-94 | 86.39 | 1.86 | 88.35 | 1.54 | 89.28 | 2.93 |

*Contd...*

| *District/ Year* | *Rice* | *Wheat* | *Total cereals* | *Total Pulses* | *Total food grains* | *Til* |
|---|---|---|---|---|---|---|
| 1996-97 | 86.43 | 1.29 | 88.2 | 0.74 | 88.93 | 1.67 |
| 2001-02 | 87.43 | 2.12 | 89.81 | 0.26 | 90 | 2.62 |
| Howrah | | | | | | |
| 1972-73 | 79.08 | 2.37 | 81.45 | 6.69 | 88.14 | 0.76 |
| 1977-78 | 77.12 | 3.07 | 80.19 | 9.86 | 90.06 | 2.51 |
| 1982-83 | 82.31 | 0.56 | 82.87 | 6.61 | 89.48 | 3.72 |
| 1987-88 | 83.02 | 0.43 | 83.45 | 4.5 | 87.95 | 5.08 |
| 1993-94 | 91.96 | 0.13 | 91.5 | 0.98 | 93.07 | 1.89 |
| 1996-97 | 89.45 | 0.31 | 89.76 | 0.38 | 90.15 | 2.03 |
| 2001-02 | 74.15 | 0.3 | 74.51 | 0.24 | 76.75 | 1.39 |
| Jalpaiguri | | | | | | |
| 1972-73 | 66.76 | 0.39 | 68.22 | 1.56 | 69.79 | 0.03 |
| 1977-78 | 65.6 | 6.65 | 73.68 | 2.06 | 75.74 | 0.06 |
| 1982-83 | 61.99 | 2.69 | 65.68 | 2.34 | 68.07 | 0.2 |
| 1987-88 | 71.48 | 5.8 | 80.13 | 1.12 | 89.25 | 0.82 |
| 1993-94 | 60.46 | 3.58 | 64.64 | 0.74 | 65.38 | 0.26 |
| 1996-97 | 57.83 | 4.99 | 63.68 | 0.14 | 64.85 | 0.3 |
| 2001-02 | 45.65 | 4.75 | 50.85 | 1.17 | 52.53 | 0.23 |
| West Dinajpur | | | | | | |
| 1972-73 | 64.48 | 5.96 | 70.44 | 7.44 | 80.11 | 0.03 |
| 1977-78 | 65.91 | 6.95 | 73.07 | 6.07 | 80.03 | 0.11 |
| 1982-83 | 65.94 | 5.79 | 71.5 | 6.64 | 79.5 | 1 |
| 1987-88 | 76.27 | 7.51 | 83.2 | 4.41 | 88.72 | 1.43 |
| 1993-94 | 71.04 | 4.9 | 75.6 | 3.4 | 79.47 | 0.18 |
| 1996-97 | 70.97 | 4.88 | 75 | 3.11 | 79.12 | 0.19 |
| 2001-02 | 66.36 | 1.2 | 67.73 | 1.65 | 70 | 1 |
| Malda | | | | | | |
| 1972-73 | 47.41 | 9.58 | 65.73 | 23.12 | 88.85 | 0.05 |
| 1977-78 | 60.18 | 7.91 | 73.94 | 13.95 | 87.89 | 0.08 |
| 1982-83 | 60.94 | 5.33 | 72.36 | 15.03 | 87.39 | 0.06 |
| 1987-88 | 61.18 | 9.66 | 73.64 | 14.28 | 87.93 | 0.07 |
| 1993-94 | 59.83 | 12.15 | 75.01 | 12.96 | 87.97 | 0.1 |
| 1996-97 | 60.98 | 12.8 | 77.14 | 7.1 | 84.24 | 0.16 |
| 2001-02 | 46.36 | 10.73 | 58.45 | 7.84 | 70.08 | 0.06 |
| Nadia | | | | | | |
| 1972-73 | 48.27 | 9.7 | 58.79 | 18.46 | 77.25 | 0.13 |
| 1977-78 | 44.82 | 8.33 | 53.99 | 19.25 | 73.24 | 0.14 |
| 1982-83 | 43.86 | 8.83 | 52.95 | 15.83 | 68.78 | 0.99 |
| 1987-88 | 49.4 | 9.1 | 58.88 | 14.24 | 73.12 | 0.95 |
| 1993-94 | 49.53 | 8.76 | 58.72 | 9.48 | 68.2 | 1.66 |
| 1996-97 | 47.62 | 8.16 | 55.81 | 7.62 | 63.49 | 1.93 |
| 2001-02 | 47.31 | 8.07 | 55.51 | 7.89 | 70.76 | 2.11 |

*Contd...*

| *District/ Year* | *Rice* | *Wheat* | *Total cereals* | *Total Pulses* | *Total food grains* | *Til* |
|---|---|---|---|---|---|---|
| Murshidabad | | | | | | |
| 1972-73 | 48.19 | 15.31 | 65.97 | 17.89 | 83.87 | 0.11 |
| 1977-78 | 46.02 | 15.07 | 62.98 | 18.51 | 81.5 | 0.44 |
| 1982-83 | 46.52 | 13.25 | 62.18 | 16.58 | 78.76 | 1.07 |
| 1987-88 | 52.62 | 14.05 | 67.73 | 8.75 | 76.49 | 1.65 |
| 1993-94 | 53.08 | 13.61 | 67.39 | 6.48 | 73.87 | 0.18 |
| 1996-97 | 52 | 14 | 67 | 6 | 72 | 0.2 |
| 2001-02 | 52.33 | 18.23 | 70.75 | 7.85 | 80.57 | 0.62 |
| Midnapore | | | | | | |
| 1972-73 | 86.84 | 1.7 | 88.77 | 4.58 | 93.34 | 0.2 |
| 1977-78 | 87.47 | 2.16 | 89.75 | 5.56 | 95.41 | 0.67 |
| 1982-83 | 91.65 | 0.75 | 92.49 | 2.51 | 95.01 | 1.14 |
| 1987-88 | 87.37 | 0.33 | 87.91 | 3.07 | 90.98 | 1.91 |
| 1993-94 | 86.11 | 0.82 | 87.02 | 1.56 | 88.59 | 2.16 |
| 1996-97 | 84.55 | 0.72 | 85.32 | 1.42 | 86.74 | 1.19 |
| 2001-02 | 74.83 | 1.18 | 76.09 | 1.44 | 80.14 | 2.18 |
| Coochbehar | | | | | | |
| 1972-73 | 69.58 | 1.53 | 71.76 | 4.31 | 76.07 | 0.05 |
| 1977-78 | 72.08 | 9.79 | 82.39 | 1.93 | 84.32 | 0.07 |
| 1982-83 | 71.3 | 3.27 | 75.05 | 1.8 | 76.85 | 0.25 |
| 1987-88 | 67.55 | 8.57 | 76.96 | 2.26 | 79.22 | 0.27 |
| 1993-94 | 69.66 | 4.45 | 74.43 | 2.69 | 77.12 | 0.55 |
| 1996-97 | 68.85 | 4.99 | 74.4 | 2.25 | 76.35 | 0.61 |
| 2001-02 | 64.96 | 5.03 | 70.15 | 2.06 | 70.38 | 0.11 |
| Purulia | | | | | | |
| 1972-73 | 83.42 | 1.44 | 88.16 | 6.95 | 95.1 | 0.13 |
| 1977-78 | 86.54 | 2.92 | 90.25 | 6.88 | 97.14 | 0.25 |
| 1982-83 | 88.52 | 0.3 | 92.57 | 6.1 | 98.39 | 0.13 |
| 1987-88 | 78.77 | 0.66 | 85.55 | 7.17 | 92.78 | 0.33 |
| 1993-94 | 85.87 | 0.67 | 91.15 | 5.6 | 96.76 | 0.47 |
| 1996-97 | 88.07 | 0.49 | 92.3 | 5.13 | 97.13 | 0.49 |
| 2001-02 | 83.67 | 0.41 | 86.18 | 5.75 | 94.49 | 0.3 |
| Darjeeling | | | | | | |
| 1972-73 | 39.77 | 0.34 | 88.16 | 6.95 | 95.1 | 0.01 |
| 1977-78 | 36.64 | 2.7 | 90.25 | 6.88 | 97.14 | 0.11 |
| 1982-83 | 32.77 | 2.56 | 92.57 | 6.1 | 98.39 | 0.32 |
| 1987-88 | 52.92 | 0.88 | 85.55 | 7.17 | 92.78 | 0.22 |
| 1993-94 | 37.88 | 2.33 | 91.15 | 5.6 | 96.76 | 0.34 |
| 1996-97 | 37.72 | 2.47 | 59.96 | 1.71 | 61.67 | 0.32 |
| 2001-02 | 30.35 | 2.66 | 50.05 | 1.02 | 55.69 | 0.36 |

*Contd...*

| District/ Year | Jute | Sugar cane | Potato | Total oilseeds | Total N.F. | Non superior cereals & NF |
|---|---|---|---|---|---|---|
| Burdwan | | | | | | |
| 1972-73 | 1.11 | 0.5 | 2.43 | 0.62 | 8.78 | 11.61 |
| 1977-78 | 2.39 | 0.34 | 3.55 | 2.07 | 8.66 | 12.73 |
| 1982-83 | 2.21 | 0.24 | 4.59 | 8.65 | 15.87 | 17.69 |
| 1987-88 | 1.34 | 0.21 | 5.26 | 9.99 | 16.92 | 17.6 |
| 1993-94 | 1.03 | 0.08 | 4.51 | 7.85 | 13.71 | 13.96 |
| 1996-97 | 1.43 | 0.06 | 4.01 | 5.48 | 13.16 | 14 |
| 2001-02 | 1.33 | 0.05 | 5.33 | 7.12 | 15.05 | 16 |
| Birbhum | | | | | | |
| 1972-73 | 0.07 | 0.86 | 1.18 | 1.32 | 5.59 | 12.84 |
| 1977-78 | 0.12 | 0.8 | 1.55 | 2.31 | 5.05 | 12.92 |
| 1982-83 | 0.05 | 0.87 | 1.44 | 6.36 | 9.06 | 14.73 |
| 1987-88 | 0.06 | 0.36 | 2.46 | 12.01 | 15.19 | 20.15 |
| 1993-94 | 0.07 | 0.34 | 1.91 | 7.34 | 9.9 | 12.65 |
| 1996-97 | 0.06 | 0.32 | 1.08 | 5.18 | 8.58 | 13.01 |
| 2001-02 | 0.02 | 0.17 | 2.79 | 8.15 | 10 | 12 |
| Hooghly | | | | | | |
| 1972-73 | 5.59 | 0.28 | 6.2 | 1.14 | 15.67 | 18.09 |
| 1977-78 | 11.73 | 0.09 | 8.59 | 3.39 | 23.88 | 26.15 |
| 1982-83 | 7.93 | 0.03 | 10.98 | 10.08 | 29.21 | 31.39 |
| 1987-88 | 5.74 | 0.02 | 12.74 | 12.78 | 31.26 | 31.67 |
| 1993-94 | 7.15 | 0.02 | 15.85 | 9.82 | 33.02 | 33.11 |
| 1996-97 | 6.34 | 0.03 | 17.84 | 10.7 | 42.76 | 34.03 |
| 2001-02 | 7.55 | 0.01 | 15.35 | 6.48 | 19.07 | 22.03 |
| 24-Paraganas | | | | | | |
| 1972-73 | 4.28 | 0.16 | 0.46 | 0.86 | 13.22 | 18.49 |
| 1977-78 | 7.06 | 0.15 | 0.68 | 1.33 | 10.52 | 16.02 |
| 1982-83 | 5.84 | 0.15 | 0.62 | 2.44 | 10.51 | 13.92 |
| 1987-88 | 6.49 | 0.03 | 0.62 | 5.92 | 13.54 | 18.71 |
| 1993-94 | 4.46 | 0.05 | 0.85 | 5.96 | 13.23 | 15.9 |
| 1996-97 | 5.89 | 0.07 | 0.75 | 4.89 | 13.76 | 16.34 |
| 2001-02 | 5.47 | 0.09 | 1.16 | 5.14 | 21.33 | 23.05 |
| Bankura | | | | | | |
| 1972-73 | 0.32 | 0.39 | 0.67 | 1.09 | 5.08 | 7.51 |
| 1977-78 | 0.37 | 0.3 | 1.21 | 4 | 6.79 | 4.57 |
| 1982-83 | 0.1 | 0.19 | 1.1 | 5.43 | 7.51 | 9.74 |
| 1987-88 | 0.07 | 0.04 | 2.38 | 8.62 | 11.68 | 13.49 |
| 1993-94 | 0.12 | 0.02 | 13.26 | 7.1 | 10.72 | 11.75 |
| 1996-97 | 0.25 | 0.09 | 12.75 | 5.96 | 11.07 | 12.01 |
| 2001-02 | 0.08 | 0.03 | 14.3 | 5.5 | 10 | 13 |

*Contd...*

| *District/ Year* | *Jute* | *Sugar cane* | *Potato* | *Total oilseeds* | *Total N.F.* | *Non superior cereals & NF* |
|---|---|---|---|---|---|---|
| Howrah | | | | | | |
| 1972-73 | 1.75 | 0.51 | 1.86 | 3.47 | 11.86 | 18.55 |
| 1977-78 | 3.56 | 0.32 | 3.07 | 2.91 | 9.04 | 19.81 |
| 1982-83 | 3.16 | 0.18 | 2.89 | 4.38 | 10.52 | 17.13 |
| 1987-88 | 1.74 | 0.15 | 3.55 | 7.16 | 12.05 | 16.55 |
| 1993-94 | 1.37 | 0.06 | 1.96 | 3.39 | 10.85 | 7.91 |
| 1996-97 | 1.43 | 0.09 | 2.34 | 0.62 | 9.85 | 8.76 |
| 2001-02 | 4.11 | 0.06 | 5.56 | 3.2 | 23.25 | 25.22 |
| Jalpaiguri | | | | | | |
| 1972-73 | 10.33 | 0.05 | 0.26 | 0.26 | 30.21 | 32.85 |
| 1977-78 | 7.42 | 0.06 | 0.94 | 0.94 | 24.26 | 27.75 |
| 1982-83 | 10.71 | 0.02 | 0.42 | 0.42 | 31.93 | 35.82 |
| 1987-88 | 12.12 | 0.01 | 0.93 | 0.93 | 10.75 | 22.72 |
| 1993-94 | 8.98 | 0.02 | 1.79 | 1.79 | 34.62 | 35.96 |
| 1996-97 | 9.31 | 0.05 | 1.08 | 2.68 | 35.15 | 34.34 |
| 2001-02 | 7.89 | 0.05 | 1.23 | 2.21 | 27.47 | 29.05 |
| West Dinajpur | | | | | | |
| 1972-73 | 8.89 | 0.27 | 0.68 | 2.43 | 19.09 | 29.56 |
| 1977-78 | 9.97 | 0.23 | 1.18 | 1.22 | 19.97 | 27.14 |
| 1982-83 | 8.88 | 0.15 | 0.78 | 3.08 | 20.5 | 28.27 |
| 1987-88 | 7.33 | 0.03 | 0.79 | 3.74 | 11.28 | 16.22 |
| 1993-94 | 0.03 | 0.13 | 0.98 | 2.6 | 20.53 | 4.06 |
| 1996-97 | 0.91 | 0.23 | 1.02 | 7.04 | 20.88 | 6.05 |
| 2001-02 | 0.26 | 0.23 | 1.43 | 8.5 | 30 | 32 |
| Malda | | | | | | |
| 1972-73 | 2.81 | 0.6 | 0.2 | 4.9 | 11.15 | 13.01 |
| 1977-78 | 5.95 | 0.63 | 0.56 | 4.31 | 12.11 | 31.91 |
| 1982-83 | 5.85 | 1.24 | 0.34 | 4.71 | 12.61 | 33.73 |
| 1987-88 | 2.29 | 0.58 | 0.53 | 8.29 | 12.07 | 29.16 |
| 1993-94 | 5.48 | 0.49 | 0.49 | 4.99 | 12.03 | 28.02 |
| 1996-97 | 5.84 | 0.56 | 0.43 | 5.86 | 15.73 | 27.07 |
| 2001-02 | 5.21 | 0.17 | 0.52 | 7.76 | 29.92 | 31.23 |
| Nadia | | | | | | |
| 1972-73 | 14.89 | 0.97 | 0.17 | 3.43 | 22.78 | 42.03 |
| 1977-78 | 18.52 | 0.96 | 0.66 | 5.14 | 26.76 | 36.4 |
| 1982-83 | 18.73 | 1.84 | 0.55 | 9.34 | 31.22 | 47.31 |
| 1987-88 | 16.37 | 0.05 | 0.22 | 9.52 | 26.88 | 41.5 |
| 1993-94 | 17.5 | 0.37 | 0.53 | 12.6 | 31.8 | 41.71 |
| 1996-97 | 18.43 | 0.54 | 0.63 | 12.85 | 34.79 | 41.05 |
| 2001-02 | 18.04 | 0.24 | 0.48 | 15.08 | 29.24 | 32.01 |

*Contd...*

| District/ Year | Jute | Sugar cane | Potato | Total oilseeds | Total N.F. | Non superior cereals & NF |
|---|---|---|---|---|---|---|
| Murshidabad | | | | | | |
| 1972-73 | 8.28 | 1.12 | 0.64 | 3.7 | 16.13 | 36.5 |
| 1977-78 | 11.62 | 1.12 | 0.9 | 4.47 | 18.5 | 33.91 |
| 1982-83 | 11.01 | 1.29 | 1.02 | 7.38 | 21.24 | 40.33 |
| 1987-88 | 11.02 | 0.29 | 0.97 | 9.97 | 23.51 | 33.33 |
| 1993-94 | 15.42 | 0.36 | 0.96 | 8.1 | 26.13 | 33.31 |
| 1996-97 | 16 | 0.31 | 0.98 | 8.9 | 28 | 33 |
| 2001-02 | 15.67 | 0.86 | 1.17 | 9.41 | 19.43 | 21.06 |
| Midnapore | | | | | | |
| 1972-73 | 1.16 | 0.35 | 1.17 | 0.61 | 6.66 | 11.76 |
| 1977-78 | 1.14 | 0.25 | 1.45 | 1.12 | 4.59 | 10.37 |
| 1982-83 | 1.07 | 0.21 | 1.54 | 1.53 | 4.99 | 7.6 |
| 1987-88 | 0.94 | 0.2 | 3.01 | 3.96 | 9.02 | 12.3 |
| 1993-94 | 0.7 | 0.07 | 4.42 | 5.32 | 11.41 | 13.07 |
| 1996-97 | 0.85 | 0.23 | 5.42 | 5.68 | 13.26 | 15.02 |
| 2001-02 | 0.53 | 0.31 | 4.19 | 5.05 | 19.86 | 23.02 |
| Coochbehar | | | | | | |
| 1972-73 | 15.8 | 0.11 | 0.43 | 2.54 | 23.93 | 28.89 |
| 1977-78 | 9.69 | 0.05 | 1.13 | 1.36 | 15.68 | 18.13 |
| 1982-83 | 14.61 | 0.03 | 0.52 | 3.55 | 28.15 | 25.43 |
| 1987-88 | 13.98 | 0.03 | 1.01 | 2.75 | 20.78 | 23.88 |
| 1993-94 | 14.22 | 0.04 | 1.99 | 3.1 | 22.88 | 25.89 |
| 1996-97 | 13.94 | 0.08 | 2.03 | 1.2 | 23.65 | 26.44 |
| 2001-02 | 16.55 | 0.08 | 2.4 | 2.48 | 29.62 | 30.23 |
| Purulia | | | | | | |
| 1972-73 | | 0.42 | 0.13 | 1.34 | 4.9 | 15.14 |
| 1977-78 | | 0.53 | 0.44 | 0.98 | 2.86 | 10.54 |
| 1982-83 | | 0.08 | 0.08 | 0.63 | 1.61 | 11.16 |
| 1987-88 | 0.06 | 0.3 | 0.36 | 2.76 | 7.22 | 20.57 |
| 1993-94 | | 0.3 | 0.3 | 2.4 | 3.24 | 13.46 |
| 1996-97 | 0.07 | 0.17 | 0.27 | 1.65 | 2.87 | 12.69 |
| 2001-02 | | 0.18 | 0.27 | 1.86 | 5.51 | 7.72 |
| Darjeeling | | | | | | |
| 1972-73 | 2.71 | | 1.94 | 1.34 | 4.9 | 59.89 |
| 1977-78 | 2.21 | | 3.28 | 0.98 | 2.86 | 60.66 |
| 1982-83 | 2.66 | | 3.52 | 0.63 | 1.61 | 64.67 |
| 1987-88 | 3.31 | | 5.57 | 2.76 | 7.22 | 46.2 |
| 1993-94 | 0.16 | | 5.87 | 2.4 | 3.24 | 59.79 |
| 1996-97 | 1.13 | | 6.03 | 1.14 | 38.33 | 57.43 |
| 2001-02 | 1.18 | | 5.81 | 1.21 | 44.31 | 46.58 |

*Notes:* N.F = Non-foodgrains
*Source:* Statistical Abstract, Government of West Bengal.

1972-73 to 2001-02, while that under wheat, total pulses and sugarcane has declined. The fluctuations has been observed in the case of til and total oilseeds. On the whole, the percentage share of total foodgrains in the GCA has shown a declining trend over years during the same period.

In Darjeeling, the percentage of area under cultivation of wheat, til and potato has shown a rising trend during 1972-73 to 2001-02. In case of total cereals, rice, total pulses, total oilseeds and jute the percentage of area cultivated has fluctuated over time with a decreasing trend. Thus, the percentage share of total foodgrains in the GCA has,in general, shown a decreasing trend during 1972-73 to 2001-02.

In case of Coochbehar, the percentage of area under cultivation of rice, wheat, total cereals, til, potato, has almost increased during 1972-73 to 2001-02 ,while that under rice ,total pulses, total oilseeds, jute, and sugarcane decreased over years. In general, the percentage share of total foodgrains in the GCA has shown an increasing trend during 1972-73 to 1996-97 ,but has shown an opposite trend during 1996-97 to 2001-02.

In Purulia, it is observed that the percentage of area under cultivation of rice, cereals, til ,total oilseeds and potato has shown an increasing trend during 1972-73 to 2001-02, while that under wheat, pulses, sugarcane has declined over years. Thus, what we see in the district is that the percentage share of total foodgrains in the GCA has shown a rising trend during 1972-73 to 2001-02.

In Midnapore, the percentage area under cultivation of total oilseeds, til and potato has shown an increasing trend over years during 1972-73 to 2001-02. In case of cultivation of wheat, rice, total cereals, pulses, jute and sugarcane, the percentage of area has decreased during 1972-73 to 2001-02. As a whole, the percentage share of total foodgrains in the GCA has shown a declining trend in the state during the same period.

In case of Burdwan, the percentage area under cultivation of rice, total oilseeds, pulses, til, jute and potato has shown an increasing trend over years, while that under wheat, total cereals and sugarcane has fluctuated with a declining trend during 1972-73 to 2001-02. Thus, it is quite evident from the above analysis that the percentage share of total foodgrains to GCA has declined over time during the same period.

In case of Murshidabad, the percentage of area under cultivation of wheat, til, potato, total oilseeds and jute has shown a rising trend, though not significant, while in case of rice, total cereals, pulses and sugarcane the opposite trend has been observed. In general, the percentage share of total foodgrains in the GCA has declined over time during 1972-73 to 2001-02 in the district.

In case of Nadia, the area under cultivation of til, total oilseeds, jute, and potato has increased during 1972-73 to 2001-02, while that under rice, wheat, total cereals, sugarcane and pulses has moved in opposite derection. In general, it may be said that the percentage share of total foodgrains to GCA has almost declined over years during 1972-73 to 2001-02.

From the above analysis, it can be said that the change in CP has moved in favour of cultivation of non-foodgrains in the districts of Midnapore, Darjeeling, Howrah, Burdwan, Nadia, Mursidabad, West Dinajpur, 24-Paraganas, Jalpaiguri, Hoogly, Bankura, Birbhum, Malda ,but CP has changed in favour of cultivation of foodgrains in two districts only, namely, Purulia and Coochbehar.

In other words, it can said that there has taken place a change in CP towards superior cereals and change has been found in case of Malda, Nadia, 24-Paraganas, Jalpaiguri, Hoogly, Bankura, Burdwan, Birbhum, and Midnapore. Among the foodgrains and non-foodgrains there has been a discernible shift in the CP towards HYV crops in all the districts of W.B. (Table 2). Due to the incidence of green revolution characterised by extensive cultivation of HYV crops in W.B. during 1980s, the pro-HYV CP change is a natural outcome in the districts of the state (Table-2). The percentage of area under the HYV crops in the GCA has increased in all the districts in W.B. The explanation behind the pro-non-foodgrains CP change is rapid population growth, changing composition of demand due to urbanisation and industrialisation, and the introduction of HYV technology in agriculture which has caused agriculture much more profitable than earlier. The high demand for food in this very densely populated state is the basic reason behind the pro-HYV foodgrains CP change. Along side this a pro-commercial CP change has also taken place in almost all the districts of W.B.

**Table 2**

**Percentage of HYV Area in the G.C.A in West Bengal by Districts**

| | *1976-1977* | *1980-1981* | *1983-1984* | *1987-1988* | *1990-1991* | *1995-1996* | *2001-2002* |
|---|---|---|---|---|---|---|---|
| Bankura | 16.63 | 24.46 | 30.88 | 40.9 | 43 | 92.67 | 98.87 |
| Birbhum | 24.19 | 41.92 | 45.58 | 46.9 | 51.46 | 95.11 | 95.03 |
| Burdwan | 31.46 | 44 | 40.92 | 61.1 | 64.06 | 91.63 | 96.11 |
| Coochbehar | 19.55 | 22.47 | 20.18 | 28.9 | 31.74 | 87.44 | 86.55 |
| Darjeeling | 15.9 | 15.46 | 13.33 | 16.11 | 17.41 | 48.36 | 41.49 |
| Hoogly | 42.34 | 44.93 | 54.56 | 66.6 | 57.65 | 90.03 | 92.67 |
| Howrah | 22.61 | 28.76 | 27.67 | 47.8 | 51.26 | 85.08 | 84.9 |
| Jalpaiguri | 13.55 | 15.29 | 12.98 | 26 | 37.29 | 71.78 | 64.78 |
| Malda | 19.03 | 25.49 | 26.9 | 35.3 | 42.6 | 82.69 | 82.28 |
| Midnapore | 17.82 | 23.52 | 26.77 | 37.35 | 49.26 | 94.54 | 93.78 |
| Nadia | 16.45 | 15.55 | 15.57 | 39.9 | 42.5 | 98.75 | 91.34 |
| 24-Paraganas | 19.59 | 24.6 | 24.91 | 37.5 | 41.96 | 84.34 | 85.16 |
| Purulia | 10.29 | 22.17 | 24.51 | 36.6 | 30.31 | 97.34 | 96.5 |
| West Dinajpur | 18.47 | 16.91 | 22.69 | 25.2 | 40.41 | 77.25 | 80.36 |
| Murshidabad | 25.74 | 35.16 | 32.92 | 37.06 | 43.07 | 94.79 | 98.59 |

*Source:* Ministry of Agriculture, Government of West Bengal.

### Change in CP Among HYV crops in the Districts of W.B.

Let us now analyse the change in CP among all HYV crops in the districts of W. B. during pre and post reform period (Table 3).

### Pre Reform Period (1976-77 to 1990-91)

In Darjeeling, among HYV crops, the percentage of area under cultivation of aman paddy, boro paddy, total paddy has increased with some fluctuations over years during 1976-77 to 1990-91. On the other side, the fluctuations have been observed for the cultivation of HYV aus paddy and HYV wheat.

In Coochbehar, the percentage of area under the cultivation of all varieties of HYV paddy has risen, though at a slow rate, over time during 1976-77 to 1990-91, while that under HYV wheat has fluctuated over years.

**Table 3**
**Percentage of Area Under Different HYV Crops in the Total HYV Cropped Area in the West Bengal by Districts**

| | 1976-1977 | 1981-1982 | 1985-1986 | 1990-1991 | 1991-1992 | 1995-1996 | 2001-2002 |
|---|---|---|---|---|---|---|---|
| *Hooghly* | | | | | | | |
| Aus paddy | 8.74 | 10.52 | 7.33 | 6.43 | 2.03 | 2.62 | 1.41 |
| Aman paddy | 47.56 | 56.01 | 59.74 | 59.14 | 46.31 | 43.4 | 42.71 |
| Boro paddy | 25.13 | 25.11 | 29.66 | 34.04 | 16.27 | 22.01 | 23.53 |
| Total paddy | 81.43 | 91.64 | 96.73 | 99.61 | 64.62 | 68.05 | 67.67 |
| Wheat | 18.57 | 8.36 | 3.27 | 0.39 | 0.49 | 0.24 | 0.24 |
| Total Food Grains | 100 | 100 | 100 | 100 | 65.28 | 68.44 | 68.01 |
| Total Oil Seeds | NA | NA | NA | NA | 10 | 7.71 | 7.3 |
| Potato | NA | NA | NA | NA | 16.25 | 18.49 | 17.3 |
| Jute | NA | NA | NA | NA | 8.27 | 5.19 | 7.11 |
| Rape and Mastard | NA | NA | NA | NA | 3.14 | 1.085 | 3.07 |
| Ground Nut | NA | NA | NA | NA | 0.4 | 1.12 | 1.01 |
| Til | NA | NA | NA | NA | 6.44 | 4.73 | 3.21 |
| *Burdwan* | | | | | | | |
| Aus paddy | 10.97 | 10.35 | 5.96 | 7.36 | 4.61 | 4.75 | 5.5 |
| Aman paddy | 54.53 | 69.05 | 64.99 | 65.01 | 59.07 | 59.85 | 60.06 |
| Boro paddy | 17.56 | 13.62 | 24.41 | 27.02 | 18.61 | 25.01 | 30.03 |
| Total paddy | 83.36 | 93.02 | 97.36 | 99.39 | 82.3 | 89.62 | 95.59 |
| Wheat | 24.32 | 19.87 | 8.27 | 3.63 | 0.36 | 0.58 | 0.68 |
| Total Food Grains | 100 | 100 | 100 | 100 | 82.94 | 87.81 | 96.27 |
| Total Oil Seeds | NA | NA | NA | NA | 9.28 | 5.51 | 0.73 |
| Potato | NA | NA | NA | NA | 5.19 | 4.94 | 0.5 |
| Jute | NA | NA | NA | NA | 2.13 | 1.43 | 0.25 |
| Rape and Mastard | NA | NA | NA | NA | 7.2 | 4.65 | 1.75 |
| Ground Nut | NA | NA | NA | NA | 0.11 | 0.21 | 0.75 |
| Til | NA | NA | NA | NA | 1.95 | 0.63 | 0.25 |
| *Bankura* | | | | | | | |
| Aus paddy | 20.23 | 9.63 | 13.64 | 7.93 | 7.58 | 7.61 | 5.49 |
| Aman paddy | 52.9 | 65.53 | 68.18 | 69.71 | 69.33 | 68.43 | 71.36 |
| Boro paddy | 2.55 | 4.97 | 9.91 | 18.73 | 7.36 | 11.35 | 11.55 |
| Total paddy | 75.68 | 80.13 | 91.73 | 96.37 | 84.27 | 87.4 | 88.42 |
| Wheat | 24.33 | 19.58 | 8.26 | 3.54 | 1.79 | 1.33 | 2.14 |
| Total Food Grains | 100 | 100 | 100 | 100 | 87.3 | 89.57 | 91 |
| Total Oil Seeds | NA | NA | NA | NA | 7.969 | 5.55 | 5.05 |
| Potato | NA | NA | NA | NA | 3.8 | 4.11 | 3.86 |
| Jute | NA | NA | NA | NA | 0.1 | 0.05 | 0.02 |
| Rape and Mastard | NA | NA | NA | NA | 4.08 | 2.4 | 2.53 |
| Ground Nut | NA | NA | NA | NA | 0.06 | 0.16 | 0.09 |
| Til | NA | NA | NA | NA | 3.77 | 2.91 | 2.34 |

*Contd...*

| | *1976-1977* | *1981-1982* | *1985-1986* | *1990-1991* | *1991-1992* | *1995-1996* | *2001-2002* |
|---|---|---|---|---|---|---|---|
| *Midnapore (W)* | | | | | | | |
| Aus paddy | 15.32 | 3.95 | 13.32 | 13.3 | 8.21 | 9.82 | 7.04 |
| Aman paddy | 58.03 | 67.21 | 61.09 | 60.77 | 65.86 | 59.33 | 60.87 |
| Boro paddy | 12.53 | 15.8 | 21.28 | 24.19 | 10.87 | 13.15 | 12.62 |
| Total paddy | 85.79 | 86.96 | 95.69 | 98.26 | 84.96 | 82.31 | 80.53 |
| Wheat | 14.11 | 13.04 | 4.31 | 1.74 | 0.6 | 1.59 | 2.34 |
| Total Food Grains | 100 | 100 | 100 | 100 | 86.93 | 84.47 | 82.87 |
| Total Oil Seeds | NA | NA | NA | NA | 6.27 | 7.63 | 7.85 |
| Potato | NA | NA | NA | NA | 4.63 | 5..00 | 6.7 |
| Jute | NA | NA | NA | NA | 0.65 | 0.56 | 0.39 |
| Rape and Mastard | NA | NA | NA | NA | 2.79 | 3.33 | 2.62 |
| Ground Nut | NA | NA | NA | NA | 0.39 | 0.55 | 0.79 |
| Til | NA | NA | NA | NA | 3 | 3.64 | 4.41 |
| *Midnapore(E)* | | | | | | | |
| Aus paddy | 9.71 | 3.24 | 3.26 | 1.77 | 1.72 | 3.12 | 4.51 |
| Aman paddy | 41.76 | 50.51 | 43.41 | 52.06 | 60.74 | 61.2 | 58.28 |
| Boro paddy | 41.51 | 41.82 | 51.05 | 47.08 | 24.02 | 23.19 | 26.33 |
| Total paddy | 92.44 | 95.57 | 97.72 | 96.87 | 86.49 | 87.52 | 89.13 |
| Wheat | 7.56 | 4.43 | 2.28 | 3.13 | 0.06 | 0.06 | 0.1 |
| Total Food Grains | 100 | 100 | 100 | 100 | 88.72 | 88.74 | 90.54 |
| Total Oil Seeds | NA | NA | NA | NA | 3.99 | 5.09 | 5.83 |
| Potato | NA | NA | NA | NA | 3.89 | 3.64 | 3.85 |
| Jute | NA | NA | NA | NA | 1.21 | 1.08 | 0.93 |
| Rape and Mastard | NA | NA | NA | NA | 1.2 | 1.6 | 0.73 |
| Ground Nut | NA | NA | NA | NA | 1.21 | 1.46 | 1.52 |
| Til | NA | NA | NA | NA | 1.56 | 2.03 | 1.57 |
| *Birbhum* | | | | | | | |
| Aus paddy | 6.34 | 3.31 | 2.04 | 7.58 | 3.38 | 2.52 | 5.39 |
| Aman paddy | 54.69 | 61.96 | 76.43 | 70.98 | 70.65 | 66.6 | 60.78 |
| Boro paddy | 2.21 | 7.43 | 11 | 15.51 | 6.56 | 14.36 | 4.08 |
| Total paddy | 63.24 | 72.7 | 89.52 | 94.07 | 80.6 | 83.5 | 69.25 |
| Wheat | 36.76 | 27.3 | 10.48 | 5.93 | 2.4 | 4.08 | 2.13 |
| Total Food Grains | 100 | 100 | 100 | 100 | 85.71 | 90.34 | 71.38 |
| Total Oil Seeds | NA | NA | NA | NA | 9.7 | 4.66 | 2.96 |
| Potato | NA | NA | NA | NA | 1.68 | 2.01 | 1.01 |
| Jute | NA | NA | NA | NA | 0.06 | 0.04 | 0 |
| Rape and Mastard | NA | NA | NA | NA | 9.09 | 4.34 | 2.83 |
| Ground Nut | NA | NA | NA | NA | 0.03 | 0.05 | 0.01 |
| Til | NA | NA | NA | NA | 0.5 | 0.16 | 0.08 |
| *Purulia* | | | | | | | |
| Aus paddy | 10.46 | 7.22 | 2.49 | 4.38 | 1.2 | 1.14 | 0.91 |
| Aman paddy | 72.62 | 82.44 | 93.91 | 91.59 | 79.32 | 81.56 | 84.52 |
| Boro paddy | 1.23 | 0.91 | 0.29 | 2.83 | 0.67 | 0.72 | 0.35 |

*Contd...*

| | 1976-1977 | 1981-1982 | 1985-1986 | 1990-1991 | 1991-1992 | 1995-1996 | 2001-2002 |
|---|---|---|---|---|---|---|---|
| Total paddy | 84.31 | 98.76 | 96.69 | 98.8 | 81.19 | 83.43 | 85.78 |
| Wheat | 15.69 | 1.24 | 3.31 | 1.2 | 0.29 | 1.31 | 1.33 |
| Total Food Grains | 100 | 100 | 100 | 100 | 91.2 | 93.3 | 95 |
| Total Oil Seeds | NA | NA | NA | NA | 2.48 | 1.77 | 0.97 |
| Potato | NA | NA | NA | NA | 0.24 | 0.44 | 0.27 |
| Jute | NA | NA | NA | NA | 0 | 0 | 0 |
| Rape and Mastard | NA | NA | NA | NA | 0.51 | 0.21 | 0.12 |
| Ground Nut | NA | NA | NA | NA | 0.93 | 0.01 | 0.3 |
| Til | NA | NA | NA | NA | 0.43 | 0.24 | 0.27 |
| *Darjeeling* | | | | | | | |
| Aus paddy | 23.22 | 14.48 | 17.42 | 23.21 | 12.38 | 6.62 | 7.93 |
| Aman paddy | 60.19 | 64.14 | 71.22 | 58.03 | 33.31 | 40.31 | 37.38 |
| Boro paddy | | 0.02 | | 0.87 | 0 | 0.12 | 1.32 |
| Total paddy | 83.41 | 78.64 | 94.64 | 82.11 | 45.69 | 47.05 | 46.63 |
| Wheat | 16.59 | 21.36 | 5.36 | 17.89 | 3.4 | 3.21 | 3.95 |
| Total Food Grains | 100 | 100 | 100 | 100 | 48.73 | 50.26 | 50.58 |
| Total Oil Seeds | NA | NA | NA | NA | 1.44 | 1.18 | 0.57 |
| Potato | NA | NA | NA | NA | 8.42 | 5.85 | 9.4 |
| Jute | NA | NA | NA | NA | 3.49 | 1.93 | 3.71 |
| Rape and Mastard | NA | NA | NA | NA | 0.71 | 0.27 | 0.12 |
| Ground Nut | NA | NA | NA | NA | 0 | 0 | 0 |
| Til | NA | NA | NA | NA | 0.15 | 0.49 | 0.32 |
| *Jalpaiguri* | | | | | | | |
| Aus paddy | 24.03 | 21.74 | 27.26 | 33.73 | 17.77 | 18.45 | 16.16 |
| Aman paddy | 40.68 | 40.59 | 62.02 | 57.93 | 57.69 | 56.66 | 48.46 |
| Boro paddy | | | 0.15 | 0.54 | 0.47 | 0.98 | 5.84 |
| Total paddy | 64.71 | 62.33 | 89.43 | 92.2 | 74.93 | 76.09 | 70.46 |
| Wheat | 35.29 | 37.67 | 10.57 | 7.8 | 3.72 | 6.15 | 7.33 |
| Total Food Grains | 100 | 100 | 100 | 100 | 81.17 | 82.24 | 79.52 |
| Total Oil Seeds | NA | NA | NA | NA | 3.04 | 3.86 | 3.13 |
| Potato | NA | NA | NA | NA | 1.49 | 0 | 3.93 |
| Jute | NA | NA | NA | NA | 12.63 | 11.4 | 12.74 |
| Rape and Mastard | NA | NA | NA | NA | 2.31 | 2.57 | 1.97 |
| Ground Nut | NA | NA | NA | NA | 0.03 | 0.05 | 0.05 |
| Til | NA | NA | NA | NA | 0.21 | 1.5 | 1.27 |
| *Coochbehar* | | | | | | | |
| Aus paddy | 7.75 | 10.19 | 23.6 | 30.57 | 18.21 | 12.33 | 4.67 |
| Aman paddy | 47.34 | 47.36 | 59.43 | 55.19 | 48.8 | 55.27 | 50.58 |
| Boro paddy | 0.35 | 0.82 | 0.46 | 5.12 | 2.13 | 3.99 | 6.41 |
| Total paddy | 55.44 | 58.37 | 83.49 | 90.88 | 69.14 | 71.59 | 61.67 |
| Wheat | 44.56 | 41.63 | 16.51 | 9.12 | 2.96 | 4.17 | 5.82 |
| Total Food Grains | 100 | 100 | 100 | 100 | 72.1 | 75.76 | 72.07 |
| Total Oil Seeds | NA | NA | NA | NA | 2.59 | 3.04 | 3.13 |

*Contd...*

| | *1976-1977* | *1981-1982* | *1985-1986* | *1990-1991* | *1991-1992* | *1995-1996* | *2001-2002* |
|---|---|---|---|---|---|---|---|
| Potato | NA | NA | NA | NA | 1.4 | 1.99 | 2.55 |
| Jute | NA | NA | NA | NA | 17.02 | 14.24 | 19.47 |
| Rape and Mastard | NA | NA | NA | NA | 1.34 | 1.72 | 2.09 |
| Ground Nut | NA | NA | NA | NA | 0.04 | 0 | 0.08 |
| Til | NA | NA | NA | NA | 0.38 | 0.21 | 0.07 |
| *West Dinajpur* | | | | | | | |
| Aus paddy | 17.45 | 16.19 | 15.74 | 16.59 | 10 | 10 | 8 |
| Aman paddy | 27.47 | 38.46 | 49.2 | 53.55 | 50 | 45 | 47 |
| Boro paddy | 4.83 | 5.71 | 9.72 | 19.62 | 25 | 30 | 32 |
| Total paddy | 49.75 | 60.36 | 74.66 | 89.76 | 85 | 85 | 87 |
| Wheat | 50.25 | 39.64 | 25.34 | 10.24 | 10 | 5 | 2 |
| Total Food Grains | 100 | 100 | 100 | 100 | 95 | 90 | 89 |
| Total Oil Seeds | NA | NA | NA | NA | 1 | 2 | 2 |
| Potato | NA | NA | NA | NA | 0 | 0 | 0 |
| Jute | NA | NA | NA | NA | 1 | 3 | 3 |
| Rape and Mastard | NA | NA | NA | NA | 1 | 2 | 2 |
| Ground Nut | NA | NA | NA | NA | 0 | 1 | 2 |
| Til | NA | NA | NA | NA | 2 | 2 | 2 |
| *Malda* | | | | | | | |
| Aus paddy | 13.47 | 7.94 | 9.39 | 11.69 | 9.36 | 7.79 | 3.53 |
| Aman paddy | 15.45 | 30.21 | 37.55 | 38.18 | 34.18 | 46.48 | 36.27 |
| Boro paddy | 26.6 | 26.59 | 28.41 | 30 | 14.28 | 16.59 | 18.75 |
| Total paddy | 55.52 | 64.74 | 75.35 | 79.87 | 57.82 | 70.86 | 58.55 |
| Wheat | 44.48 | 35.26 | 24.65 | 20.13 | 7.78 | 11.7 | 13.05 |
| Total Food Grains | 100 | 100 | 100 | 100 | 65.6 | 82.03 | 71.6 |
| Total Oil Seeds | NA | NA | NA | NA | 5.73 | 6.89 | 11 |
| Potato | NA | NA | NA | NA | 0.4 | 0.51 | 0.65 |
| Jute | NA | NA | NA | NA | 0 | 0 | 0 |
| Rape and Mastard | NA | NA | NA | NA | 5.45 | 5.63 | 9.21 |
| Ground Nut | NA | NA | NA | NA | 0 | 0 | 0 |
| Til | NA | NA | NA | NA | 0.17 | 0.14 | 0.06 |
| *Murshidabad* | | | | | | | |
| Aus paddy | 14.34 | 12.19 | 5.36 | 10.06 | 11 | 7.01 | 5.62 |
| Aman paddy | 21.49 | 32.36 | 31.28 | 39.24 | 30.33 | 29.29 | 27.13 |
| Boro paddy | 5.27 | 11.46 | 16.61 | 19.06 | 8.87 | 10.19 | 13.03 |
| Total paddy | 41.1 | 56.01 | 53.25 | 68.36 | 50.21 | 46.49 | 45.78 |
| Wheat | 58.9 | 43.99 | 46.75 | 31.64 | 12.26 | 15.5 | 15.96 |
| Total Food Grains | 100 | 100 | 100 | 100 | 68.51 | 61.99 | 61.74 |
| Total Oil Seeds | NA | NA | NA | NA | 10.16 | 7.15 | 7.95 |
| Potato | NA | NA | NA | NA | 0.89 | 0.94 | 0.88 |
| Jute | NA | NA | NA | NA | 13.98 | 17.09 | 16.57 |
| Rape and Mastard | NA | NA | NA | NA | 9.56 | 6.34 | 7.3 |

*Contd...*

| | *1976-1977* | *1981-1982* | *1985-1986* | *1990-1991* | *1991-1992* | *1995-1996* | *2001-2002* |
|---|---|---|---|---|---|---|---|
| Ground Nut | NA | NA | NA | NA | 0 | 0.02 | 0.08 |
| Til | NA | NA | NA | NA | 0.57 | 0.53 | 0.47 |
| *Nadia* | | | | | | | |
| Aus paddy | 28.63 | 19.24 | 13.68 | 15.8 | 10 | 10 | 5 |
| Aman paddy | 21.98 | 31.21 | 24.62 | 30.61 | 30 | 35 | 40 |
| Boro paddy | 11.45 | 26.61 | 31.21 | 37.14 | 40 | 35 | 40 |
| Total paddy | 62.06 | 77.06 | 69.51 | 83.55 | 80 | 80 | 85 |
| Wheat | 37.94 | 22.94 | 30.49 | 16.45 | 2 | 1 | 1 |
| Total Food Grains | 100 | 100 | 100 | 100 | 82 | 81 | 86 |
| Total Oil Seeds | NA | NA | NA | NA | 1.5 | 1 | 1 |
| Potato | NA | NA | NA | NA | 1.5 | 2 | 1 |
| Jute | NA | NA | NA | NA | 0 | 1 | 0 |
| Rape and Mastard | NA | NA | NA | NA | 2.5 | 2 | 0 |
| Ground Nut | NA | NA | NA | NA | 2.5 | 3 | 1 |
| Til | NA | NA | NA | NA | 0 | 0 | 1 |
| *24-Paraganas(N)* | | | | | | | |
| Aus paddy | 15.28 | 10.37 | 11.06 | 10.08 | 5.16 | 5.31 | 7.52 |
| Aman paddy | 45.08 | 56.49 | 54.74 | 51.2 | 43.66 | 47.86 | 36.1 |
| Boro paddy | 21.39 | 25.58 | 30.51 | 34.72 | 18.51 | 23.63 | 28.86 |
| Total paddy | 84.75 | 95.44 | 96.31 | 96 | 67.43 | 76.8 | 72.48 |
| Wheat | 15.25 | 4.55 | 3.69 | 4 | 2.1 | 1.88 | 3.58 |
| Total Food Grains | 100 | 100 | 100 | 100 | 69.53 | 78.68 | 76.06 |
| Total Oil Seeds | NA | NA | NA | NA | 8.97 | 8.29 | 12.82 |
| Potato | NA | NA | NA | NA | 0.94 | 0.9 | 2.23 |
| Jute | NA | NA | NA | NA | 9.14 | 10.03 | 14.77 |
| Rape and Mastard | NA | NA | NA | NA | 7.34 | 6.44 | 10.65 |
| Ground Nut | NA | NA | NA | NA | 0.24 | 0.58 | 1.05 |
| Til | NA | NA | NA | NA | 1.42 | 1.15 | 1.1 |
| *24-Paraganas(S)* | | | | | | | |
| Aus paddy | 6.41 | 4.17 | 2.89 | 1.54 | 0.32 | 0.54 | 2.49 |
| Aman paddy | 71.15 | 64.64 | 75.14 | 68.91 | 82 | 83.92 | 74.28 |
| Boro paddy | 15.38 | 27.44 | 20.92 | 29.05 | 10.51 | 11.75 | 13.75 |
| Total paddy | 92.94 | 96.25 | 98.95 | 99.5 | 92.83 | 95.21 | 90.52 |
| Wheat | 7.06 | 3.75 | 1.05 | 0.5 | 0.04 | 0.04 | 0.27 |
| Total Food Grains | 100 | 100 | 100 | 100 | 92.87 | 95.25 | 90.79 |
| Total Oil Seeds | NA | NA | NA | NA | 1.29 | 0.72 | 1 |
| Potato | NA | NA | NA | NA | 0.27 | 0.27 | 0.94 |
| Jute | NA | NA | NA | NA | 0.55 | 0.22 | 0.25 |
| Rape and Mastard | NA | NA | NA | NA | 0.82 | 0.45 | 0.53 |
| Ground Nut | NA | NA | NA | NA | 0.01 | 0 | 0.07 |
| Til | NA | NA | NA | NA | 0.37 | 0.18 | 0.21 |

*Contd...*

| | *1976-1977* | *1981-1982* | *1985-1986* | *1990-1991* | *1991-1992* | *1995-1996* | *2001-2002* |
|---|---|---|---|---|---|---|---|
| *Howrah* | | | | | | | |
| Aus paddy | 4.84 | 3.19 | 4.43 | 2.2 | 0.52 | 0.52 | 0.89 |
| Aman paddy | 50.52 | 47.87 | 36.69 | 38.74 | 52.07 | 55.06 | 52.71 |
| Boro paddy | 29.07 | 40.77 | 56.63 | 57.12 | 36.53 | 34.49 | 33.75 |
| Total paddy | 84.43 | 91.83 | 97.75 | 98.06 | 89.12 | 90 | 87.35 |
| Wheat | 15.57 | 8.17 | 2.25 | 1.94 | 0.06 | 1.07 | 0.33 |
| Total Food Grains | 100 | 100 | 100 | 100 | 89.18 | 91.07 | 87.68 |
| Total Oil Seeds | NA | NA | NA | NA | 3.76 | 2.37 | 3.15 |
| Potato | NA | NA | NA | NA | 2.22 | 6.61 | 5.22 |
| Jute | NA | NA | NA | NA | 2.22 | 2.43 | 3.33 |
| Rape and Mastard | NA | NA | NA | NA | 0.65 | 0.97 | 1.1 |
| Ground Nut | NA | NA | NA | NA | 0.07 | 0.21 | 1.3 |
| Til | NA | NA | NA | NA | 3.01 | 1.12 | 0.74 |

*Source:* Ministry of Agriculture, Government of West Bengal.

In Malda, it is observed that the percentage of area under HYV cultivation of aus paddy and wheat has shown a declining trend during 1976-77 to 1990-91 ,while an opposite trend is observed in case of the other two varieties of HYV paddy, namely aman and boro.

In Jalpaiguri, the percentage of area under cultivation of all varieties of HYV paddy has shown a rising trend during 1976-77 to 1990-91, while that of HYV wheat has fluctuated over time.

It is seen in case of West Dinajpur that the percentage of area under cultivation of all varieties of HYV paddy has slowly increased over time during 1976-77 to 1990-91, while that under HYV wheat has declined with some fluctuations. On the other hand, the percentage of area has fluctuated in case of the cultivation of HYV aus paddy.

In Murshidabad, the percentage of area under HYV cultivation of aus paddy and wheat has shown a declining trend during 1976-77 to 1990-91, but the opposite trend is found in case of cultivation of other two varieties of HYV paddy.

In Nadia, the percentage of area under cultivation of HYV aus paddy has remained more or less constant over time during 1976-77 to 1990-91, while that of other two varieties, namely aman and boro of HYV paddy, increased roughly during the same period. On the other hand, the percentage of area under HYV wheat has fluctuated over years in the district.

In 24-Paraganas(S), the percentage of area under HYV cultivation of aus paddy and wheat has shown a declining trend, though not a significant one, during 1976-77 to 1990-91, while that under boro paddy and total paddy increased over years. The area under HYV aman paddy has fluctuated over years.

In 24-Paraganas(N), the percentage of area under cultivation of HYV aus paddy and HYV wheat has decreased during 1976-77 to 1990-91, while that under other varieties of HYV paddy increased over years. In Howrah, the percentage of area under HYV cultivation of aman paddy and wheat has declined with some fluctuations during 1976-77 to 1990-91, while that under boro paddy and total paddy increased roughly. On the other side, the percentage area under cultivation of HYV aus paddy has fluctuated over years.

In both Burdwan and Hoogly, the percentage of area under different HYV crops reveals that the percentage area under aman paddy, boro paddy and total paddy has an increasing trend, while that under wheat has declined sharply during the period 1976-77 to 1990-91. The percentage of area under aus paddy has fluctuated over years during the same period.

In Birbhum, the percentage of area under aus, aman and boro paddy has very slowly risen over years, while an opposite trend is observed for wheat, during 1976-77 to 1990-91. But the percentage of area under total HYV paddy has fluctuated over years during the same period.

Again, the percentage of area under different HYV crops in Bankura, Purulia and Midnapore shows the same trend. In these districts, the percentage of area under aman and boro and total paddy has increased over years, while that under wheat and aus paddy has shown a declining trend during 1976-77 to 1990-91.

### Post Reform Period (1991-92 to 2001-02)

The percentage of area under aus paddy has risen over years for seven districts and that under aman and boro paddy has increased over time in case of 8 and 14 districts respectively during 1991-92 to 2001-02 (Table 3). The percentage of area under total paddy has increased over years in case of 11 districts during the same period. It is found that the change in CP has moved in favour of cultivation of HYV foodgrains for 10 districts during the post reform period.

In sum, it is observed that among the HYV crops, the paddy, in particular the aman and boro varieties, has the most domineering crops, in almost all the districts in W.B. It is a very natural outcome in these districts because the HYV technology has been applied most intensively for the expansion of paddy cultivation to meet the high demand for food.

## Change in CP in Blocks of Midnapore

In this section, an analysis of change in CP during 1992-2003 for different blocks of Midnapore can be done. For the purpose of analysis, the triennia average of percentage area under each crop to GCA has been calculated for the periods 1992-1995, 1996-1999 and 2000-2003. It is evident from Table 4 that the percentage area under total foodgrains has decereased over years for most (viz., 43) of the blocks in Midnapur district. That is, the CP has moved in favour of NF cultivation in most of the blocks. On the other hand, only in case of eleven blocks, viz, Binpur-I, Gopiballabpur-I, Debra, Garbeta-I, II & III, Mohanpur, Chandrakona-I & II, Midnapur, Ramnagar-I, the CP has moved in favour of foodgrains cultivation. Paddy is the principal crop among the foodgrains. It is also found that the percentage area under paddy has declined over years for most (viz., 42) of the blocks in the district. Only in case of 12 blocks, viz., Binpur-I, Gopiballabpur-I, Debra, Mohanpur, Garbeta-I,II,&III, Midnapur, Chandrakona-I&II, Nandigram-II, & Ramnagar-I, the percentage area under paddy cultivation has increased over years during the period under study. Side by side, the percentage area under other foodgrains, viz, wheat has declined over years for 19 blocks, viz., Binpur-I, Jamboni, Gopiballabpur-I, Pingla, Keshiary, Dantan-II, Narayangarh, Mohanpur, Garbeta-I, II & III, Kharagpur-II, Chandrakona-I&II, Nandigram-I&II, Sutahata, Ramnagar-I, in the district.

**Table 4**
**Percentage Share of Each Crop to Gross Cropped Area in Blocks of Midnapur During 1992-95 To 2000-03**

| *Block/Period* | *Paddy* | *Wheat* | *Potato* | *Jute* | *Mustard* | *Til* | *Total Food-grains* |
|---|---|---|---|---|---|---|---|
| *Jhargram* | | | | | | | |
| 1992-95 | 92.35 | 2.12 | 0.00 | 0.05 | 0.39 | 0.74 | 94.47 |
| 1996-99 | 94.74 | 0.89 | 0.00 | 0.00 | 0.71 | 0.59 | 95.63 |
| 2000-03 | 84.07 | 3.71 | 3.38 | 2.00 | 2.00 | 3.00 | 87.78 |
| *Binpur-1* | | | | | | | |
| 1992-95 | 66.21 | 14.56 | 5.73 | 0.02 | 1.80 | 11.66 | 80.77 |
| 1996-99 | 84.83 | 0.99 | 8.23 | 0.40 | 1.46 | 4.07 | 85.83 |
| 2000-03 | 89.63 | 1.02 | 5.42 | 0.77 | 1.24 | 1.88 | 90.65 |
| *Binpur-2* | | | | | | | |
| 1992-95 | 91.67 | 3.03 | 1.00 | 0.76 | 1.59 | 1.52 | 94.71 |
| 1996-99 | 91.04 | 2.76 | 1.24 | 0.06 | 1.91 | 1.69 | 93.80 |
| 2000-03 | 50.12 | 6.15 | 10.57 | 7.05 | 8.37 | 7.70 | 56.27 |
| *Jamboni* | | | | | | | |
| 1992-95 | 79.94 | 16.79 | 0.63 | 0.00 | 2.18 | 0.46 | 96.73 |
| 1996-99 | 92.17 | 2.23 | 1.00 | 0.00 | 3.00 | 1.60 | 94.40 |
| 2000-03 | 40.71 | 12.00 | 15.21 | 5.79 | 13.95 | 10.00 | 52.71 |
| *Nayagram* | | | | | | | |
| 1992-95 | 95.04 | 1.01 | 1.07 | 0.00 | 1.44 | 0.68 | 96.04 |
| 1996-99 | 95.43 | 0.68 | 0.89 | 0.00 | 1.68 | 0.54 | 96.11 |
| 2000-03 | 55.23 | 5.50 | 10.82 | 0.00 | 12.00 | 9.00 | 60.73 |
| *Sankrail* | | | | | | | |
| 1992-95 | 92.60 | 2.09 | 0.80 | 0.00 | 2.02 | 2.33 | 94.69 |
| 1996-99 | 92.08 | 2.50 | 0.79 | 0.05 | 3.01 | 1.44 | 94.59 |
| 2000-03 | 65.88 | 7.35 | 6.12 | 0.05 | 10.00 | 8.21 | 73.23 |
| *Gopiballabpur-1* | | | | | | | |
| 1992-95 | 85.77 | 5.56 | 1.63 | 0.00 | 5.44 | 0.54 | 91.34 |
| 1996-99 | 86.66 | 4.91 | 1.34 | 0.00 | 6.56 | 0.41 | 91.57 |
| 2000-03 | 88.00 | 4.94 | 1.41 | 0.00 | 4.85 | 0.19 | 92.94 |
| *Gopiballabpur-2* | | | | | | | |
| 1992-95 | 88.62 | 1.78 | 2.73 | 0.00 | 4.13 | 2.64 | 90.40 |
| 1996-99 | 90.57 | 0.82 | 3.10 | 0.00 | 4.27 | 1.14 | 91.39 |
| 2000-03 | 45.00 | 12.39 | 14.00 | 0.00 | 15.00 | 12.00 | 57.39 |
| *Debra* | | | | | | | |
| 1992-95 | 92.70 | 0.66 | 5.64 | 0.84 | 0.09 | 0.07 | 93.36 |
| 1996-99 | 92.92 | 0.71 | 5.64 | 0.30 | 0.39 | 0.03 | 93.64 |
| 2000-03 | 93.06 | 0.81 | 5.65 | 0.09 | 0.38 | 0.01 | 93.87 |

*Contd...*

| *Block/Period* | *Paddy* | *Wheat* | *Potato* | *Jute* | *Mustard* | *Til* | *Total Food-grains* |
|---|---|---|---|---|---|---|---|
| *Pingla* | | | | | | | |
| 1992-95 | 77.41 | 18.26 | 0.43 | 0.13 | 3.74 | 0.03 | 95.67 |
| 1996-99 | 92.36 | 1.30 | 0.57 | 1.13 | 4.50 | 0.13 | 93.67 |
| 2000-03 | 62.83 | 1.20 | 10.21 | 6.10 | 9.23 | 8.00 | 64.03 |
| *Keshiary* | | | | | | | |
| 1992-95 | 86.91 | 10.09 | 0.76 | 0.00 | 0.88 | 1.36 | 97.00 |
| 1996-99 | 95.38 | 0.97 | 0.89 | 0.00 | 0.85 | 1.92 | 96.35 |
| 2000-03 | 75.63 | 2.97 | 10.78 | 0.00 | 6.00 | 4.27 | 78.60 |
| *Dantan-1* | | | | | | | |
| 1992-95 | 92.92 | 0.69 | 1.30 | 0.01 | 3.03 | 1.48 | 93.61 |
| 1996-99 | 92.15 | 0.75 | 1.31 | 0.03 | 3.00 | 2.20 | 92.90 |
| 2000-03 | 55.00 | 5.01 | 17.21 | 0.03 | 12.00 | 10.00 | 60.01 |
| *Dantan-2* | | | | | | | |
| 1992-95 | 86.19 | 12.39 | 0.15 | 0.00 | 1.17 | 0.10 | 98.58 |
| 1996-99 | 97.88 | 0.08 | 0.29 | 0.00 | 1.43 | 0.32 | 97.96 |
| 2000-03 | 57.00 | 10.15 | 11.00 | 0.00 | 12.00 | 7.61 | 67.15 |
| *Narayangarh* | | | | | | | |
| 1992-95 | 93.93 | 1.29 | 0.12 | 0.04 | 3.29 | 0.92 | 95.22 |
| 1996-99 | 94.89 | 1.32 | 0.38 | 0.03 | 1.94 | 1.00 | 96.21 |
| 2000-03 | 61.41 | 2.00 | 8.21 | 0.01 | 7.12 | 8.69 | 63.41 |
| *Mohanpur* | | | | | | | |
| 1992-95 | 84.49 | 14.91 | 0.04 | 0.00 | 0.56 | 0.00 | 99.39 |
| 1996-99 | 97.84 | 0.06 | 0.09 | 0.00 | 1.02 | 0.99 | 97.91 |
| 2000-03 | 90.53 | 0.23 | 7.13 | 0.00 | 1.06 | 1.05 | 90.76 |
| *Salboni* | | | | | | | |
| 1992-95 | 69.03 | 14.05 | 11.44 | 0.00 | 1.86 | 2.76 | 83.08 |
| 1996-99 | 75.23 | 3.32 | 14.15 | 0.00 | 1.50 | 3.62 | 78.55 |
| 2000-03 | 60.00 | 5.84 | 25.00 | 0.00 | 2.41 | 3.93 | 65.84 |
| *Keshpur* | | | | | | | |
| 1992-95 | 73.62 | 2.89 | 11.75 | 0.52 | 6.71 | 4.47 | 76.51 |
| 1996-99 | 74.03 | 3.05 | 13.87 | 0.51 | 4.33 | 4.17 | 77.08 |
| 2000-03 | 60.12 | 9.14 | 20.00 | 0.41 | 5.43 | 4.14 | 69.26 |
| *Garbeta-1* | | | | | | | |
| 1992-95 | 53.87 | 13.87 | 22.16 | 0.02 | 2.39 | 2.39 | 67.74 |
| 1996-99 | 58.12 | 0.59 | 28.14 | 0.03 | 3.17 | 3.17 | 58.71 |
| 2000-03 | 69.00 | 5.00 | 20.00 | 0.03 | 2.72 | 2.72 | 74.00 |
| *Garbeta-2* | | | | | | | |
| 1992-95 | 57.55 | 6.60 | 24.70 | 0.36 | 1.54 | 9.08 | 64.15 |
| 1996-99 | 60.49 | 4.78 | 22.65 | 0.09 | 2.00 | 9.52 | 65.27 |
| 2000-03 | 77.00 | 3.79 | 12.00 | 0.09 | 1.79 | 5.00 | 80.79 |

*Contd...*

| *Block/Period* | *Paddy* | *Wheat* | *Potato* | *Jute* | *Mustard* | *Til* | *Total Food-grains* |
|---|---|---|---|---|---|---|---|
| *Garbeta-3* | | | | | | | |
| 1992-95 | 55.74 | 2.20 | 29.98 | 0.00 | 2.04 | 0.00 | 57.94 |
| 1996-99 | 52.53 | 1.65 | 30.30 | 0.00 | 1.84 | 0.00 | 54.18 |
| 2000-03 | 65.50 | 1.63 | 25.81 | 0.00 | 2.33 | 0.00 | 67.13 |
| *Midnapur* | | | | | | | |
| 1992-95 | 87.48 | 2.21 | 6.82 | 0.00 | 3.91 | 0.00 | 89.68 |
| 1996-99 | 88.52 | 2.09 | 4.98 | 0.00 | 4.17 | 0.00 | 90.60 |
| 2000-03 | 88.00 | 2.70 | 4.84 | 0.00 | 3.70 | 0.00 | 90.70 |
| *Sabang* | | | | | | | |
| 1992-95 | 97.24 | 0.28 | 0.37 | 0.00 | 1.81 | 0.30 | 97.52 |
| 1996-99 | 96.82 | 0.39 | 0.46 | 0.00 | 2.14 | 0.20 | 97.21 |
| 2000-03 | 47.00 | 0.39 | 10.53 | 12.00 | 12.00 | 12.62 | 47.39 |
| *Kharagpur-1* | | | | | | | |
| 1992-95 | 96.63 | 0.00 | 1.04 | 0.00 | 2.14 | 0.19 | 96.63 |
| 1996-99 | 95.93 | 0.38 | 1.05 | 0.00 | 2.00 | 0.54 | 96.31 |
| 2000-03 | 65.21 | 0.78 | 10.00 | 0.00 | 11.23 | 12.36 | 65.99 |
| *Kharagpur-2* | | | | | | | |
| 1992-95 | 87.24 | 10.12 | 0.47 | 0.00 | 2.11 | 0.06 | 97.36 |
| 1996-99 | 93.33 | 2.98 | 0.51 | 0.00 | 2.64 | 0.54 | 96.31 |
| 2000-03 | 78.23 | 2.92 | 8.75 | 0.00 | 6.13 | 0.98 | 81.15 |
| *Chandrakona-1* | | | | | | | |
| 1992-95 | 58.60 | 12.40 | 15.43 | 0.41 | 4.51 | 8.62 | 70.99 |
| 1996-99 | 64.89 | 0.12 | 14.44 | 0.29 | 7.50 | 12.42 | 65.01 |
| 2000-03 | 85.28 | 0.14 | 12.80 | 0.48 | 0.50 | 0.62 | 85.42 |
| *Chandrakona-2* | | | | | | | |
| 1992-95 | 56.76 | 11.37 | 16.50 | 1.08 | 1.40 | 12.85 | 68.13 |
| 1996-99 | 64.18 | 0.08 | 17.72 | 1.11 | 2.08 | 14.84 | 64.26 |
| 2000-03 | 70.00 | 5.00 | 17.01 | 1.23 | 1.90 | 4.00 | 75.00 |
| *Ghantal* | | | | | | | |
| 1992-95 | 92.00 | 0.70 | 3.60 | 2.45 | 0.55 | 0.70 | 92.70 |
| 1996-99 | 93.00 | 0.00 | 4.34 | 2.47 | 0.00 | 0.00 | 93.00 |
| 2000-03 | 65.00 | 11.44 | 16.49 | 3.00 | 1.00 | 1.00 | 76.44 |
| *Daspur-1* | | | | | | | |
| 1992-95 | 94.60 | 0.26 | 3.87 | 0.65 | 0.00 | 0.61 | 94.86 |
| 1996-99 | 95.67 | 0.00 | 4.33 | 0.00 | 0.00 | 0.00 | 95.67 |
| 2000-03 | 58.00 | 3.00 | 14.00 | 5.00 | 10.00 | 8.00 | 61.00 |
| *Daspur-2* | | | | | | | |
| 1992-95 | 95.89 | 0.39 | 2.82 | 0.49 | 0.39 | 0.00 | 96.29 |
| 1996-99 | 97.01 | 0.00 | 2.99 | 0.00 | 0.00 | 0.00 | 97.01 |
| 2000-03 | 51.00 | 1.00 | 13.00 | 6.00 | 7.00 | 6.00 | 52.00 |

*Contd...*

| *Block/Period* | *Paddy* | *Wheat* | *Potato* | *Jute* | *Mustard* | *Til* | *Total Food-grains* |
|---|---|---|---|---|---|---|---|
| *Tamluk-1 (Tamluk)* | | | | | | | |
| 1992-95 | 99.56 | 0.07 | 0.20 | 0.07 | 0.10 | 0.00 | 99.63 |
| 1996-99 | 99.74 | 0.00 | 0.26 | 0.00 | 0.00 | 0.00 | 99.74 |
| 2000-03 | 87.76 | 11.50 | 0.42 | 0.01 | 0.30 | 0.00 | 99.26 |
| *Tamluk-2 (Sahid Matangini)* | | | | | | | |
| 1992-95 | 99.31 | 0.00 | 0.45 | 0.24 | 0.00 | 0.00 | 99.31 |
| 1996-99 | 99.31 | 0.00 | 0.45 | 0.24 | 0.00 | 0.00 | 99.31 |
| 2000-03 | 87.81 | 11.36 | 0.55 | 0.20 | 0.05 | 0.00 | 99.18 |
| *Panskura-1* | | | | | | | |
| 1992-95 | 96.66 | 0.43 | 1.00 | 1.11 | 0.55 | 0.23 | 97.09 |
| 1996-99 | 97.67 | 0.00 | 1.00 | 1.30 | 0.00 | 0.00 | 97.67 |
| 2000-03 | 70.00 | 3.85 | 6.30 | 6.00 | 7.00 | 5.00 | 73.85 |
| *Panskura-2* | | | | | | | |
| 1992-95 | 97.00 | 0.34 | 1.10 | 1.09 | 0.23 | 0.00 | 97.34 |
| 1996-99 | 97.00 | 0.00 | 1.26 | 1.16 | 0.00 | 0.00 | 97.00 |
| 2000-03 | 55.00 | 2.00 | 6.00 | 6.00 | 4.58 | 12.00 | 57.00 |
| *Moyna* | | | | | | | |
| 1992-95 | 94.96 | 0.61 | 1.34 | 1.00 | 1.00 | 1.10 | 95.56 |
| 1996-99 | 95.61 | 0.94 | 1.05 | 1.06 | 0.50 | 0.81 | 96.55 |
| 2000-03 | 80.00 | 1.31 | 5.43 | 5.32 | 3.00 | 2.30 | 81.31 |
| *Mahisadal-1 (Nandakumar)* | | | | | | | |
| 1992-95 | 97.42 | 0.60 | 0.60 | 0.19 | 0.98 | 0.00 | 98.02 |
| 1996-99 | 97.01 | 0.54 | 0.54 | 0.08 | 0.99 | 0.00 | 97.55 |
| 2000-03 | 80.00 | 3.00 | 5.50 | 5.20 | 5.30 | 0.00 | 83.00 |
| *Mahisadal-2 (Mahisadal)* | | | | | | | |
| 1992-95 | 99.25 | 0.09 | 0.53 | 0.00 | 0.12 | 0.00 | 99.35 |
| 1996-99 | 99.40 | 0.00 | 0.60 | 0.00 | 0.00 | 0.00 | 99.40 |
| 2000-03 | 60.21 | 7.21 | 6.30 | 0.00 | 5.21 | 7.00 | 67.42 |
| *Nandigram-1* | | | | | | | |
| 1992-95 | 94.58 | 1.58 | 2.82 | 0.27 | 0.75 | 0.00 | 96.16 |
| 1996-99 | 94.03 | 1.72 | 3.25 | 0.22 | 0.77 | 0.00 | 95.75 |
| 2000-03 | 70.00 | 1.04 | 10.00 | 8.00 | 7.21 | 0.00 | 71.04 |
| *Nandigram-2* | | | | | | | |
| 1992-95 | 79.84 | 14.63 | 3.00 | 0.56 | 1.69 | 0.00 | 94.47 |
| 1996-99 | 80.74 | 14.63 | 3.02 | 0.60 | 1.00 | 0.00 | 95.37 |
| 2000-03 | 94.37 | 0.00 | 3.00 | 0.49 | 1.78 | 0.00 | 94.37 |

*Contd...*

| *Block/Period* | *Paddy* | *Wheat* | *Potato* | *Jute* | *Mustard* | *Til* | *Total Food-grains* |
|---|---|---|---|---|---|---|---|
| *Nandigram-3 (Chandipur)* | | | | | | | |
| 1992-95 | 97.45 | 0.03 | 1.35 | 0.51 | 0.60 | 0.00 | 97.48 |
| 1996-99 | 97.89 | 0.00 | 1.50 | 0.49 | 0.00 | 0.00 | 97.89 |
| 2000-03 | 81.97 | 0.75 | 10.23 | 2.34 | 2.45 | 0.00 | 82.72 |
| *Sutahata-1 (Sutahata)* | | | | | | | |
| 1992-95 | 99.40 | 0.08 | 0.39 | 0.00 | 0.12 | 0.00 | 99.49 |
| 1996-99 | 99.22 | 0.00 | 0.78 | 0.00 | 0.00 | 0.00 | 99.22 |
| 2000-03 | 55.21 | 0.00 | 8.32 | 8.32 | 9.23 | 2.30 | 55.21 |
| *Sutahata-2 (Haldia)* | | | | | | | |
| 1992-95 | 99.36 | 0.04 | 0.33 | 0.00 | 0.09 | 0.00 | 99.41 |
| 1996-99 | 99.27 | 0.00 | 0.73 | 0.00 | 0.00 | 0.00 | 99.27 |
| 2000-03 | 83.11 | 15.95 | 0.80 | 0.01 | 0.12 | 0.00 | 99.05 |
| *Potashpur-1* | | | | | | | |
| 1992-95 | 98.70 | 0.08 | 0.30 | 0.36 | 0.51 | 0.05 | 98.79 |
| 1996-99 | 99.04 | 0.00 | 0.33 | 0.33 | 0.00 | 0.00 | 99.04 |
| 2000-03 | 85.93 | 0.12 | 3.02 | 3.33 | 3.19 | 0.01 | 86.05 |
| *Potashpur-2* | | | | | | | |
| 1992-95 | 85.85 | 13.03 | 1.00 | 0.00 | 0.12 | 0.00 | 98.88 |
| 1996-99 | 85.36 | 13.03 | 1.31 | 0.00 | 0.30 | 0.00 | 98.40 |
| 2000-03 | 65.00 | 13.03 | 11.52 | 2.31 | 5.23 | 2.31 | 78.03 |
| *Khejuri-1* | | | | | | | |
| 1992-95 | 97.33 | 2.67 | 0.00 | 0.00 | 0.00 | 0.00 | 100.00 |
| 1996-99 | 95.92 | 3.77 | 0.00 | 0.00 | 0.31 | 0.00 | 99.69 |
| 2000-03 | 93.00 | 3.08 | 0.00 | 1.23 | 1.62 | 0.00 | 96.08 |
| *Khejuri-2* | | | | | | | |
| 1992-95 | 99.43 | 0.00 | 0.57 | 0.00 | 0.00 | 0.00 | 99.43 |
| 1996-99 | 83.06 | 16.15 | 0.73 | 0.00 | 0.07 | 0.00 | 99.20 |
| 2000-03 | 82.49 | 16.16 | 1.25 | 0.00 | 0.10 | 0.00 | 98.65 |
| *Bhagawanpur-1* | | | | | | | |
| 1992-95 | 98.34 | 0.09 | 1.15 | 0.30 | 0.02 | 0.00 | 98.43 |
| 1996-99 | 98.75 | 0.00 | 1.25 | 0.00 | 0.00 | 0.00 | 98.75 |
| 2000-03 | 84.03 | 14.28 | 1.57 | 0.05 | 0.05 | 0.00 | 98.31 |
| *Bhagawanpur-2* | | | | | | | |
| 1992-95 | 97.78 | 0.10 | 1.45 | 0.65 | 0.00 | 0.02 | 97.87 |
| 1996-99 | 98.13 | 0.00 | 1.02 | 0.83 | 0.00 | 0.00 | 98.13 |
| 2000-03 | 84.10 | 13.22 | 1.50 | 0.78 | 0.00 | 0.21 | 97.31 |

*Contd...*

| *Block/Period* | *Paddy* | *Wheat* | *Potato* | *Jute* | *Mustard* | *Til* | *Total Food-grains* |
|---|---|---|---|---|---|---|---|
| *Ramnagar-1* | | | | | | | |
| 1992-95 | 95.69 | 0.30 | 2.12 | 0.32 | 0.71 | 0.85 | 95.99 |
| 1996-99 | 97.64 | 0.00 | 2.36 | 0.00 | 0.00 | 0.00 | 97.64 |
| 2000-03 | 97.33 | 0.00 | 2.48 | 0.16 | 0.02 | 0.01 | 97.33 |
| *Ramnagar-2* | | | | | | | |
| 1992-95 | 97.01 | 0.23 | 1.44 | 0.22 | 0.14 | 0.89 | 97.24 |
| 1996-99 | 98.47 | 0.00 | 1.53 | 0.00 | 0.00 | 0.00 | 98.47 |
| 2000-03 | 65.00 | 7.12 | 12.00 | 5.21 | 5.00 | 2.10 | 72.12 |
| *Egra-1* | | | | | | | |
| 1992-95 | 99.51 | 0.03 | 0.34 | 0.00 | 0.12 | 0.00 | 99.54 |
| 1996-99 | 99.61 | 0.00 | 0.39 | 0.00 | 0.00 | 0.00 | 99.61 |
| 2000-03 | 70.00 | 1.90 | 13.00 | 3.00 | 2.00 | 2.00 | 71.90 |
| *Egra-2* | | | | | | | |
| 1992-95 | 99.58 | 0.09 | 0.25 | 0.00 | 0.09 | 0.00 | 99.66 |
| 1996-99 | 99.68 | 0.00 | 0.32 | 0.00 | 0.00 | 0.00 | 99.68 |
| 2000-03 | 67.21 | 2.11 | 10.32 | 4.21 | 5.06 | 4.21 | 69.32 |
| *Contai-1* | | | | | | | |
| 1992-95 | 98.06 | 0.05 | 1.71 | 0.13 | 0.04 | 0.00 | 98.11 |
| 1996-99 | 98.22 | 0.00 | 1.78 | 0.00 | 0.00 | 0.00 | 98.22 |
| 2000-03 | 62.42 | 10.21 | 10.21 | 3.21 | 5.00 | 3.21 | 72.63 |
| *Contai-2* | | | | | | | |
| 1992-95 | 98.59 | 0.04 | 1.10 | 0.10 | 0.10 | 0.00 | 98.63 |
| 1996-99 | 98.55 | 0.00 | 1.32 | 0.09 | 0.00 | 0.00 | 98.55 |
| 2000-03 | 73.25 | 4.50 | 11.00 | 5.03 | 5.01 | 1.21 | 77.75 |
| *Contai-3* | | | | | | | |
| 1992-95 | 97.77 | 0.20 | 1.30 | 0.08 | 0.40 | 0.20 | 97.97 |
| 1996-99 | 98.21 | 0.00 | 1.60 | 0.09 | 0.00 | 0.00 | 98.21 |
| 2000-03 | 82.00 | 15.90 | 2.00 | 0.10 | 0.00 | 0.00 | 97.90 |

*Source:* (1) District Statistical Handbook,Midnapur (Various Issues).
(2) P.A.O. Office, Purba Midnapore &Paschim Midnapore.

Concentrating on the area allocation among different non-food crops it is observed that the percentage area under potato cultivation has increased over years for most (viz. ,49) of the blocks in the district. The blocks experienced the decline in percentage area under potato cultivation during the period under study are Binpur-I, Gopiballabpur-I, Garbeta-3, Midnapur, Chandrakona-1. Again, the percentage area under mustard has decreased only for 5 blocks, viz., Binpur-I,

Chandrakona-I, Keshpur, Contai-III, and Ramnagar-I and increased for other blocks. Side by side, the percentage area under til cultivation has decreased over years during 1992-2003 for eight blocks, viz, Binpur-I, Gopiballabpur-I, Debra, Keshpur, Garbeta-2, Patashpur-I, Ramnager-1, Contai-3, but increased for other blocks .

The percentage area under jute cultivation has increased over years during 1992-2003 for most (viz., 45) of the blocks ,but declined only for nine blocks, viz , Debra, Narayangarh, Garbeta-II, Keshpur, Tamluk, Sahid Matangini, Bhagwanpur-I, Ramanagar-I, and Nandigram-II.

From the above analysis it becomes evident that the CP has moved in favour of non-foodgrains cultivation for most of the blocks during 1992-2003. This indicates a move towards commercialisation of agriculture in most of the blocks in Midnapore district during the post- reform period.

## A Summing-Up

In W.B. the pro-non-foodgrain CP change has taken place in 13 districts . Due to the incidence of green revolution in W.B. during 1980s, pro-HYV CP change has become a natural outcome in the districts of the state. Again, the CP has moved in favour of non-foodgrains cultivation for most of the blocks in Midnapore.

# CHAPTER–8

## TREND OF INSTITUTIONAL CREDIT IN INDIA: STATE, DISTRICT AND BLOCK LEVEL ANALYSIS

The previous two chapters 6 and 7 have highlighted the nature of change in CP at the state, district and block levels. The present chapter focuses on the trend of IC as one of the factors influencing CP change in India. Agricultural development bears a close positive correlation with credit and finance. The relationship between credit and agricultural development is high but indirect. Agricultural credit is one of the most crucial inputs in all agricultural development programmes in India. At the time of independence, the most important source of agricultural credit was the money lenders. The predominant position of the money lenders was due to the reason that there was no other source worth the name and the farmers were forced to borrow from them . This almost total dependence of farmers on the money lenders enabled the latter to dictate terms and exploit the former in a number of ways. On account of this reason, they can be termed anti-social elements. The government has, therefore, undertaken various steps to regulate the activities of the money lenders. The most important move to free the agriculturists from the clutches of the money lenders was the expansion of institutional credit to agriculture.

The change in CP is influenced by several natural, technological, economic and institutional factors including availability of credit. But the credit must be accessible, cheap, safe and productive. In this perspective, the provision of sufficient amount of IC to the farmers is an appropriate one for changing CP in the desired direction. It is to be noted that the nationalisation of commercial banks in India has made a structural change in the banking operations-a change from private banking into mass banking. In this regard, the priority sectors, in particular agriculture, have been given a special emphasis with a provision of larger amount of IC to the farmers. The fact is evident from All India Debt and Investment Survey (AIDIS) conducted decennially since 1951-52. From the four surveys conducted so far it is observed that the institutional sources of rural credit have increased their participation considerably. From a meagre 7.2% in 1951 the share of institutional sources increased to 61.2% in1981 and to 66.30% in 1991-commendable performances by any measure. The steep decline in the importance of non-institutional sources in rural credit from 70.8% in 1971 to 38.8% in 1981 and 33.70% in 1991 was mainly due to steep rise in the participation of banks in rural credit The rising participation of institutional agencies in rural credit is provided by the fact that the total direct institutional finances to agriculture which amounted to Rs. 1798 crores in 1970-71 increased to Rs. 2928 crores in 1979-80 (by 16.20%) and to Rs. 13,000 crores ( by 43.99%) in 1992-93. We shall now highlight the trend, direction/ composition of IC for agriculture in all India, sample states, districts of W.B. and blocks of Midnapore.

### Agricultural Credit in India by States: Trend ,and Direction/ Composition

In this section , we shall examine the trend, direction/ composition of flow of IC in all India as well as 12 sample states , viz., A.P., Gujarat, Haryana, H.P., M.P., Maharashtra, Orissa, Punjab, Rajasthan, Tamil Nadu, U.P., and W.B. It is to note that IC means the credit provided by both commercial banks including RRBs and Co-operatives. In this analysis the commercial bank credit includes credit of both commercial banks and RRBs. On the other hand, the co-operative credit is the sum of credit of different co-operative organisations, viz., Primary Agricultural Credit Societies (PACS), Farmers' Service Societies (FSS), Large Size Agricultural Multi-Purpose Society (LAMPS), Land Development Banks(LDBs).

We shall analyse the trend, direction/composition of flow of agricultural credit in the following manner.

### Institutional Credit-Trend in Sample States vis-à-vis All India

Let us discuss first the trend of IC in the sample states as well as all-India. The IC comprises the commercial bank (including RRBs) credit and co-operative credit. So far as the trend of IC is concerned, it is observed that in all-India, the IC per hectare has increased over years from Rs. 88.27 in June 1973 to Rs.5262.13 in March 2000 (Table 1). In case of A.P., the IC per hectare has shown a rising trend during June 1973–June 1983 from Rs. 155.31 to Rs. 914.41 and declined to Rs. 206 in June 1988 ,but increased again to Rs. 5228.04 in March 2000. In case of Gujarat, the IC per hectare has increased from Rs. 181.81 in June 1973 to Rs. 2614.75 in March 2000. In case of Haryana, H.P. , Punjab , Maharashtra, Rajasthan, Tamil Nadu, W.B. and M.P., the IC per hectare has a rising trend over years during June 1973 to March 2000. In Orissa, the IC per hectare has shown a rising trend during June 1973–June 1983 from Rs. 69.53 to Rs. 503.30. But it declined to Rs. 376.01 in June 1988, but again it increased to Rs. 3960.07 in March 2000. It is observed from Table 3(a) that the CAGRs of IC have been significant for All India and sample states.

### Commercial Bank Credit–Trend in Sample States vis-à-vis All India

Focusing on the trend of commercial bank credit (CBC) it is observed that in India as a whole, the CBC per hectare has risen from Rs. 35.25 in June 1973 to Rs.1570.89 in March 2002 (Table 2). In case of A.P., the CBC per hectare has shown a rising trend during June 1973–June 1983 from Rs. 36.76 to Rs. 514.99 . But it declined to Rs. 134.89 in 1988. However, it again increased to Rs. 5652.62 during March 2002. The CBC per hectare has risen over years during June 1973 to March

2002 for both Gujarat and Haryana. For other sample states the CBC per hectare has also shown a rising trend during the same period. It is observed from Table 3(a) that the CAGRs of CBC have been significant for All India and sample states.

**Table 1**
**Institutional Credit/Hectare in India by States**

*(in Rs.)*

| *States* | *June 1973* | *June 1978* | *June 1983* | *June 1988* | *March 1994* | *March 1997* | *March 2000* |
|---|---|---|---|---|---|---|---|
| Andhra Pradesh | 155.31 | 299.55 | 914.41 | 206 | 3271.3 | 3677.14 | 5228.04 |
| Gujarat | 181.81 | 305.68 | 850.5 | 1491.04 | 2370.16 | 7220.93 | 2614.75 |
| Haryana | 125.32 | 386.32 | 953.79 | 2853.82 | 3489.81 | 4099.71 | 23472.31 |
| H.P. | 107.29 | 212.48 | 557.28 | 1224.41 | 2006.82 | 2806.52 | 6067.05 |
| Madhya Pradesh | 72.8 | 133.18 | 307.95 | 762.95 | 1679.51 | 2379.62 | 1506.36 |
| Maharashtra | 150.72 | 267.85 | 703.03 | 961.99 | 2089.62 | 3589.72 | 11594.14 |
| Orissa | 69.53 | 146.43 | 503.3 | 376.07 | 1200.58 | 1923.1 | 3960.07 |
| Punjab | 229.4 | 429.81 | 1134.48 | 2334.36 | 3499.02 | 4999.15 | 28357.38 |
| Rajasthan | 35.51 | 109.76 | 295.25 | 994.14 | 1058.77 | 1960.78 | 3409.88 |
| Tamil Nadu | 332.3 | 657.13 | 1465.65 | 2623.11 | 5735.24 | 7756.36 | 23052.33 |
| Uttar Pradesh | 29.29 | 159.81 | 575.3 | 819.85 | 1415.15 | 1965.26 | 4373.22 |
| West Bengal | 107.69 | 265.64 | 491.33 | 887.58 | 1032.98 | 1692.09 | 3897.05 |
| All India | 88.27 | 148.22 | 320.73 | 603.63 | 837.44 | 2206.45 | 5262.13 |

*Source:* (i) Banking Statistics, R.B.I.
(ii) Statistical Statements Relating to Co-oprative Movement in India, R.B.I.
(iii) Area and Production of Principal Crops in India, Government of India.

## Co-operative Credit-Trend in Sample States vis-à-vis All India

Let us analyse the trend of Co-operative Credit (COC) per hectare in the sample states in all India perspective. In India, the COC per hectare has increased from Rs. 53.02 in June 1973 to Rs. 4459.35 in March 2000 (Table 3). In case of A.P., the COC per hectare showed rising trend over years during June 1973 to June 1983 from Rs. 118.55 to Rs. 399.42, but has declined to Rs. 71.11 in June 1988 and again increased to Rs. 1119.20 in March 1994, but declined to Rs. 984.52 in March 2000. The states like Gujarat, Haryana, H.P., M.P., Maharastra, Orissa, Punjab, Tamil Nadu and W.B. have shown a rising trend of COC per hectare during June 1973 to March 2000. On the other hand,

**Table 2**
**All Scheduled Commercial Banks**
**Agricultural Credit per hectare in India by States**

*(in Rs.)*

| *States* | *June 1973* | *June 1978* | *June 1983* | *June 1988* | *March 1994* | *March 1997* | *March 2002* |
|---|---|---|---|---|---|---|---|
| Andhra Pradesh | 36.76 | 176.85 | 514.99 | 134.89 | 2152.1 | 3485.61 | 5652.62 |
| Gujarat | 44.28 | 95.62 | 256.66 | 798.68 | 1253.12 | 4573.49 | 2883.27 |
| Haryana | 15.71 | 142.91 | 405.08 | 1252.48 | 1802.55 | 1854.7 | 7365.82 |
| H.P. | 6.25 | 67.95 | 323.84 | 716.54 | 1122.1 | 1206.77 | 9401.6 |
| Madhya Pradesh | 6.52 | 38.13 | 123.82 | 340.9 | 556.75 | 264.75 | 1487.25 |
| Maharashtra | 48.35 | 119.35 | 314.19 | 430.87 | 865.15 | 1370.93 | 2993.11 |
| Orissa | 3.43 | 43.53 | 199.1 | 208.87 | 872.5 | 1177.36 | 1439.43 |
| Punjab | 21.98 | 205.81 | 640.39 | 1200.93 | 2110.7 | 2488.47 | 1080.78 |
| Rajasthan | 8.6 | 39.33 | 146.53 | 462.49 | 676.63 | 682.54 | 1979.6 |
| Tamil Nadu | 83.98 | 94.48 | 853.1 | 1898.29 | 3341.4 | 4573.49 | 7681.85 |
| Uttar Pradesh | 24.54 | 77.61 | 260.6 | 480.77 | 1078.8 | 1212.91 | 2714.46 |
| West Bengal | 73.19 | 142.77 | 284.06 | 570.59 | 969.44 | 1224.9 | 2153.11 |
| All India | 35.25 | 113.86 | 162.77 | 332.61 | 253.65 | 1767.68 | 1570.89 |

*Source:* (i) Banking Statistics, R.B.I.
(ii) Area and Production of Principal Crop in India, Government of India.

**Table 3**
**Co-operative Agricultural Credit/hectare in India by States**

*(in Rs.)*

| *States* | *June 1973* | *June 1978* | *June 1983* | *June 1988* | *March 1994* | *March 1997* | *March 2000* |
|---|---|---|---|---|---|---|---|
| Andhra Pradesh | 118.55 | 122.7 | 399.42 | 71.11 | 1119.2 | 191.53 | 984.52 |
| Gujarat | 137.53 | 210.06 | 593.9 | 692.36 | 1117.04 | 2647.44 | 790.01 |
| Haryana | 109.61 | 243.41 | 548.71 | 1602.34 | 1687.26 | 2245.01 | 18823.77 |
| H.P. | 101.04 | 144.53 | 233.44 | 7.87 | 884.72 | 1599.75 | 20968.58 |
| Madhya Pradesh | 66.28 | 95.05 | 184.13 | 422.05 | 1122.76 | 2114.87 | 2564.86 |
| Maharashtra | 102.37 | 148.5 | 388.84 | 531.12 | 1224.47 | 2218.79 | 9415.48 |
| Orissa | 66.1 | 102.9 | 304.2 | 167.2 | 328.08 | 745.75 | 2875.04 |
| Punjab | 207.42 | 224 | 94.09 | 1133.43 | 1388.32 | 2510.68 | 20037.1 |
| Rajasthan | 26.91 | 70.43 | 148.72 | 531.65 | 382.14 | 1278.24 | 2100.31 |
| Tamil Nadu | 248.32 | 562.65 | 612.54 | 727.82 | 1793.84 | 3183.07 | 16479.31 |
| Uttar Pradesh | 4.75 | 82.2 | 314.7 | 339.08 | 336.57 | 752.35 | 2406.3 |
| West Bengal | 34.5 | 122.87 | 207.27 | 316.99 | 463.54 | 467.19 | 2639.53 |
| All India | 53.02 | 34.36 | 157.96 | 271.02 | 583.79 | 438.77 | 4459.35 |

*Source:* (i) Statistical Statements Relating to Co-operative Movement in India R.B.I./NABARD
(ii) Area and Production of Principal Crops in India, Government of India.

for the two states, viz. , Rajasthan and U.P., the COC per hectare has increased over years during June 1973 to June 1988, but declined in March 1994 and again increased in March 2000. It is observed from Table 3(a) that the CAGRs of COC have been significant for All India and sample states.

**Table 3(a)**
**Compound Annual Growth Rates (CAGRs) of Credit Per Hectare in India by States**

| *States* | *Institutional Credit/Ha* | *Commercial Bank Credit/Ha* | *Co-operative Credit/ Ha* |
|---|---|---|---|
| All India | 16.98* | 18.93* | 19.21* |
| Punjab | 18.35* | 22.92* | 21.61* |
| Haryana | 18.87* | 24.91* | 21.32* |
| Rajasthan | 19.92* | 27.86* | 25.31* |
| Maharastra | 13.98* | 18.18* | 20.31* |
| Gujarat | 18.88* | 21.92* | 19.32* |
| Tamil Nadu | 15.63* | 23.92* | 21.61* |
| A.P | 16.72* | 19.53* | 15.23* |
| M.P | 21.92* | 23.77* | 19.95* |
| U.P | 16.98* | 19.92* | 16.98* |
| H.P | 19.92* | 29.99* | 19.32* |
| Orissa | 16.78* | 23.77* | 18.53* |
| West Bengal | 10.72* | 13.89* | 15.31* |

*Notes:* *indicates value significant at 1% level.

## Commercial Bank Credit by Size Classes of Land Holdings in Sample States vis-à-vis All India-An Analysis of Direction/ Composition

Let us highlight the direction/composition of CBC across the different size classes of land holdings. In A.P. it is observed that the percentage share of a size class CBC in total CBC has shown slowly rising trend over years during 1980-95 for all the size classes of land holdings (Table 4). In Tamil Nadu, the percentage share of a size class CBC to total CBC (PCBC) has risen over years for all size classes of land holdings except the size class of upto 2.5 acre. In M.P., the PCBC has shown a rising trend for size class of 'above 2.5 to 5 acres' and the declining trends for the size class of 'above 10 acres' and no discernible trend for the size class of upto 2.5 acres. In the case of U.P., Gujarat, Maharashtra, Punjab, and all-India, the

**Table 4**
**Percentage Share of Disbursed Credit from Scheduled Comercial Banks of a Size Class in the States in India**

| *States* | *Year* | *Upto 2.5 Acre* | *Above 2.5- 5 Acre* | *Above 5 Acres* | *Above 10 Acres* |
|---|---|---|---|---|---|
| Andhra Pradesh | 1980 | 29.18 | 26.57 | 21.8 | 22.44 |
| | 1981 | 26.42 | 26.59 | 32.61 | 23.38 |
| | 1986 | 31.1 | 40.43 | 28.47 | |
| | 1991 | 31.39 | 32.61 | 36.07 | |
| | 1995 | 32.2 | 30.73 | 37.07 | |
| Gujarat | 1980 | 9.02 | 11.19 | 20.94 | 58.85 |
| | 1981 | 9.85 | 10.1 | 19.04 | 61 |
| | 1986 | 8.29 | 14.5 | 77.21 | |
| | 1991 | 8.32 | 16.59 | 75.09 | |
| | 1995 | 12.39 | 17.26 | 70.42 | |
| Haryana | 1980 | 3.94 | 3.94 | 12.82 | 79.29 |
| | 1981 | 4.83 | 4.91 | 14.42 | 75.84 |
| | 1986 | 39.42 | 60.58 | | |
| | 1991 | 6.51 | 13.75 | 79.74 | |
| | 1995 | 5.85 | 11.14 | 83.01 | |
| H. Pradesh | 1980 | 36.66 | 27.76 | 17.48 | 18.1 |
| | 1981 | 42.98 | 26.3 | 12.9 | 17.83 |
| | 1986 | 53.45 | 15.6 | 30.94 | |
| | 1991 | 31.26 | 16.83 | 57.9 | |
| | 1995 | 33.69 | 20.66 | 45.65 | |
| M. Pradesh | 1980 | 9.98 | 5.39 | 31.72 | 52 |
| | 1981 | 11.22 | 17.57 | 25.47 | 43 |
| | 1986 | 12.77 | 23.79 | 63.92 | |
| | 1991 | 26.63 | 30 | 43.39 | |
| | 1995 | 11.02 | 20.02 | 68.96 | |
| Maharashtra | 1980 | 10.01 | 12.57 | 25.1 | 52.3 |
| | 1981 | 8.7 | 13.11 | 24.61 | 53.5 |
| | 1986 | 11.88 | 19.68 | 68.44 | |
| | 1991 | 16.1 | 22.67 | 61.21 | |
| | 1995 | 19.29 | 22.61 | 58.1 | |
| Orissa | 1980 | 53.58 | 21.82 | 14.78 | 9.81 |
| | 1981 | 44.15 | 22 | 17.24 | 16.59 |
| | 1986 | 42.9 | 34.19 | 22.9 | |
| | 1991 | 37.25 | 30.74 | 32 | |
| | 1995 | 35.21 | 33.17 | 31.6 | |
| Punjab | 1980 | 4.8 | 5.73 | 19.9 | 69.57 |
| | 1981 | 3.83 | 6.46 | 17.67 | 72.04 |
| | 1986 | 20.15 | 15.27 | 64.59 | |

*Contd...*

| States | Year | Upto 2.5 Acre | Above 2.5-5 Acre | Above 5 Acres | Above 10 Acres |
|---|---|---|---|---|---|
| | 1991 | 5.88 | 15.21 | 78.93 | |
| | 1995 | 6.14 | 15.49 | 78.37 | |
| Rajasthan | 1980 | 8.16 | 11.37 | 16.47 | 63.55 |
| | 1981 | 7.9 | 12 | 17.11 | 62.98 |
| | 1986 | 8.87 | 13.64 | 77.49 | |
| | 1991 | 10.98 | 17.41 | 71.6 | |
| | 1995 | 11.92 | 13.38 | 74.7 | |
| Tamil Nadu | 1980 | 47.32 | 19.6 | 12.83 | 20.25 |
| | 1981 | 44.77 | 21.03 | 12.93 | 21.27 |
| | 1986 | 38.9 | 29.85 | 31.25 | |
| | 1991 | 40.58 | 24.61 | 34.82 | |
| | 1995 | 37.86 | 29.02 | 33.12 | |
| Uttar Pradesh | 1980 | 13.44 | 16.54 | 10.19 | 61.83 |
| | 1981 | 11.53 | 17.2 | 16.58 | 54.69 |
| | 1986 | 18.44 | 22.32 | 59.24 | |
| | 1991 | 20.44 | 23.75 | 56.01 | |
| | 1995 | 19.57 | 22.74 | 57.75 | |
| West Bengal | 1980 | 31.16 | 29.67 | 13.45 | 26.8 |
| | 1981 | 32.67 | 25.75 | 8.39 | 33.05 |
| | 1986 | 27.16 | 16.91 | 55.93 | |
| | 1991 | 36.44 | 19.32 | 44.24 | |
| | 1995 | 43.46 | 19.04 | 37.5 | |
| All India | 1980 | 21.44 | 16.84 | 16.97 | 44.74 |
| | 1981 | 21.41 | 13.28 | 18.57 | 46.74 |
| | 1986 | 22.82 | 22.18 | 55 | |
| | 1991 | 12.76 | 33.59 | 53.66 | |
| | 1995 | 39.25 | 36.93 | 23.82 | |
| | 2000 | 22.61 | 23.56 | 53.81 | |
| | 2004 | 26 | 25 | 49 | |

*Source:* Report on Currency and Finance, R.B.I.

PCBC has risen over years with some fluctuations for the size classes of 'upto 2.5 acres', 'above 2.5 to 5 acres' and 'above 5 acres'. In Orissa, the PCBC has shown a rising trend during 1980-95 for the size classes of 'above 2.5 acres', 'above 5 acres' and 'above10 acres' ,but declining trend for the size class of 'upto 2.5 acres'. In Haryana, the PCBC has remained more or less constant for the size classes of 'upto 2.5 acres' and 'above 2.5 to 5 acres', has risen for the size classes of 'above 5 acres' and has fallen for the size class, 'above 10 acres'. In H.P., the PCBC has shown fluctuation for the size classes of 'upto 2.5 acres' and 'above 5 acres', but declining trend for the size classes of '2.5 to 5 acres'. In W.B., the PCBC has slowly risen over years during 1980-95 for all the size classes except the size class of 'above 2.5 to 5 acres'. In Rajasthan, the PCBC has remained almost constant for all the size classes, except for the size class of 'above 10 acres' where it has declined over years during 1980-95.

The analysis of data on the credit availability for India and W.B. (Tables 4a and 4b) shows that weaker sections (i.e., relatively smaller size classes of land holdings) has received considerable amount of loans per hectare in relation to the amount of loans received by the rich farmers (land holding size greater than 4 hectares). It is observed that the weaker sections have sometimes got larger amount of credit as compared to the rich farmers.

**Table 4a**
**Commercial Bank's Agricultural Credit Per Hectare by Size Classes of Land Holdings (L.H.s) in India**

| | *Agricultural Credit Per Hectare (in Rs.)* | | |
|---|---|---|---|
| *L. H.s (in Hectare)* | *1980-81* | *1985-86* | *1990-91* |
| Below 1 Hectare | 208.58 | 565.99 | 512.63 |
| 1.0-4.0 Hectare | 117.18 | 511 | 1255.25 |
| Above 4 Hectare | 170.07 | 460.88 | 573.24 |

*Source:* (1) Report on Currency and Finance, R.B.I.
(2) Agricultural Census, Government of India

**Table 4b**
**Commercial Bank's Agricultural Credit per Hectare by Size Classes of Land Holdings (L.H.s) in West Bengal**

| | *Agricultural Credit Per Hectare (in Rs.)* | | |
|---|---|---|---|
| *L. H.s (in Hectare)* | *1980-81* | *1985-86* | *1990-91* |
| Below 1 Hectare | 211.91 | 463.24 | 625.24 |
| 1.0-4.0 Hectare | 156.7 | 300 | 404.14 |
| Above 4 Hectare | 199.27 | 868.1 | 823.16 |

*Source:* (1) Report on Currency and Finance, R.B.I.
(2) Agricultural Census, Government of West Bengal.

## PACS Credit and Co-operative Credit by Size Classes of Land Holdings in Sample States and India: An Analysis of Direction/ Composition

Concentrating on the percentage of a size class PACS credit to total PACS credit (PPACC) it is observed that in all India, there exists no definite trend for any size class of land holdings during 1976-77 to 1997-98 (Table 5). In Gujarat, the PPACC has increased, though very slowly, over years for three size classes, namely, upto 1 hectare, 1-2 hectares and 2-4 hectares, but declined for other size classes of land holdings. In case of H.P., the PPACC has decreased over years for all the size classes of land holdings. In Punjab, the PPACC has increased, though at a very slow rate, over years during 1976-77 to 1997-98 for the size classes 1 to 2 hectares, 2 to 4 hectares, and 4 to 8 hectares, but declined for other size classes. In Rajasthan, the PPACC has slowly declined over years during 1976-77 to 1997-98 for the size classes of 'upto 1 hectare', '4 to 8 hectare', and 'above 8 hectare', but declined for other classes. In case of Tamil Nadu, the PPACC has shown fluctuations for all the size classes of land holdings during 1976-77 to 1997-98. In U.P., the PPACC has no trend for all size classes.

**Table 5**
**Percentage Share of PACS Credit of a Size Class of Land Holding in India by Statee**

| *State/ Year* | *Total* | *Upto 1 ha* | *1-2 ha* | *2-4 ha* | *4-8 ha* | *above 8 ha* |
|---|---|---|---|---|---|---|
| Gujarat | | | | | | |
| 1976-77 | 100 | 4.58 | 9.67 | 17.43 | 34.34 | 33.85 |
| 1983-84 | 100 | 7.51 | 13.42 | 21.76 | 26.59 | 30.52 |
| 1991-92 | 100 | 4.48 | 12.72 | 21.05 | 25.62 | 38.7 |
| 1994-95 | 100 | 9.31 | 16.6 | 26.15 | 28.99 | 18.94 |
| 1997-98 | 100 | 12.56 | 14.53 | 20.29 | 25.8 | 26.82 |
| H.P | | | | | | |
| 1977-78 | 100 | 17.94 | 39.06 | 25.8 | 14.12 | 3.05 |
| 1983-84 | 100 | 27.33 | 44.15 | 3.1 | 15.89 | 9.51 |
| 1989-90 | 100 | 33.67 | 37.01 | 15.73 | 6.61 | 6.96 |
| 1993-94 | 100 | 54.8 | 26.66 | 11.85 | 6.3 | 0.39 |
| 1997-98 | 100 | 54.36 | 20.18 | 10.47 | 7.99 | 2.99 |

Contd...

| *State/ Year* | *Total* | *Upto 1 ha* | *1-2 ha* | *2-4 ha* | *4-8 ha* | *above 8 ha* |
|---|---|---|---|---|---|---|
| Punjab | | | | | | |
| 1976-77 | 100 | 19.02 | 15.43 | 25.16 | 17.93 | 22.45 |
| 1983-84 | 100 | 9.83 | 25.21 | 27.09 | 21.04 | 16.81 |
| 1989-90 | 100 | 22.19 | 13.23 | 37.37 | 14.7 | 12.48 |
| 1993-94 | 100 | 10.53 | 26.73 | 36.39 | 24 | 2.32 |
| 1997-98 | 100 | 10.74 | 26.08 | 43.59 | 16.57 | 3.02 |
| Rajasthan | | | | | | |
| 1977-78 | 100 | 6.48 | 13.86 | 21.66 | 31.62 | 25.94 |
| 1983-84 | 100 | 13.22 | 20.01 | 23.41 | 24.1 | 19.25 |
| 1991-92 | 100 | 11.09 | 37.6 | 20.56 | 20.26 | 10.4 |
| 1994-95 | 100 | 7.07 | 17.67 | 27.23 | 27.45 | 20.55 |
| 1997-98 | 100 | 7.01 | 17.01 | 25.05 | 26.97 | 23.96 |
| Tamil Nadu | | | | | | |
| 1976-77 | 100 | 31.15 | 18.9 | 23.41 | 18.63 | 7.89 |
| 1980-81 | 100 | 14.1 | 31.26 | 27.93 | 20.74 | 10.59 |
| 1988-89 | 100 | 19.65 | 20.44 | 42.63 | 9.72 | 7.54 |
| 1994-95 | 100 | 44.27 | 18.13 | 24.79 | 8.51 | 4.36 |
| 1997-98 | 100 | 25 | 20.37 | 18.67 | 22.26 | 13.7 |
| U.P. | | | | | | |
| 1976-77 | 100 | 19.32 | 27.64 | 23.21 | 15.37 | 19.46 |
| 1980-81 | 100 | 27.39 | 27.8 | 23.56 | 17.56 | 3.68 |
| 1988-89 | 100 | 56.72 | 17.22 | 12.3 | 10.68 | 3.03 |
| 1997-98 | 100 | 18.08 | | | 14.42 | 4.11 |
| Haryana | | | | | | |
| 1976-77 | 100 | 10.36 | 27.08 | 29.59 | 19.06 | 13.88 |
| 1983-84 | 100 | 9.6 | 22.56 | 29.7 | 28.91 | 9.21 |
| 1989-90 | 100 | 4.91 | 18.16 | 29.9 | 40.56 | 6.43 |
| 1994-95 | 100 | 8.03 | 21.14 | 29.82 | 40.72 | |
| M.P. | | | | | | |
| 1976-77 | 100 | 4.77 | 11.67 | 20.62 | 29.95 | 32.97 |
| 1983-84 | 100 | 12.1 | 17.83 | 27.6 | 24.05 | 18.43 |
| 1989-90 | 100 | 11.39 | 17.91 | 26.25 | 26.79 | 17.69 |
| 1997-98 | 100 | 18.86 | 24.45 | 20.66 | 19.26 | 16.77 |
| Maharastra | | | | | | |
| 1976-77 | 100 | 7.34 | 12.28 | 20.7 | 30.02 | 29.65 |
| 1983-84 | 100 | 10.38 | 17.9 | 24.71 | 24.14 | 22.84 |
| 1991-92 | 100 | 7.16 | 13.4 | 21.55 | 25.35 | 32.16 |
| 1994-95 | 100 | 7.68 | 14.87 | 19.27 | 23.87 | 34.29 |
| 1997-98 | 100 | 13.56 | 18.13 | 24.11 | 21.28 | 22.92 |
| Orissa | | | | | | |
| 1976-77 | 100 | 19.9 | 28.01 | 26.47 | 16.58 | 19.03 |
| 1983-84 | 100 | 30.2 | 34.75 | 23.64 | 9.75 | 1.65 |

*Contd...*

| *State/ Year* | *Total* | *Upto 1 ha* | *1-2 ha* | *2-4 ha* | *4-8 ha* | *above 8 ha* |
|---|---|---|---|---|---|---|
| 1989-90 | 100 | 21.03 | 24.6 | 46.82 | 26.3 | 1.55 |
| 1994-95 | 100 | 10.01 | 27.1 | 32.12 | 11.45 | 19.21 |
| 1997-98 | 100 | 24.9 | 34.17 | 21.82 | 8.11 | 11 |
| West Bengal | | | | | | |
| 1980-81 | 100 | 32.81 | 35.55 | 17.08 | 10.83 | 3.71 |
| 1983-84 | 100 | 38.58 | 37.24 | 19.29 | 3.55 | 1.31 |
| 1987-88 | 100 | 41.43 | 34.93 | 18.55 | 4.07 | 1 |
| 1997-98 | 100 | 34.95 | 29.47 | 15.65 | 9.44 | 10.49 |
| A.P. | | | | | | |
| 1976-77 | 100 | 21.16 | 21.45 | 21.1 | 19.35 | 16.91 |
| 1978-79 | 100 | 21.28 | 22.91 | 22.85 | 18.46 | 14.49 |
| 1987-88 | 100 | 30.61 | 24.35 | 23.77 | 12.94 | 8.38 |
| All India | | | | | | |
| 1976-77 | 100 | 15.05 | 18.31 | 22.1 | 23.73 | 20.81 |
| 1983-84 | 100 | 24.12 | 2.92 | 31.23 | 24.56 | 17.71 |
| 1989-90 | 100 | 20.43 | 20.88 | 26.05 | 19.6 | 13.04 |
| 1994-95 | 100 | 18.76 | 20.4 | 23.43 | 20.89 | 16.51 |
| 1997-98 | 100 | 20.32 | 21.33 | 22.19 | 10.36 | 17.8 |

*Source:* Statistical Statements Relating to Co-operative Movements in India, NABARD/R.B.I.

In Haryana, the PPACC has shown fluctuations for the size classes of 'upto 1 hectare', '1 to 2 hectare' and '2 to 4 hectare', but rising and declining trends for the size classes of '4 to 8 hectare' and 'above 8 hectare' respectively. In the case of M.P., the PPACC has fluctuated for all the size classes showing no discernible trend, except the size class of above 8 hectare for which it has declined over years during 1976-77 to 1997-98. In both Maharastra and Orissa, the PPACC has fluctuated for all size classes of land holdings during 1976-77 to 1997-98. In both A.P. and W.B., almost similar trends are observed for all the size classes of land holdings except the size class of 1-2 hectare.

The analysis of data on the credit availability for India and W.B. (Tables 5a and 5b) shows that the weaker sections (smaller size class of land holdings) have received considerable amount of loans per hectare in relation to the amount of loans received by rich farmers (land holding size greater than 4 hectares). It is observed that the weaker sections have sometimes got larger amount of credit as compared to the rich farmers.

**Table 5(a)**
**PACS Credit by Size Classes of Land Holdings (L.H.s) in India**

*(in Rs. Per hectare)*

| *Size of L.H.s* | *1980-81* | *1985-86* | *1990-91* |
|---|---|---|---|
| Upto 2.0 Hectare | 203.75 | 135.55 | 270.03 |
| 2.0-4.0 Hectare | 168.54 | 219.89 | 254.6 |
| Above 4.0 Hectare | 512.49 | 154.46 | 186.7 |

*Source:* Statistical Statements Relating to Cooperative Movements in India, NABARD and Agricultural Census, Government of India.

**Table 5(b)**
**PACS Credit by Size Classes of Land Holdings (L.H.s) in West Bengal**

*(in Rs. Per hectare)*

| *Size of L.H.s* | *1985-86* | *1990-91* |
|---|---|---|
| Upto 2.0 Hectare | 182.67 | 126.99 |
| 2.0-4.0 Hectare | 119.01 | 129.41 |
| Above 4.0 Hectare | 77.65 | 119.39 |

*Source:* Statistical Statements Relating to Cooperative Movements in India, NABARD and Agricultural Census, Government of India.

Focusing on the percentage of COC of a size class in the total COC (PCOC) in all-India, it is observed that it has fluctuated over years for all size of classes of land holdings during 1976-77 to 1994-95 (Table 6). The PCOC has increased over years for smaller land holdings, viz. ,up to 1 hectare, 1-2 hectare, and 2-4 hectare and decreased over years for larger land holdings like 4-8 hectare, and above 8 hectare.

**Table 6**
**Percentage Share of Co-operative Credit of a Size Class of Land Holding in India**

| *Year* | *Total* | *upto 1 ha* | *1-2 ha* | *2-4 ha* | *4-8 ha* | *above 8 ha* |
|---|---|---|---|---|---|---|
| 1976-77 | 100 | 15.89 | 18.11 | 21.75 | 23.14 | 21.58 |
| 1980-81 | 100 | 18.55 | 21.18 | 24.28 | 21.6 | 14.39 |
| 1989-90 | 100 | 21.08 | 21.85 | 24.67 | 19.1 | 13.29 |
| 1994-95 | 100 | 19.41 | 19.56 | 22.95 | 20.37 | 17.7 |

*Source:* As in Table 5.

**Table 7**
**Cropwise PACS Short Term Credit Per Hectare in India by States**

(in Rs.)

| State/ Year | Food grains | Wheat | Paddy | Pulses | Cotton | Oil Seeds | Sugar-cane | Jute |
|---|---|---|---|---|---|---|---|---|
| H.P. | | | | | | | | |
| 1980-81 | 15.84 | 15.34 | | | | | 86.77 | |
| 1983-84 | 33.53 | 28.71 | | | | | 27.02 | |
| 1989-90 | 39.04 | 31.16 | | | | | 236 | |
| 1992-93 | 46.08 | 43.8 | | | | | | |
| W.B. | | | | | | | | |
| 1980-81 | 63.02 | 44.16 | | | | | 704.07 | 71.08 |
| 1982-83 | 71.27 | 49.95 | | | | | 294.65 | 76.68 |
| U.P. | | | | | | | | |
| 1980-81 | 78.4 | 86.55 | | | 160.14 | | 154.12 | |
| 1983-84 | 97.58 | 107.42 | | | 248.16 | | 147.95 | |
| 1987-88 | 104.06 | 118.78 | | | 124.98 | | 159.93 | |
| 1988-89 | 100.18 | 115.82 | | | 119 | | 163.62 | |
| Tamil Nadu | | | | | | | | |
| 1980-81 | 52.51 | | | | 36.54 | | 319.32 | |
| 1983-84 | 160.06 | | | | 473.33 | | 1010.25 | |
| 1989-90 | 538.27 | | | | 728.33 | | 950.4 | |
| 1992-93 | 360.45 | | | | 477.37 | | | |
| M.P. | | | | | | | | |
| 1980-81 | 64.26 | 112.34 | 52.49 | 12.94 | 187.87 | | 1513.38 | |
| 1983-84 | 77.63 | 128.66 | 66.2 | 12.5 | 241.52 | | 2441.57 | |
| 1989-90 | 156.11 | 256.3 | 137.6 | 30.28 | 268.03 | 79.43 | 3958.7 | |

Contd...

| State/ Year | Food grains | Wheat | Paddy | Pulses | Cotton | Oil Seeds | Sugar-cane | Jute |
|---|---|---|---|---|---|---|---|---|
| Haryana | | | | | | | | |
| 1980-81 | 282.1 | 134.17 | 462.98 | 19.22 | 270.46 | | 796.97 | |
| 1983-84 | 351.01 | 385.48 | 594.29 | 49.1 | 318.45 | | 1059.76 | |
| 1989-90 | 575.97 | 510.51 | 758.87 | 43.64 | 317.93 | 140.83 | 2979.88 | |
| 1992-93 | 659.09 | 583.93 | 717.27 | 31.75 | 371.21 | 256.71 | 3233.9 | |
| A.P. | | | | | | | | |
| 1980-81 | 108.16 | | 143.94 | 9.52 | 110.52 | | 1379.06 | 35.82 |
| 1983-84 | 185.22 | | 265.66 | 12.99 | 96.06 | | | 3006.65 |
| 1989-90 | 412.47 | | 463.91 | 9.21 | | | | |
| 1992-93 | 478.43 | | 460.78 | 8.29 | 450.44 | 34.77 | 4415.86 | 4.18 |
| 1993-94 | 31898.03 | | 20142.54 | 11083.85 | | | | |
| Gujarat | | | | | | | | |
| 1980-81 | 263.19 | 302.72 | 194.32 | 21.04 | 252.44 | | 1122.16 | |
| 1983-84 | 326.49 | 453.53 | 167.54 | 34.13 | 327 | | 1266.99 | |
| 1988-89 | 573.6 | 577.64 | 359.45 | 58.61 | 45.45 | 284.52 | 10694.73 | |
| 1993-94 | 1355.81 | 1447.98 | 571.78 | 108.81 | 91.73 | 675.9 | 4853.04 | |
| 1994-95 | 1310.23 | 988.11 | 641 | 108.95 | 92.73 | 78.81 | 1549.98 | |
| Punjab | | | | | | | | |
| 1980-81 | 397.73 | 328.18 | 473.07 | | 207.47 | | 450.53 | |
| 1983-84 | 540.84 | 430.74 | 624.46 | | 299.62 | | 189.41 | |
| 1988-89 | 551.2 | 1475.48 | 574.79 | | 217.33 | | 504.73 | |
| 1992-93 | 1000.34 | 842.57 | 1276.78 | 10.88 | 137.47 | | 1382.38 | |
| 1994-95 | 1428.06 | 1266.75 | 1668.04 | 12.67 | 19.81 | | 577.05 | |

Contd...

| *State/ Year* | *Food grains* | *Wheat* | *Paddy* | *Pulses* | *Cotton* | *Oil Seeds* | *Sugar-cane* | *Jute* |
|---|---|---|---|---|---|---|---|---|
| Rajasthan | | | | | | | | |
| 1980-81 | 59.49 | 134.36 | | 7.43 | 190.54 | | 279.26 | |
| 1983-84 | 85.03 | 178.54 | | 7.39 | 202.53 | | 337.35 | |
| 1988-89 | 103.28 | 195.64 | | 11.6 | 271.36 | | 1341.5 | |
| Maharastra | | | | | | | | |
| 1980-81 | 153.43 | 55.13 | 279.29 | 4.29 | 172.36 | | 2132.22 | 2.2 |
| 1983-84 | 184.33 | 55.44 | 119.83 | 8.35 | 165.42 | | 3553.06 | 108.53 |
| 1988-89 | 461.88 | 127.21 | 154.5 | 28.47 | 480.82 | | 7290.55 | |
| 1992-93 | 217.81 | 63.64 | 52.14 | 12.29 | 145.74 | 88.91 | 3154.4 | 66.57 |
| 1994-95 | 868.74 | 208.22 | 354.52 | 29.64 | 628.62 | 260.61 | 9523.96 | |
| Orissa | | | | | | | | |
| 1980-81 | 78.39 | | 100.22 | 2.15 | | | 1231.24 | 25.08 |
| 1983-84 | 102.42 | | 125.6 | 3 | | | 1989.57 | 36.1 |
| 1988-89 | 87.77 | | 107.05 | 4.77 | | | 1538.34 | 47.04 |
| 1992-93 | 65.66 | 135.01 | 80.15 | 1.06 | 286.11 | 20.98 | 954.35 | 21.58 |
| 1994-95 | 174.07 | 2643.33 | 172.03 | 12.77 | 221.06 | 132.69 | 5964.38 | 360.08 |
| All India | | | | | | | | |
| 1980-81 | 968.56 | 154.03 | | | 177.59 | | 676.23 | 54.72 |
| 1983-84 | 151.25 | 172.93 | | | 237.42 | | 775.75 | 89.72 |
| 1988-89 | 244.11 | 270.89 | | | 441.92 | | 1443.56 | 174.07 |
| 1992-93 | 1270.63 | 226.62 | | | 323.29 | | 1078.94 | 4153.04 |
| 1993-94 | 363.56 | 323.39 | | | 542.48 | | 2187.07 | 6541.12 |

*Source:* As in table 8.6

### Trend of Cropwise PACS Credit in Sample States vis-à-vis All India

Let us have a look at the trend of cropwise PACS credit for different size classes of land holdings in India as well as sample states (Table 7). In the case of both foodgrains and wheat, the PACS credit has increased over years in all the sample states and India. In case of paddy, the PACS credit has shown a rising trend for the sample states, other than Orissa. The PACS credit for other crops, like pulses, cotton, oilseeds and sugarcane, has shown a rising trend in most of the sample states and all India.

### Agricultural Credit in the Districts of W.B.: Trend and Direction/ Composition

In this section, we shall examine the trend, direction/ composition of flow of IC in the districts of W.B.

We shall highlight the trend, direction/composition of flow of agricultural credit in the following manner.

### Trend of IC in the Districts of W.B.

Let us examine the trend of IC in different districts of W.B. It is observed that there exists a rising trend of IC during June 1981 to March 2001 in seven districts in the state, namely, Bankura, Birbhum, Burdwan, Hooghly, Nadia, Purulia and West Dinajpur (Table 8). On the other hand, the IC per hectare has fluctuated over years during June 1981 to March 2001 in the other districts of W.B. For example,in case of Coochbehar the IC per hectare has increased over years during June 1981 to June 1985 from Rs. 249.57 to Rs. 329.71, but declined to Rs. 275.07 in June 1988 and increased over years since June 1988. It is observed from Table 10(a) that the CAGRs of IC have been significant for W.B. and its districts.

### Trend of Commercial Bank Credit in the Districts of W.B.

As far as trend of CBC per hectare is concerned, it is observed that it has increased over years during June 1973 to March 2002 in the 13 districts, viz., Bankura, Birbhum, Burdwan, Coochbehar, Hooghly, Howrah, Malda, Mursidabad, Purulia, 24-Parganas, West Dinajpur, Darjeeling and Midnapore ,but fluctuated in other districts, viz., Jalpaiguri, Nadia during the same period (Table 9). It is observed from Table 10(a) that the CAGRs of CBC have been significant for W.B. and its districts.

**Table 8**
**Institutional Credit per Hectre in West Bengal by Districts**

*(in Rs.)*

| | *June 1981* | *June 1985* | *June 1988* | *June 1991* | *March 1994* | *March 1997* | *March 2001* |
|---|---|---|---|---|---|---|---|
| Bankura | 240.18 | 313.66 | 339.26 | 456.16 | 793.87 | 932.66 | 1392.42 |
| Birbhum | 342.46 | 403.65 | 435.26 | 777.72 | 1239.86 | 1582.34 | 1859.64 |
| Burdwan | 403.32 | 522.66 | 591.79 | 1243.65 | 1486.52 | 1842.73 | 2682.45 |
| Coochbehar | 249.57 | 329.71 | 275.07 | 337.02 | 528.75 | 742.16 | 889.75 |
| Darjeeling | 1157.06 | 938.24 | 2021.04 | 2752.82 | 4242.05 | 3689.07 | 1746.37 |
| Hoogly | 548.75 | 915.97 | 729.7 | 1107.24 | 1608.04 | 2342.55 | 4212.06 |
| Howrah | 781.7 | 947.4 | 1575.04 | 1652.08 | 1631.38 | 1426.68 | 2629.24 |
| Jalpaiguri | 552.83 | 224.48 | 289.9 | 689.18 | 1099.39 | 1266.44 | 766.16 |
| Malda | 279.5 | 911.68 | 555.21 | 587.54 | 1020.15 | 1198.82 | 1089.99 |
| Midnapore | 402 | 892.11 | 830.35 | 1249.52 | 1578.05 | 1529.48 | 2797.87 |
| Nadia | 443.15 | 562.96 | 595.96 | 795.9 | 816 | 1185.69 | 1572.63 |
| 24-Parganas | 571.62 | 577.25 | 750.91 | 1451.05 | 1538 | 1265.58 | 2575.47 |
| Purulia | 165.72 | 179.21 | 296.86 | 425.89 | 673.44 | 685.07 | 741.68 |
| West Dinajpur | 150.72 | 215.8 | 375.1 | 464.42 | 585 | 873.44 | 1095.96 |
| Mursidabad | 442.85 | 375.66 | 556.27 | 449.87 | 802.89 | 879.46 | 1572.63 |

*Source:* (i) Banking Statistics, R.B.I.
(ii) Statistical Statements Relating to Co-operative Movement in India, R.B.I./NABARD
(iii) Statistical Abstract, Government of West Bengal
(iv) Co-operative Statistical Cell, Government of West Bengal

## Trend of Co-operative Credit in the Districts of W.B.

Highlighting on the trend of COC per hectare it is found that for most of the districts (i.e., 13 districts) the COC per hectare has fluctuated during 1980-81 to 2000-01 (Table 10). However, despite the fluctuations the COC per hectare has shown rising trends in all these 13 districts, namely, Bankura, Birbhum, Burdwan, Coochbehar, Hooghly, Howrah, Jalpaiguri, Malda, Mursidabad, 24-Parganas, Purulia, West Dinajpur and Midnapore. In the other two districts, namely, Nadia and Darjeeling the COC per hectare has increased over years during the same period. It is observed from Table 10(a) that the CAGRs of COC have been significant for W.B. and its districts.

**Table 9**
**Commercial Bank's Agricultural Credit per Hectare in West Bengal by Districts**

*(in Rs.)*

| | *June 1973* | *June 1978* | *June 1983* | *June 1988* | *March 1994* | *March 1997* | *March 2002* |
|---|---|---|---|---|---|---|---|
| Bankura | 4.85 | 42 | 137.33 | 231.38 | 536 | 736 | 910.15 |
| Birbhum | 12.17 | 54.46 | 210.4 | 352.8 | 996 | 1196 | 1177.7 |
| Burdwan | 19.34 | 117.05 | 267.49 | 376.65 | 1143.25 | 1343 | 2114.71 |
| Coochbehar | 1.39 | 38.45 | 142.8 | 173.06 | 463 | 663 | 644.56 |
| Darjeeling | 752.25 | 764.07 | 820.58 | 1531.94 | 2655 | 2481.49 | 2845.56 |
| Hoogly | 46.76 | 293.39 | 579.26 | 587.38 | 1300.56 | 1500.56 | 2675.83 |
| Howrah | 77.91 | 111.44 | 286.39 | 1075.92 | 1086 | 1286 | 1117.33 |
| Jalpaiguri | 272.34 | 343.41 | 366.61 | 227.71 | 984 | 1184 | 1020.67 |
| Malda | 7.04 | 55.47 | 208.52 | 443.98 | 730 | 930 | 849.42 |
| Midnapore | 9.79 | 41.8 | 98.85 | 519.5 | 941.46 | 1141 | 1310.75 |
| Nadia | 27.46 | 121.06 | 298.11 | 297.5 | 501.55 | 781.55 | 891.12 |
| 24-Parganas | 50.06 | 144.4 | 315.9 | 575.56 | 1202.05 | 1402 | 1449.6 |
| Purulia | 13.17 | 36.29 | 82.39 | 173.81 | 427.53 | 627.53 | 472.33 |
| West Dinajpur | 1.41 | 36.75 | 107.24 | 183.7 | 294.05 | 496.85 | 753.23 |
| Mursidabad | 7.49 | 62.66 | 176.62 | 370.37 | 516 | 717 | 760.66 |

*Source:* (i) Banking Statistics,R.B.I.
(ii) Statistical Abstract, Government of West Bengal.

**Table 10**
**Cooperative Credit per Hectare in West Bengal by Districts**

*(in Rs.)*

| | *1980-1981* | *1984-1985* | *1987-1988* | *1990-1991* | *1993-1994* | *1996-1997* | *2000-2001* |
|---|---|---|---|---|---|---|---|
| Bankura | 169.44 | 138 | 107.88 | 123.57 | 257.87 | 196.66 | 574.57 |
| Birbhum | 134.61 | 168 | 82.46 | 178.87 | 243.12 | 386.34 | 717.74 |
| Burdwan | 206.55 | 260 | 215.14 | 276.52 | 343.27 | 499.73 | 922.81 |
| Coochbehar | 182.07 | 179 | 102.01 | 97.32 | 135.75 | 79.16 | 230.7 |
| Darjeeling | 152.73 | 295.65 | 489.1 | 1249.52 | 1587.05 | 1207.58 | 660.02 |
| Hoogly | 102.64 | 248 | 142.32 | 101.25 | 307.48 | 841.99 | 738.72 |
| Howrah | 547 | 640 | 499.12 | 690.35 | 545.38 | 140.68 | 584.09 |
| Jalpaiguri | 93 | 79.49 | 62.11 | 45.73 | 115.39 | 82.44 | 3663.15 |
| Malda | 110.23 | 116 | 111.23 | 259.02 | 290.15 | 268.62 | 343.36 |
| Midnapore | 337 | 388 | 310.76 | 604.31 | 645.59 | 388.48 | 1770.03 |
| Nadia | 284.86 | 279.01 | 283.76 | 298.38 | 234.45 | 404.14 | 812.71 |
| 24-Parganas | 351.41 | 147.21 | 175.35 | 255.32 | 335.95 | 638.05 | 1343.54 |

*Contd...*

| | 1980-1981 | 1984-1985 | 1987-1988 | 1990-1991 | 1993-1994 | 1996-1997 | 2000-2001 |
|---|---|---|---|---|---|---|---|
| Purulia | 104.01 | 115.61 | 123.05 | 105.5 | 345.91 | 57.54 | 228.74 |
| West Dinajpur | 150.62 | 175.21 | 191.31 | 190.57 | 290.15 | 376.59 | 561.77 |
| Mursidabad | 179 | 186.89 | 185.9 | 312.06 | 286.89 | 162.46 | 768.38 |

*Source:* (i) Statistical Abstract, Government of West Bengal.
(ii) Co-operative Statistical Cell, Government of West Bengal.

**Table 10(a)**
**Compound Annual Growth Rates (CAGRs) of Credit Per Hectare in West Bengal by Districts**

| *Districts* | *Institutional Credit/Ha* | *Commercial Bank Credit/Ha* | *Co-operative Credit/ Ha* |
|---|---|---|---|
| West Bengal | 19.88* | 19.85* | 13.05* |
| Burdwan | 19.72* | 19.53* | 12.29* |
| Birbhum | 23.37* | 19.31* | 12.02* |
| Bankura | 25.86* | 17.36* | 21.92* |
| Coochbehar | 22.54* | 19.72* | 16.18* |
| Hoogly | 19 58* | 18.65* | 23.66* |
| Howrah | 19.72* | 12.64* | 16.18* |
| Jalpaiguri | 15.55* | 19.84* | 16.18* |
| Malda | 25.86* | 13.77* | 16.39* |
| Midnapore | 28.45* | 11.29* | 11.21* |
| Mursidabad | 22.14* | 10.41* | 12.33* |
| Nadia | 16.18* | 10.74* | 13.04* |
| 24-Parganas | 19.72* | 17.47* | 14.08* |
| Purulia | 19.72* | 15.83* | 16.18* |
| West Dinajpur | 29.69* | 19.35* | 18.33* |
| Darjeeling | 16.72* | 18.41* | 11.10* |

*Notes:* (i) *indicates value significant at 1% level.

## PACS Credit by Size Classes of Land Holdings in the Districts of W.B.: An Analysis of Direction/ Composition

Focusing on the PPACC in the districts of W.B. it is observed that in Birbhum, Bankura and Midnapore, there exists a slow rising trend for the size classes of 'up to 1 hectare', '1 to 2 hectare', but a declining trend for the size class of 4 to 8 hectare. The PPACC has not shown any definite trend during 1976-77 to 1986-87 for different size classes of land holding in Malda ,Jalpaiguri, 24-Parganas, Howrah, Nadia, Mursidabad, Hooghly, Burdwan, Coochbehar, Purulia and Darjeeling (Table 11). It has declined over years during 1982-83 to 1984-85 in West Dinajpur.

**Table 11**
**Percentage Share of PACS Credit of a Size Class of Land Holding in West Bengal by Districts**

| *Districts/ year* | *upto 1 ha* | *1-2 ha* | *2-4 ha* | *4-8 ha* | *above 8 ha* |
|---|---|---|---|---|---|
| 24-Parganas | | | | | |
| 1980-81 | 30.31 | 45.29 | 22.62 | 1.76 | |
| 1982-83 | 33.42 | 42.18 | 23.56 | 0.84 | |
| 1984-85 | 39.67 | 37.78 | 20.29 | 2.26 | |
| 1985-86 | 44.05 | 36.32 | 17.31 | 2.32 | |
| 1986-87 | 33.89 | 24.34 | 36.98 | 9.78 | |
| Howrah | | | | | |
| 1982-83 | 53.15 | 33.71 | 13.14 | | |
| 1984-85 | 55.34 | 27.83 | 16.84 | | |
| 1985-86 | 55.32 | 30.06 | 14.62 | | |
| Nadia | | | | | |
| 1980-81 | 36.88 | 18.82 | 21.74 | 22.56 | |
| 1982-83 | 48.94 | 48.7 | 2.36 | | |
| 1984-85 | 46.43 | 46.14 | 7.42 | | |
| 1985-86 | 47.64 | 46.72 | 5.64 | | |
| 1986-87 | 30.42 | 69.58 | | | |
| Mursidabad | | | | | |
| 1982-83 | 36.72 | 32.88 | 28.4 | 2 | |
| 1984-85 | 33.64 | 30.53 | 33.55 | 2.28 | |
| 1985-86 | 33.56 | 31.48 | 32.96 | 2 | |
| Hooghly | | | | | |
| 1980-81 | 23.01 | 31.81 | 15.11 | 19.61 | 10.42 |
| 1982-83 | 58.59 | 24.79 | 12.92 | 3.7 | |
| 1984-85 | 74.09 | 20.2 | 5.71 | | |
| Burdwan | | | | | |
| 1980-81 | 55.52 | 19.8 | 23.23 | 14.1 | 7.75 |
| 1982-83 | 22.91 | 34.85 | 25.12 | 12.8 | 4.32 |
| 1984-85 | 24.58 | 36.06 | 24.72 | 11.6 | 3.08 |
| 1985-86 | 14.61 | 38.6 | 37.92 | 5.14 | 3.72 |
| 1986-87 | 19.69 | 31.25 | 35.79 | 12.54 | 0.74 |
| Bankura | | | | | |
| 1980-81 | 11.3 | 31.48 | 27.25 | 20.73 | 9.23 |
| 1982-83 | 54.52 | 29.5 | 14.46 | 1.52 | |
| 1984-85 | 57.01 | 30.88 | 14.99 | 0.002 | |
| 1985-86 | 48.81 | 25.55 | 13.18 | 12.45 | |
| 1986-87 | 43.66 | 38.65 | 15.58 | 2.11 | |
| Birbhum | | | | | |
| 1980-81 | 23.21 | 20.11 | 21.95 | 21.69 | 13.04 |
| 1982-83 | 22.99 | 33.5 | 15.69 | 3.26 | 24.31 |

*Contd...*

| *Districts/ year* | *upto 1 ha* | *1-2 ha* | *2-4 ha* | *4-8 ha* | *above 8 ha* |
|---|---|---|---|---|---|
| 1984-85 | 23.47 | 23.88 | 25.72 | 12.43 | 14.51 |
| 1985-86 | 26.78 | 25.89 | 30.56 | | 16.71 |
| Purulia | | | | | |
| 1982-83 | 39.84 | 24.56 | 35.61 | | |
| 1984-85 | 10.69 | 81.41 | 6.98 | 0.93 | |
| 1985-86 | 37.84 | 29.19 | 6.38 | 26.58 | |
| Midnapore | | | | | |
| 1980-81 | 40.33 | 25.34 | 22.73 | 10.63 | 0.98 |
| 1982-83 | 45.91 | 32.17 | 15.08 | 6.59 | 0.25 |
| 1984-85 | 56.2 | 32.29 | 10.11 | 1.39 | |
| 1985-86 | 51.3 | 34.7 | 13.39 | 0.61 | |
| 1986-87 | 50.63 | 34.36 | 11.6 | 3.41 | |
| Malda | | | | | |
| 1980-81 | 52.91 | 12.86 | 29.11 | 5.12 | |
| 1982-83 | 32.67 | 31.37 | 31.31 | 4.65 | |
| 1984-85 | 40.78 | 36.05 | 19.08 | 4.09 | |
| 1985-86 | 38.62 | 34.68 | 20.94 | 5.56 | |
| Jalpaiguri | | | | | |
| 1980-81 | 13.57 | 65.78 | 11.85 | 8.8 | |
| 1982-83 | 35.05 | 18.31 | 29.66 | 16.97 | |
| 1984-85 | 36.46 | 30.43 | 20.33 | 12.78 | |
| 1985-86 | 42.72 | 34.61 | 9.16 | 13.51 | |
| 1986-87 | 68.69 | 19.33 | 11.99 | | |
| Coochbehar | | | | | |
| 1982-83 | 33.72 | 42.75 | 21.46 | 2.07 | |
| 1984-85 | 42.8 | 31.23 | 24.87 | 1.1 | |
| 1985-86 | 34.19 | 42.8 | 20.77 | 2.23 | |
| 1986-87 | 29.32 | 42.64 | 21.07 | 6.98 | |
| Darjeeling | | | | | |
| 1980-81 | 0.85 | 98.42 | 0.54 | 0.19 | |
| 1982-83 | 15.63 | 62.36 | 12.8 | 9.23 | |
| 1984-85 | 3.98 | 62.8 | 20.01 | 13.21 | |
| 1985-86 | | 14.85 | 8.92 | 76.22 | |
| 1986-87 | 65.79 | 34.21 | | | |

*Source:* Co-operative Statistical Cell, Government of W.B.

## LDB Credit by Size Classes of Land holdings in The Districts of W.B.: An Analysis of Direction/ Composition

Let us consider the trend of the percentage share of different land holdings size classes in LDB credit in the total LDB credit (PLDBC) in the different districts of W.B. It is observed that the PLDBC has shown different trends for different size classes of land holdings in the different districts of W.B. (Table 12).

**Table 12**
**Percentage Share of Land Development Bank's Outstanding Credit of a Size Class of Land Holding in West Bengal by Districts**

| *Districts* | *Year* | *Upto 1Ha.* | *1-2 Ha.* | *2-4 Ha.* | *4-8 Ha.* | *Above 8Ha.* |
|---|---|---|---|---|---|---|
| Bankura | 1980-81 | 23.21 | 20.4 | 21.95 | 21.69 | 13.04 |
| | 1985-86 | 5.25 | 18.77 | 62.45 | 13.57 | |
| | 1987-88 | 10.88 | 25.72 | 55.67 | 7.81 | |
| Birbhum | 1980-81 | | 16.45 | 59.2 | 23.56 | |
| | 1985-86 | 23.94 | 47.84 | 24.58 | 3.58 | 0.04 |
| | 1988-89 | 26.99 | 54.42 | 7.86 | 10.93 | |
| Burdwan | 1980-81 | 16.73 | 17.51 | 28.37 | 25.33 | 12.06 |
| | 1985-86 | 26.13 | 42.72 | 19.11 | 11.37 | 0.66 |
| | 1987-88 | 34.52 | 26.77 | 19.83 | 22.17 | 1.69 |
| Coochbehar | 1980-81 | 5.81 | 89.2 | 3.7 | 1.28 | |
| | 1985-86 | 38.43 | 42.07 | 10.24 | 4.18 | 5.08 |
| | 1987-88 | 32.28 | 54.29 | 11.71 | 1.42 | 0.29 |
| Darjeeling | 1980-81 | 35.29 | 30.32 | 22.49 | 7.79 | |
| | 1985-86 | | 55.13 | 33.12 | 11.73 | |
| | 1986-87 | 43.24 | 39.01 | 11.73 | 6 | |
| Hooghly | 1985-86 | 46.79 | 39.88 | 12.91 | 0.4 | |
| | 1988-89 | 29.64 | 37.7 | 20.3 | 12.33 | 0.33 |
| Howrah | 1985-86 | 60.48 | 26.4 | 12.63 | | |
| | 1988-89 | 35.01 | 41.89 | 12.06 | 11.64 | 4.42 |
| Jalpaiguri | 1980-81 | 27.94 | 4.01 | 24.74 | 6.89 | 0.42 |
| | 1985-86 | 22.95 | 45.93 | 31.12 | | |
| | 1988-89 | 26 | 62.76 | 11.24 | | |
| Malda | 1980-81 | 30.19 | 23.85 | 26.37 | 19.58 | |
| | 1985-86 | 27.01 | 57.13 | 15.85 | | |
| | 1986-87 | 21.02 | 63.09 | | 15.9 | |
| Midnapore | 1980-81 | 46.92 | 33.84 | 14.84 | 3.67 | 0.73 |
| | 1985-86 | 49.6 | 37.72 | 10.39 | 2.18 | 0.86 |
| | 1988-89 | 15.21 | 40.37 | 33.58 | 9.39 | 1.44 |
| Nadia | 1985-86 | 33.66 | 33.37 | 26.66 | 6.29 | |
| | 1987-88 | 34.99 | 42.04 | 11.3 | 1.03 | 1.25 |
| 24-Parganas | 1980-81 | 30.31 | 45.29 | 22.02 | 1.78 | |
| | 1985-86 | 50.99 | 32.58 | 11.11 | 5.3 | |
| | 1988-89 | 63.08 | 34.5 | 2.41 | | |
| Purulia | 1985-86 | 37.84 | 29.19 | 6.38 | 26.58 | |
| West-Dinajpur | 1980-81 | 4.37 | 36.84 | 32.77 | 26.02 | |
| | 1985-86 | 25.89 | 44.57 | 26.6 | 2.76 | 0.17 |
| | 1988-89 | 67.94 | 13.76 | 8.84 | 7.56 | 1.95 |
| Murshidabad | 1980-81 | 34.78 | 17.74 | 20.5 | 26.96 | |
| | 1985-86 | 50.31 | 38.09 | 7.73 | 3.71 | 0.16 |
| | 1987-88 | 56.6 | 41.85 | 0.18 | 1.35 | |

*Source:* Co-operative Statistical Cell, Govt. of West Bengal.

In case of Birbhum, the PLDBC has shown a rising trend for two size classes-1 to 2 hectares and 'upto1 hectare', but a declining trend for other size classes. In Bankura, the PLDBC has shown an increasing trend for the size classes of 1–2 hectares and 2-4 hectares and an opposite trend for other size classes. In Nadia, Mursidabad and Darjeeling, the PLDBC has revealed the similar trend across the different size classes of land holdings. In Midnapore ,the PLDBC has shown a slowly rising trend over years for all the size classes except the size class of 'up to 1 hectare'. In Malda, the PLDBC has decreased over years for all the size classes except the size class of 1-2 hectare. In West Dinajpur, the PLDBC has declined with some fluctuations for the size classes of 1-2 hectare, 2-4 hectare, 4-8 hectare, but has increased with some fluctuations for the size class of ' above 8 hectare' and increased sharply over years for the size class of 'upto 1 hectare'. In Jalpaiguri, the PLDBC has shown a declining trend for the size classes of 'upto 1 hectare' and '2-4 hectare'. In case of 24-Parganas, the PLDBC has increased sharply over years for the size classes of 'up to 1 hectare' and '4-8 hectare'. In case of Howrah, PLDBC has risen over years for the size classes of '1-2 hectare', and '2-4 hectares'. In case of Hooghly, the PLDBC has increased over years for the size classes of '2-4 hectares' and '4-8 hectares'. In Burdwan, the PLDBC has fluctuated over years for the size classes of '1-2 hectare', '2-4 hectare' and 'above 8 hectare' ,while it has increased sharply for the size class of 'up to 1 hectare'. In case of Coochbehar, the PLDBC has declined over time for the size class of 'above 8 hectare', while for other size classes it exhibits no trend.

## Agricultural Credit in Blocks of Midnapore

Let us examine the trend of IC for agriculture in different blocks of Midnapore district. It is obvious from Table 13 that the IC per hectare has shown a rising trend for most (viz., 39) of the blocks of Midnapore district. For example, the IC per hectare in Jhargram block has increased from Rs. 344.00 in 1999-2000 to Rs. 1,105.00 in 2002-03. Again, the IC per hectare in Debra has risen from Rs. 358.00 in 1999-2000 to Rs. 1296.00 in 2002-2003. The IC per hectare in Ghatal has risen from Rs.

**Table 13**
**Institutional Credit Per Hectare in Blocks of Midnapur District During 1999-2000 To 2002-2003**

*(in Rs.)*

| *Block* | *1999-2000* | *2000-2001* | *2001-2002* | *2002-2003* |
|---|---|---|---|---|
| Jhargram | 344 | 589 | 420 | 1105 |
| Binpur-i | 486 | 677 | 821 | 1225 |
| Binpur-ii | 486 | 305 | 406 | 537 |
| Jamboni | 486 | 320 | 295 | 493 |
| Nayagram | 163 | 165 | 273 | 884 |
| Sankrail | 276 | 479 | 502 | 640 |
| Gopiballabpur-i | 358 | 493 | 639 | 1742 |
| Gopiballabpur-ii | 179 | 304 | 219 | 824 |
| Debra | 358 | 810 | 1011 | 1296 |
| Pingla | 179 | 937 | 680 | 746 |
| Keshiary | 845 | 875 | 1164 | 800 |
| Dantan-i | 831 | 699 | 414 | 322 |
| Dantan-ii | 519 | 842 | 559 | 492 |
| Narayangarh | 485 | 574 | 837 | 322 |
| Mohanpur | 685 | 1939 | 1414 | 2256 |
| Salboni | 1356 | 334 | 420 | 648 |
| Keshpur | 425 | 329 | 604 | 636 |
| Garbeta-i | 563 | 572 | 808 | 3005 |
| Garbeta-ii | 725 | 1531 | 1887 | 2051 |
| Garbeta-iii | 1188 | 742 | 560 | 1411 |
| Midnapur | 358 | 1611 | 2777 | 1617 |
| Sabang | 809 | 518 | 280 | 242 |
| Kharagpur-i | 809 | 750 | 633 | 660 |
| Kharagpur-ii | 834 | 1068 | 1489 | 1453 |
| Chandrakona-i | 929 | 2027 | 3275 | 3368 |
| Chandrakona-ii | 1058 | 993 | 5890 | 3000 |
| Ghatal | 1249 | 1655 | 1743 | 1855 |
| Daspur-i | 1616 | 1146 | 1449 | 499 |
| Daspur-ii | 958 | 685 | 816 | 1649 |
| Tamluk-i (Tamluk) | 370 | 2062 | 2326 | 3323 |
| Tamluk-ii (Sahid Matangini) | 674 | 2584 | 3808 | 4274 |
| Panskura-i | 2739 | 832 | 1356 | 4853 |
| Panskura-ii | 761 | 389 | 698 | 898 |
| Moyna | 471 | 1482 | 1062 | 2041 |
| Mahisadal-i (Nandakumar) | 717 | 1955 | 2586 | 1396 |
| Mahisadal-ii (Mahisadal) | 1603 | 1126 | 661 | 636 |

*Contd...*

| Block | 1999-2000 | 2000-2001 | 2001-2002 | 2002-2003 |
|---|---|---|---|---|
| Nandigram-i | 894 | 877 | 2211 | 1151 |
| Nandigram-ii | 1960 | 3322 | 3649 | 4627 |
| Nandigram-iii (Chandipur) | 2816 | 2120 | 2676 | 1534 |
| Sutahata-i (Sutahata) | 1955 | 766 | 376 | 759 |
| Sutahata-ii (Haldia) | 985 | 2332 | 2868 | 4085 |
| Potashpur-i | 1911 | 1878 | 1702 | 1655 |
| Potashpur-ii | 1227 | 1532 | 1567 | 1113 |
| Khejuri-i | 1551 | 2489 | 2238 | 3151 |
| Khejuri-ii | 3325 | 898 | 1202 | 1123 |
| Bhagwanpur-i | 1309 | 2807 | 3234 | 3854 |
| Bhagwanpur-ii | 3600 | 2306 | 3516 | 3863 |
| Ramnagar-i | 3149 | 3402 | 3958 | 2499 |
| Ramnagar-ii | 3602 | 1433 | 1867 | 3960 |
| Egra-i | 1509 | 1290 | 1665 | 1755 |
| Egra-ii | 1244 | 1472 | 1315 | 1427 |
| Contai-i | 1447 | 1294 | 2091 | 6617 |
| Contai-ii | 2985 | 1541 | 2032 | 1665 |
| Contai-iii | 2100 | 3400 | 3500 | 3186 |

*Source:* United Bank of India.

1249.00 in 1999-2000 to Rs. 1855.00 in 2002-2003. The IC per hectare in Mahisadal-1 has also increased from Rs. 717.00 in 1999- 2000 to Rs. 1396.00 in 2002 - 2003. The IC per hectare in Egra-I has increased from Rs. 1509.00 to Rs. 1755.00 during 1999-2000 to 2002-2003. But the declining trend of IC has been found for a number of blocks, viz, Keshiary, Dantan-II, etc. during same period.

## A Summing-Up

Agricultural credit is one of the most crucial inputs in all agricultural development programmes in India. There has been a massive flow of IC to agriculture in both the sample states of India ,districts of W.B. and blocks of Midnapore. Side by side, the flow of credit is more favourable to smaller size of land holdings i.e., the percentage share of different type of IC has increased over years for the smaller size of land holdings and sometimes they have received larger amount of agricultural credit as compared to the larger size of land holdings.

# CHAPTER–9

## IMPACT OF INSTITUTIONAL CREDIT AND OTHER FACTORS ON CHANGE IN CROPPING PATTERN IN INDIA: STATE, DISTRICT AND BLOCK LEVEL ANALYSIS

In the earlier chapters we have observed the trend, direction/ composition of flow of IC, the nature of change in CP and the significant impact of change in CP on agricultural growth in India. This chapter is primarily devoted to analyse the role of IC in the change in CP influencing agricultural growth. This change in CP is influenced by several economic, infrastructural and natural/geographical factors. Among these factors, the most crucial factor is IC which is the lifeblood of modern agricultural activities, particularly, commercial cropping. The IC acts indirectly through different factors (viz., fertilizer consumption, irrigation, etc.) influencing change in CP. So, we shall examine first whether there exists significant impact of IC, mainly bank credit (i.e., bank nationalization) on change in CP during the period from 1950-51 to 1992-93. We shall also examine the impact of the supply side factors, viz., IC and the related factors (influenced by IC), viz., fertilizer consumption per hectare, the percentage of gross irrigated area (GIA) to GCA and the demand side factor, viz., PCI or PCSDP on the change in CP in India, the sample States, the districts of W.B. and the blocks of Midnapore.

### Impact of Bank Nationalization on change in CP in India: State Level Analysis

To examine the impact of bank nationalization on change in CP in India and sample states, the following regression equations have been estimated:

| | | | |
|---|---|---|---|
| Punjab: NFA=33.53 | -0.01*X | -2.66***D; | $R^2$=0.71, F=49.41 |
| | (-6.05) | (-1.85) | |
| Haryana: NFA=15.77 | +0.001X | +6.08*D; | $R^2$=0.38, F=12.37 |
| | (0.61) | (3.24) | |
| W. B.: NFA=15.12 | +0.008*X | +1.49D; | $R^2$=0.37, F=12.004 |
| | (2.51) | (1.25) | |
| Maharashtra: NFA=31.84 | +0.0005X | -2.34***D; | $R^2$=0.098, F=2.23 |
| | (0.23) | (-1.74) | |
| Orissa: NFA=33.12 | +0.02***X | -17.35*D; | $R^2$=0.49, F=19.80 |
| | (1.73) | (-4.94) | |
| Rajasthan: NFA=29.73 | -0.0005X | -1.96D; | $R^2$=0.014, F=0.29 |
| | (-0.034) | (-0.53) | |
| Tamil Nadu : NFA=25.45 | +0.005**X | +4.49**D; | $R^2$=0.51, F=21.76 |
| | (2.58) | (2.33) | |
| H.P.: NFA=10.36 | -0.01*X | +3.25**D; | $R^2$=0.44, F=16.17 |
| | (-5.60) | (2.29) | |
| M. P.: NFA=60.98 | +0.02**X | -5.14**D; | $R^2$=0.09, F=2.05 |
| | (1.77) | (-1.96) | |
| A. P. : NFA=24.99 | +0.006*X | +1.95D; | $R^2$=0.59, F=29.70 |
| | (5.62) | (1.22) | |
| Gujarat: NFA= 48.93 | +0.01**X | +5.09*D; | $R^2$=0.58, F=28.53 |
| | (2.27) | (2.87) | |
| U. P.: NFA=68.61 | +0.02*X | -2.64***D; | $R^2$=0.27, F=7.51 |
| | (3.29) | (-1.44) | |
| India: NFA=24.59 | +0.06*X | +0.04*D; | $R^2$=0.72, F=53.49 |
| | (5.76) | (3.42) | |

*, ** and *** indicate significant at 1%, 5% and 10% level respectively.

The above results show that the bank nationalization (i.e. the larger provision of commercial banks' credit for cultivation) has made a positive and significant impact on change in CP in India and the sample states like Haryana, Tamil Nadu, H.P. and Gujarat. The timely availability of bank credit in required amount and its proper utilization may be the reasons behind this significant impact. In the states like Punjab, Orissa, Rajasthan, Maharashtra, M.P. and U.P. the bank nationalization has made adverse effect on the change in CP. On the other hand, that impact has been positive but insignificant for the states like W.B. and A.P. There may have a number of causes behind this negative or insignificant impact of bank nationalization on the change in CP in those sample states. In reality, it is observed that the farmers do not get the credit on time and/or in required amount. Often it is found that the poor Indian farmers utilize the credit for unproductive or non-agricultural purposes. These are the probable factors explaining the negative or insignificant impact of bank nationalization on the change in CP in some sample states.

## Impact of IC and Other Factors on Change in CP in India: State Level Analysis

Let us examine now the impact of factors like fertilizer consumption, irrigation, PCI/PCSDP and IC on change in CP in India and the sample states. As an index of CP change, we have used here both the CPI and NFA due to (i) the significant relationship between IC and NFA and (ii) the higher income elasticity of demand for non-food grains. Due to the commercialization of agriculture in Indian economy, the NFA as an index of CP change is also considered here to examine the impact of IC and other related factors viz., fertilizer consumption, irrigation, etc. on the emerging commercial cropping in the country. Again, in the context of commercialization of Indian agriculture, the percentage of HYV area to GCA (% HYV area) has been used as an index of CP change to examine the impact of IC on the recent commercialized agriculture in the country.

Before examining the impact of area under irrigation as percentage of the gross cropped area, IC, PCI/PCSDP and fertilizer consumption (FC) per hectare on the CP we have estimated the CAGRs of these factors. It is seen from the Tables 1, 1a and 1b that there has been significant growth of all these factors in all the sample states and India. It may be presumed that the high growth rate of these factors might have contributed to the change in the CP in the states during the period. However, to identify the factors contributing to the change in CP we regress CP upon IC, the percentage irrigated area and FC. In addition, we have also considered PCI / PCSDP as a factor on the demand side influencing the CP in the States. It is seen from Table 1c that there has been a significant growth rate of PCI /PCSDP in all sample states and all India.

**Table 1**
**Compound Annual Growth Rates (CAGRs) of Some Selected Factors Influencing Cropping Pattern in India by States during 1972-73 to 1995-96**

| *States* | *Credit per hectare* Total Credit | Commercial Bank | *Fertilizer consumption per hectare* | *Irrigation (%GIA to GCA)* |
|---|---|---|---|---|
| All India | 16.18* | 18.53* | 7.57* | 1.92* |
| | (4.20) | (16.20) | (21.12) | (23.31) |
| Punjab | 17.35* | 20.92* | 6.18* | 1.10* |
| | (10.24) | (10.66) | (9.23) | (9.54) |
| Haryana | 18.77* | 24.61* | 9.31* | 2.22* |
| | (9.58) | (13.66) | (21.74) | (9.33) |
| Rajasthan | 19.72* | 25.86* | 9.42* | 2.43* |
| | (11.16) | (17.90) | (39.31) | (7.73) |
| Maharashtra | 13.88* | 16.18* | 8.43* | 1.51* |
| | (10.12) | (8.19) | (17.48) | (5.01) |
| Gujarat | 13.88* | 20.92* | 6.61* | 3.04* |
| | (2.91) | (19.65) | (13.13) | (6.36) |
| Tamil Nadu | 15.03* | 20.92* | 5.76* | 2.72* |
| | (11.69) | (17.39) | (13.19) | (6.32) |
| A.P | 15.72* | 18.53* | 9.19* | 0.09* |
| | (5.74) | (6.65) | (18.33) | (2.96) |
| M.P | 20.92* | 23.37* | 10.07* | 4.70* |
| | (8.55) | (22.43) | (17.49) | (30.29) |
| U.P | 16.18* | 19.72* | 7.46* | 2.33* |
| | (10.04) | (32.29) | (13.74) | (26.24) |
| H.P | 19.72* | 29.69* | 6.50* | 0.40* |
| | (7.39) | (12.15) | (13.28) | (4.54) |
| Orissa | 16.18* | 23.37* | 6.61* | 1.21* |
| | (6.64) | (5.80) | (25.01) | (4.67) |
| W.B. | 10.52* | 13.88* | 9.85* | 3.05* |
| | (11.56) | (17.70) | (17.70) | (5.93) |

*Notes:* (i) Figures in the parenthesis indicate 't' value.
(ii) * indicates 't' value significant at 1% level.
(iii) ** indicates 't' value significant at 5% level.

*Source:* (i) Banking Statistics, R.B.I. ,
(ii) Fertilizer Statistics, G.O.I. ,
(iii) Area and Production of Principal Crops in India, G.O.I. ,
(iv) Statistical Statement Relating to Co-operative Movements in India, NABARD/R.B.I.

**Table 1a**
**Consumption of Fertilizer per unit of Gross Cropped Area in India by States (Kg/Hectare)**

| | *1972-1973* | *1976-1977* | *1982-1983* | *1987-1988* | *1993-1994* | *1995-1996* | *2000-2001* |
|---|---|---|---|---|---|---|---|
| Andhra Pradesh | 20.32 | 31.10 | 56.89 | 90.23 | 117.00 | 137.30 | 159.84 |
| Gujarat | 20.32 | 19.90 | 37.92 | 48.21 | 64.60 | 68.50 | 70.14 |
| Haryana | 20.32 | 25.20 | 49.61 | 78.29 | 113.50 | 1230.70 | 147.20 |
| Himachal Pradesh | 11.25 | 9.80 | 19.04 | 28.25 | 29.10 | 30.50 | 36.85 |
| Madhya Pradesh | 7.11 | 6.30 | 10.96 | 29.02 | 32.40 | 34.70 | 38.85 |
| Maharashtra | 10.16 | 14.70 | 25.34 | 40.26 | 54.60 | 65.30 | 74.35 |
| Orissa | 7.15 | 8.00 | 10.28 | 15.02 | 20.90 | 25.20 | 37.89 |
| Punjab | 50.80 | 59.30 | 128.14 | 157.02 | 159.90 | 167.30 | 166.69 |
| Rajasthan | 4.98 | 5.70 | 8.98 | 16.01 | 25.90 | 31.90 | 8.50 |
| Tamil Nadu | 40.64 | 38.40 | 77.18 | 105.80 | 117.60 | 106.90 | 145.31 |
| Uttar Pradesh | 20.32 | 31.40 | 57.77 | 79.34 | 90.00 | 101.40 | 111.31 |
| West Bengal | 13.00 | 19.20 | 37.35 | 70.81 | 83.20 | 99.30 | 113.68 |
| India | 17.07 | 20.38 | 37.06 | 49.65 | 66.6 | 74.80 | 86.34 |

*Source:* Fertilizer Statistics, G.O.I.

**Table 1b**
**Percentage of Gross Irrigated Area to Gross Cropped Area in India by States**

| | *1972-1973* | *1976-1977* | *1982-1983* | *1987-1988* | *1993-1994* | *1998-1999* |
|---|---|---|---|---|---|---|
| Andhra Pradesh | 29.00 | 35.00 | 35.40 | 35.36 | 39.50 | 44.71 |
| Gujarat | 14.80 | 17.10 | 23.90 | 22.93 | 26.40 | 35.31 |
| Haryana | 47.80 | 51.10 | 67.10 | 82.86 | 76.20 | 79.78 |
| Himachal Pradesh | 17.80 | 26.60 | 16.10 | 17.74 | 17.30 | 18.87 |
| Madhya Pradesh | 8.50 | 9.80 | 12.40 | 15.14 | 20.60 | 33.27 |
| Maharashtra | 8.60 | 11.30 | 12.50 | 11.48 | 15.20 | 15.45 |
| Orissa | 18.20 | 19.20 | 24.10 | 22.68 | 22.30 | 27.99 |
| Punjab | 76.70 | 80.80 | 88.90 | 99.74 | 94.60 | 92.24 |
| Rajasthan | 16.70 | 17.60 | 22.20 | 30.02 | 27.30 | 31.81 |
| Tamil Nadu | 47.70 | 42.00 | 45.30 | 43.77 | 47.20 | 54.85 |
| Uttar Pradesh | 37.10 | 42.10 | 49.10 | 56.98 | 61.00 | 66.43 |
| West Bengal | 21.70 | 20.20 | 26.20 | 22.81 | 40.00 | 26.81 |
| India | 24.10 | 26.00 | 29.40 | 32.52 | 34.80 | 39.22 |

*Source:* Indian Agriculture in Brief, G.O.I. & CMIE, Agriculture (Various Issues).

**Table 1c**
**Per Capita Net State Domestic Product at Constant Prices (in Rs.) in India by States**

| | *1972-1973* | *1977-1978* | *1982-1983* | *1987-1988* | *1992-1993* | *1999-2000* |
|---|---|---|---|---|---|---|
| Andhra Pradesh | 534 | 616 | 7456 | 1530 | 5718 | 9318 |
| Gujarat | 650 | 885 | 968 | 1942 | 8045 | 13022 |
| Haryana | 846 | 988 | 1083 | 2598 | 9037 | 13709 |
| Himachal Pradesh | 638 | 733 | 768 | 1818 | 6390 | 10514 |
| Madhya Pradesh | 474 | 508 | 602 | 649 | 4558 | 7947 |
| Maharashtra | 729 | 759 | 1033 | 2813 | 9795 | 15082 |
| Orissa | 520 | 619 | 729 | 1320 | 4114 | 5265 |
| Punjab | - | - | 1093 | 3310 | 11217 | 14980 |
| Rajasthan | 589 | 654 | 619 | 1241 | 5088 | 8277 |
| Tamil Nadu | 467 | 506 | 816 | 1821 | 6809 | 12314 |
| Uttar Pradesh | 467 | 599 | 1583 | 1455 | 4345 | 5682 |
| West Bengal | 1216 | 1352 | 1522 | 1828 | 5633 | 9330 |

*Source:* Rural Development Statistics, NIRD & Statistical Abstract, Govt. of W.B.

To examine the influence of IC on the change in CP, the simple regression equations have been estimated for the whole period (1972-73 to 1995-96), sub period I (1972-73 to 1982-83) and sub-period II (1983-84 to 1995-96) in India and the sample states. The specifications of these models are given in Table 2.

From the estimated equations it is observed as a whole that the IC has influenced significantly the change in CP in most of the sample states, viz., Punjab, Haryana, Orissa. Rajasthan, Tamil Nadu, H.P., Gujarat, U.P. and W.B. Here the Chow tests reveal that there have been differential influences of IC on change in CP in the sample states including all India. [where F = 41.33, 6.94, 5.05, 10.33, 5.24, 11.64, 6.99, 2.91, 3.10 and 8.23 (significant at 1% level) respectively for Punjab Haryana, Maharashtra, Rajasthan, H.P., A.P., Gujarat, M.P., W. B. and all India and 4.22 and 2.91 (significant at 5% level) for Orissa and M.P. respectively and 2.10 and 2.68 for Tamil Nadu and U.P. respectively].

Since IC acts through different supply side factors, viz., irrigation, fertiliser, etc. to change CP, it is in the fitness of

**Table 2**

**Estimates of Simple Regression Equations Concerning Cropping Pattern and Institutional Credit (IC) in India by States , Dependent Variable = Cropping Pattern Index , Number of Observations = 24 (1972-73 to 1995-1996)**

| *States* | *Constants* | *Explanatory Variables IC/hectare* | $R^2$ | *F* |
|---|---|---|---|---|
| Punjab | | | | |
| 1972-73 to 1995-96 | 94.68 | 0.03* | 0.16 | 4.19 |
| | (13.84) | (1.97) | | |
| 1972-73 to 1982-83 | 98.96 | 0.02** | 0.44 | 7.20 |
| | (78.34) | (2.68) | | |
| 1983-84 to 1995-96 | 70.23 | 0.07 | 0.14 | 1.79 |
| | (2.07) | (1.23) | | |
| Haryana | | | | |
| 1972-73 to 1995-96 | 105.86 | 0.02** | 0.49 | 21.14 |
| | (59.32) | (4.42) | | |
| 1972-73 to 1982-83 | 103.82 | 0.03** | 0.47 | 8.01 |
| | (57.17) | (2.83) | | |
| 1983-84 to 1995-96 | 116.56 | 0.0001 | 0.00001 | 0.00001 |
| | (22.76) | (0.01) | | |
| Maharastra | | | | |
| 1972-73 to 1995-96 | 81.04 | -0.02 | 0.044 | 1.01 |
| | (21.82) | (-0.96) | | |
| 1972-73 to1982-83 | 76.39 | 0.03 | 0.06 | 0.57 |
| | (13.06) | (0.75) | | |
| 1983-84 to 1995-96 | 77.59 | -0.01 | 0.03 | 0.34 |
| | (14.05) | (-0.50) | | |
| Orissa | | | | |
| 1972-73 to 1995-96 | 102.67 | -0.014 | 0.31 | 9.88 |
| | (171.47) | (-3.11) | | |
| 1972-73 to 1982-83 | 103.31 | -0.02** | 0.47 | 7.86 |
| | (161.17) | (-2.80) | | |
| 1983-84 to 1995-96 | 100.81 | -0.08 | 0.01 | 0.01 |
| | (80.07) | (-0.35) | | |
| Rajasthan | | | | |
| 1972-73 to 1995-96 | 84.41 | -0.08** | 0.38 | 13.48 |
| | (29.02) | (-3.53) | | |
| 1972-73 to 1982-83 | 72.76 | 0.11*** | 0.21 | 2.41 |
| | (15.18) | (1.55) | | |
| 1983-84 to 1995-96 | 88.18 | -0.11** | 0.56 | 14.00 |
| | (16.54) | (-3.37) | | |

*Contd...*

| *States* | *Constants* | *Explanatory Variables IC/hectare* | $R^2$ | *F* |
|---|---|---|---|---|
| Tamil Nadu | | | | |
| 1972-73 to 1995-96 | 99.60 | 0.03** | 0.45 | 18.00 |
| | (27.88) | (4.04) | | |
| 1972-73 to 1982-83 | 98.31 | 0.03 | 0.54 | 10.44 |
| | (34.24) | (3.23) | | |
| 1983-84 to 1995-96 | 107.85 | 0.02 | 0.13 | 1.64 |
| | (11.40) | (1.16) | | |
| H.P | | | | |
| 1972-73 to 1995-96 | 167.34 | -0.04* | 0.18 | 5.35 |
| | (19.25) | (-2.11) | | |
| 1972-73 to 1982-83 | 165.03 | 0.09** | 0.64 | 16.21 |
| | (63.49) | (4.03) | | |
| 1983-84 to 1995-96 | 148.27 | -0.03 | 0.08 | 0.96 |
| | (8.57) | (-0.89) | | |
| A.P | | | | |
| 1972-73 to 1995-96 | 110.91 | 0.001 | 0.06 | 1.40 |
| | (156.31) | (1.12) | | |
| 1972-73 to 1982-83 | 109.75 | -0.0009 | 0.001 | 0.009 |
| | (57.67) | (-0.096) | | |
| 1983-84 to 1995-96 | 112.99 | 0.0001 | 0.004 | 0.05 |
| | (183.69) | (0.19) | | |
| Gujarat | | | | |
| 1972-73 to 1995-96 | 84.17 | 0.05* | 0.41 | 15.29 |
| | (24.25) | (3.72) | | |
| 1972-73 to 1982-83 | 77.81 | 0.08** | 0.43 | 6.69 |
| | (12.67) | (2.59) | | |
| 1983-84 to 1995-96 | 97.34 | 0.009 | 0.13 | 1.64 |
| | (44.72) | (1.17) | | |
| U.P | | | | |
| 1972-73 to 1995-96 | 100.11 | 0.05 | 0.202 | 5.57 |
| | (28.07) | (2.25) | | |
| 1972-73 to 1982-83 | 98.89 | 0.06** | 0.504 | 9.14 |
| | (41.44) | (3.02) | | |
| 1983-84 to 1995-96 | 121.65 | -0.07 | 0.045 | 0.52 |
| | (6.68) | (-0.65) | | |
| M.P | | | | |
| 1972-73 to 1995-96 | 126.34 | 0.01 | 0.034 | 0.77 |
| | (37.33) | (0.84) | | |
| 1972-73 to 1982-83 | 123.28 | -0.0009 | 0.00009 | .00008 |
| | (15.38) | (-0.009) | | |
| 1983-84 to 1995-96 | 132.36 | 0.002 | 0.002 | 0.02 |
| | (23.46) | (0.15) | | |

*Contd...*

| *States* | *Constants* | *Explanatory Variables IC/hectare* | $R^2$ | *F* |
|---|---|---|---|---|
| W.B | | | | |
| 1972-73 to 1995-96 | 104.64 | 0.03** | 0.27 | 8.14 |
| | (45.58) | (2.71) | | |
| 1972-73 to 1982-83 | 102.90 | 0.032** | 0.68 | 19.59 |
| | (99.17) | (4.42) | | |
| 1983-84 to 1995-96 | 113.27 | 0.004 | 0.004 | 0.04 |
| | (27.42) | (0.20) | | |
| All India | | | | |
| 1972-73 to 1995-96 | 102.94 | 0.01 | 0.03 | 0.68 |
| | (59.21) | (0.77) | | |
| 1972-73 to 1982-83 | 98.22 | 0.07** | 0.56 | 11.53 |
| | (54.97) | (3.39) | | |
| 1983-84 to 1995-96 | 111.34 | -0.05 | 0.13 | 1.64 |
| | (18.54) | (-1.17) | | |

*Notes :* (i) Figures in the parenthesis indicate 't' values.
(ii) * indicates 't' value significant at 1% level.
(iii) ** indicates 't' value significant at 5% level.

*Source:* (i) Banking Statistics , R.B.I .
(ii) Fertilizer Statistics , G.O.I .
(iii) Area and Production of Pincipal Crops in India , G.O.I.
(iv) Statistical Statements Relating to Co-operative Movements in India, NABARD/R.B.I.

the things to examine the influence of these factors on CP change. As factors influencing change in CP we have considered here two supply side factors viz., fertiliser consumption per hectare and the percentage of GIA in GCA and one demand side factor, namely, PCI, or PCSDP. Let us analyse the influence of these factors on change in CP in the sample states as well as India. For this purpose we have made a regression analysis. The specifications of estimated regression equations are given in Table 3 . From the results presented in Table 3 it can be said that among the three factors, the percentage of GIA in GCA has influenced significantly the change CP in both Haryana and all India. But none of these factors has influenced significantly the change in CP in Punjab, A.P. , Gujarat, M.P. and W.B. In Orissa and Tamil Nadu, the FC per hectare has influenced significantly the change in CP. But, in Maharashtra, H.P. and U.P., both the

**Table 3**
**Estimates of Multiple Regression Equations Concerning Cropping Pattern, Fertiliser Consumption, Irrigated Area, and Income in India by States, Dependent Variable = Cropping Pattern Index, Number of Observation = 24 (1972-73 to 1995-1996)**

| *States* | *Constant* | *Explanatory variables* | | | $R^2$ | *F* |
|---|---|---|---|---|---|---|
| | | *FC/ Hectare* | *% of GIA to GCA* | *PCI/ PCSDP* | | |
| Punjab | 99.76 | 0.06 | -0.05 | 0.002 | 0.08 | 0.58 |
| | (1.44) | (0.35) | (-0.05) | (0.34) | | |
| Haryana | 94.15 | 0.05 | 0.23* | 0.004 | 0.74 | 18.97 |
| | (17.36) | (0.69) | (1.91) | (0.20) | | |
| Maharastra | 84.40 | 0.16 | -0.36* | -0.003** | 0.42 | 4.83 |
| | (27.21) | (1.01) | (-2.13) | (-1.45) | | |
| Orissa | 102.18 | -0.39* | 0.09 | 0.002 | 0.49 | 6.41 |
| | (51.28) | (-2.03) | (0.79) | (0.98) | | |
| Rajasthan | 91.96 | 0.26 | -0.36 | -0.012 | 0.52 | 7.22 |
| | (11.95) | (1.2) | (-0.78) | (-1.86) | | |
| Tamil Nadu | 95.96 | 0.26** | -0.1009 | -0.006 | 0.71 | 16.32 |
| | (5.04) | (3.27) | (-0.25) | (-0.4) | | |
| H.P. | 209.31 | -1.01 | 0.39 | -0.03** | 0.72 | 17.14 |
| | (18.48) | (-1.15) | (2.93) | (-1.58) | | |
| A.P. | 109.16 | 0.04 | -0.06 | 0.005 | 0.51 | 6.94 |
| | (11.09) | (1.14) | (-0.19) | (0.89) | | |
| Gujarat | 79.71 | 0.17 | 0.46 | -0.005 | 0.44 | 5.24 |
| | (10.61) | (0.94) | (0.86) | (-0.17) | | |
| U.P. | 67.91 | 0.12 | 0.95** | -0.014 | 0.65 | 12.38 |
| | (7.14) | (0.95) | (3.24) | (-2.47) | | |
| M.P. | 80.18 | -1.09 | 2.95 | 0.05 | 0.17 | 1.37 |
| | (2.69) | (-1.19) | (-1.19) | (0.01) | | |
| W.B. | 90.09 | 0.03 | 0.08 | 0.01 | 0.76 | 21.11 |
| | (6.56) | (0.18) | (0.64) | (0.82) | | |
| All India | 99.20 | -0.06 | 0.66** | -0.006 | 0.11 | 0.82 |
| | (13.71) | (-0.85) | (1.33) | (-0.74) | | |

*Notes:* (i) Figures in the parenthesis indicate 't' value.
(ii) * indicates 't' value significant at 1% level.
(iii) ** indicates 't' value significant at 5% level.

*Source:* (i) Banking Statistics, R.B.I.
(ii) Fertilizer Statistics, G.O.I.
(iii) Area and Production of Principal Crops in India, G.O.I.
(iv) Statistical Statement Relating to Co-operative Movements in India, NABARD/R.B.I.

percentage of GIA in GCA and PCSDP have had significant influence on the change in C.P. In Rajasthan, only the demand side factor i.e., PCSDP has significantly influenced the change in CP.

Let us analyse now the impact of supply side factors only, viz., fertiliser consumption per hectare and the percentage of GIA in GCA on the change in CP. The results of the estimated regression equations for the whole period and the sub-periods are given in Table 4 .

As a whole, in some states like Punjab, A.P., and Gujarat including all India none of these factors ,viz., FC per hectare and the percentage of GIA to GCA has any significant influence on the change in CP, but in the State of H.P. these factors are seen to have significant impacts on the change in CP. In the states, viz., W.B. and Tamil Nadu the FC has had significant influence on the CP change. In Haryana, both of these two factors have made significant impact on change in CP in the whole period ,but this is not so in two sub-periods. In Maharashtra, the percentage of GIA in GCA has made a significant effect on CP change in both the whole and the second sub-period. In Orissa the FC per hectare has made significant impact on the change in CP in all the periods, and irrigation has made significant impact only in the second sub-period. In Rajasthan, the percentage of GIA in GCA has made significant impact on change in CP during the whole period and the second sub-period, but the FC has no significant effect on CP change in any period. In M.P. the percentage of GIA in GCA only has significant influence on CP change during 1972-73 to 1995-96. In U.P. the irrigation has significant effect on change in CP in all the periods, but the FC has made significant effect only in second sub-period. The Chow tests reveal that there has been differential influences of FC per hectare and the percentage of GIA in GCA on change in CP in most of the sample states including all India [where F = 0.60, 3.15*, 4.23*, 5.38**, 1.94***, 8.97**, 3.03*, 2.19*, 3.96**, 5.24**, 1.81***, 0.13 and 28.55** respectively for Punjab, Haryana, Maharashtra, Orissa, Rajasthan, Tamil Nadu, H.P., A.P., Gujarat, U.P. M.P. , W.B. and all India ( *, ** , *** indicate significant at 5%, 1%

**Table 4**
**Estimates of Multiple Regression Equations concerning Cropping Pattern , Fertiliser Consumption, and Irrigated Area in India by States, Dependent Variable = Cropping Pattern Index, Number of Observation = 24 (1972-73 to 1995-96)**

| *States* | *Constant* | *Explanatory variables* FC/ Hectare | % of GIA to GCA | $R^2$ | *F* |
|---|---|---|---|---|---|
| Punjab | | | | | |
| 1972-73 to 1995-96 | 87.32 | 0.78 | 0.11 | 0.072 | 0.81 |
| | (1.46) | (0.51) | (0.34) | | |
| 1972-73 to 1982-83 | 79.75 | 0.04 | 0.24 | 0.75 | 11.92 |
| | (4.48) | (1.12) | (0.97) | | |
| 1983-84 to 1995-96 | 210.49 | -0.79 | 0.26 | 0.05 | 0.26 |
| | (1.11) | (-0.63) | (0.19) | | |
| Haryana | | | | | |
| 1972-73 to 1995-96 | 94.19 | 0.065** | 0.23* | 0.703 | 24.85 |
| | (17.84) | (1.41) | (1.98) | | |
| 1972-73 to 1982-83 | 105.72 | 0.33* | -0.13 | 0.52 | 4.27 |
| | (10.02) | (2.12) | (-0.22) | | |
| 1983-84 to 1995-96 | 95.20 | -0.034 | 0.33** | 0.63 | 8.51 |
| | (16.33) | (-0.55) | (2.99) | | |
| Maharastra | | | | | |
| 1972-73 to 1995-96 | 85.67 | -0.048 | -0.47 | 0.36 | 5.91 |
| | (27.96) | (-0.71) | (-3.05) | | |
| 1972-73 to 1982-83 | 57.35 | -0.033 | 2.11 | 0.20 | 1.02 |
| | (2.95) | (-0.05) | (0.84) | | |
| 1983-84 to 1995-96 | 81.39 | 0.02 | -0.45 | 0.86 | 30.71 |
| | (28.40) | (0.34) | (-6.99) | | |
| Orissa | | | | | |
| 1972-73 to 1995-96 | 102.15 | -0.21 | 0.08 | 0.46 | 8.94 |
| | (51.33) | (-3.27) | (0.68) | | |
| 1972-73 to 1982-83 | 107.41 | -1.06** | 0.16 | 0.54 | 4.74 |
| | (36.18) | (-2.72) | (0.79) | | |
| 1983-84 to 1995-96 | 101.39 | -0.30** | 0.18*** | 0.59 | 7.19 |
| | (38.19) | (-3.38) | (-1.15) | | |
| Rajasthan | | | | | |
| 1972-73 to 1995-96 | 100.16 | -0.08 | -1.07 | 0.43 | 7.92 |
| | (14.94) | (-0.59) | (-3.68) | | |
| 1973-74 to 1982-83 | 77.94 | -0.04 | 0.11 | 0.29 | 0.01 |
| | (4.98) | (-0.09) | (0.14) | | |
| 1983-84 to 1995-96 | 100.31 | -0.02 | -1.16 | 0.55 | 6.11 |
| | (7.95) | (-0.10) | (-2.77) | | |

*Contd...*

| *States* | *Constant* | *Explanatory variables* FC/ Hectare | % of GIA to GCA | $R^2$ | *F* |
|---|---|---|---|---|---|
| Tamil Nadu | | | | | |
| 1972-73 to 1995-96 | 95.68 | 0.26** | -0.094 | 0.72 | 27.00 |
| | (5.49) | (6.82) | (-0.27) | | |
| 1972-73 to 1982-83 | 103.43 | 0.30** | -0.28 | 0.64 | 7.02 |
| | (7.28) | (3.58) | (-0.84) | | |
| 1983-84 to 1995-96 | 57.17 | 0.56** | -0.012 | 0.82 | 22.78 |
| | (1.88) | (5.91) | (-0.02) | | |
| H.P | | | | | |
| 1972-73 to 1995-96 | 213.56 | -3.12** | 0.35** | 0.68 | 22.31 |
| | (18.70) | (-6.25) | (2.57) | | |
| 1972-73 to 1982-83 | 267.04 | 0.89** | -6.19 | 0.73 | 10.86 |
| | (6.78) | (3.05) | (-2.71) | | |
| 1983-84 to 1995-96 | 308.49 | -6.32** | 0.28* | 0.83 | 24.41 |
| | (9.28) | (-5.64) | (2.16) | | |
| A.P | | | | | |
| 1972-73 to 1995-96 | 109.16 | 0.04 | -0.06 | 0.51 | 10.93 |
| | (11.09) | (1.14) | (-0.19) | | |
| 1972-73 to 1982-83 | 118.14 | 0.06 | -0.32 | 0.11 | 0.47 |
| | (6.64) | (0.97) | (-0.56) | | |
| 1983-84 to 1995-96 | 100.93 | 0.03 | 0.25 | 0.58 | 6.90 |
| | (12.45) | (1.03) | (0.96) | | |
| M.P. | | | | | |
| 1972-73 to 1995-96 | 99.38 | -0.82 | 3.20*** | 0.143 | 1.75 |
| | (5.42) | (-0.97) | (1.95) | | |
| 1972-73 to 1982-83 | 108.35 | -3.82 | 4.41 | 0.15 | 0.70 |
| | (4.14) | (-1.19) | (1.013) | | |
| 1983-84 to 1995-96 | 110.78 | -1.42 | 3.59 | 0.16 | 0.95 |
| | (3.61) | (-1.23) | (1.09) | | |
| Gujarat | | | | | |
| 1972-73 to 1995-96 | 79.61 | 0.15 | 0.47 | 0.43 | 7.92 |
| | (10.92) | (1.14) | (0.91) | | |
| 1972-73 to 1982-83 | 73.43 | 0.62 | 0.14 | 0.45 | 3.32 |
| | (5.85) | (1.7) | (0.15) | | |
| 1983-84 to 1995-96 | 96.47 | 0.07 | -0.012 | 0.19 | 1.17 |
| | (14.83) | (1.17) | (-0.04) | | |
| U.P. | | | | | |
| 1972-73 to 1995-96 | 74.13 | -0.08 | 0.77* | 0.53 | 11.84 |
| | (7.18) | (-0.73) | (2.402) | | |
| 1972-73 to 1982-83 | 52.72 | -0.09 | 1.30** | 0.66 | 7.65 |
| | (2.02) | (-0.40) | (1.66) | | |
| 1983-84 to 1995-96 | 53.60 | 0.34*** | 0.52*** | 0.66 | 9.71 |
| | (3.69) | (1.41) | (1.45) | | |

*Contd...*

| States | Constant | FC/ Hectare | % of GIA to GCA | $R^2$ | F |
|---|---|---|---|---|---|
| | | *Explanatory variables* | | | |
| West Bengal | | | | | |
| 1972-73 to 1995-96 | 101.20 | 0.15** | 0.09 | 0.75 | 31.50 |
| | (41.07) | (5.16) | (0.70) | | |
| 1972-73 to 1982-83 | 102.90 | 0.17* | -0.1004 | 0.55 | 4.82 |
| | (19.05) | (2.42) | (-0.03) | | |
| 1983-84 to 1995-96 | 99.12 | 0.18** | 0.08 | 0.45 | 4.09 |
| | (16.44) | (2.00) | (0.42) | | |
| All India | | | | | |
| 1972-73 to 1995-96 | 96.16 | -0.079 | 0.36 | 0.084 | 0.96 |
| | (15.46) | (-1.28) | (1.29) | | |
| 1972-73 to 1982-83 | 84.68 | 0.06 | 0.67*** | 0.71 | 9.83 |
| | (9.35) | (0.61) | (1.63) | | |
| 1983-84 to 1995-96 | 115.39 | -0.23 | 0.06 | 0.06 | 0.34 |
| | (25.53) | (-5.45) | (0.37) | | |

*Notes:* (i) Figures in the parentheses indicate 't' value.
(ii) * indicates 't' value significant at 1% level.
(iii) ** indicates 't' value significant at 5% level.

*Source:* (i) Banking Statistics, R.B.I.
(ii) Fertilizer Statistics, G.O.I.
(iii) Area and Production of Principal Crops in India, G.O.I.
(iv) Statistical Statements Relating to Co-operative Movement in India, NABARD/R.B.I.

and 10% respectively)] during the two sub-periods. It is to note that the IC has also similar differential impact on CP change in those sample states. Thus, it may be said that it is the difference in the levels of IC among the states accounts for the differential impact of the FC per hectare and the percentage of GIA in GCA on change in CP in these sample states.

Let us examine now the impact of FC or % irrigated area separately on change in CP in the sample states and India respectively. To examine the impact of FC per hectare on the change in CP the simple regression equations are estimated and the results of the regression models are given in Table 5.

The FC per hectare has made significant influence on CP change in Punjab, Haryana, Orissa, Tamil Nadu, H.P., A.P., U.P. , W.B. , Gujarat and all India during the whole period as

**Table 5**
**Estimates of Simple Regression Equations Concerning Cropping Pattern, and Fertiliser Consumption in India by States, Dependent Variable = Cropping Pattern Index, Number of Observation = 24 (1972-73 to 1995-96)**

| *States/ Periods* | *Constants* | *Explanatory variable F.C/hectare* | $R^2$ | *F* |
|---|---|---|---|---|
| Punjab | | | | |
| 1972-73 to 1995-96 | 95.18 | 0.09 | 0.07 | 1.66 |
| | (9.70) | (1.23) | | |
| 1972-73 to 1982-83 | 96.93 | 0.06 | 0.71 | 3.07 |
| | (85.98) | (4.80) | | |
| 1983-84 to 1995-96 | 221.24 | -0.70 | 0.04 | 0.46 |
| | (1.29) | (-0.64) | | |
| Haryana | | | | |
| 1972-73 to 1995-96 | 104.25 | 0.14** | 0.64 | 39.11 |
| | (65.16) | (5.98) | | |
| 1972-73 to 1982-83 | 100.51 | 0.26** | 0.50 | 9.01 |
| | (34.30) | (3.00) | | |
| 1983-84 to 1995-96 | 108.42 | 0.11 | 0.22 | 3.10 |
| | (20.78) | (1.61) | | |
| Maharastra | | | | |
| 1972-73 to 1995-96 | 80.08 | -0.07 | 0.04 | 0.92 |
| | (27.40) | (-0.91) | | |
| 1972-73 to 1982-83 | 72.58 | 0.44 | 0.13 | 1.36 |
| | (10.27) | (1.17) | | |
| 1983-84 to 1995-96 | 70.15 | 0.11 | 0.08 | 0.96 |
| | (12.75) | (0.91) | | |
| Orissa | | | | |
| 1972-73 to 1995-96 | 103.43 | -0.18** | 0.45 | 18.00 |
| | (163.84) | (-4.03) | | |
| 1972-73 to 1982-83 | 108.80 | -0.85** | 0.51 | 9.22 |
| | (46.46) | (-3.04) | | |
| 1983-84 to 1995-96 | 104.73 | -0.24** | 0.47 | 9.75 |
| | (66.22) | (-2.83) | | |
| Rajasthan | | | | |
| 1972-73 to 1995-96 | 76.92 | -0.11 | 0.02 | 0.45 |
| | (26.52) | (-0.67) | | |
| 1972-73 to 1982-83 | 79.99 | -0.04 | 0.001 | 0.009 |
| | (21.61) | (0.11) | | |
| 1983-84 to 1995-96 | 66.76 | 0.21 | 0.11 | 1.36 |
| | (14.61) | (1.08) | | |

*Contd...*

| *States/ Periods* | *Constants* | *Explanatory variable F.C/hectare* | $R^2$ | *F* |
|---|---|---|---|---|
| Tamil Nadu | | | | |
| 1972-73 to 1995-96 | 91.12 | 0.26** | 0.72 | 56.57 |
| | (28.22) | (7.11) | | |
| 1972-73 to 1982-83 | 92.26 | 0.27** | 0.61 | 13.80 |
| | (22.38) | (3.71) | | |
| 1983-84 to 1995-96 | 56.53 | 0.57** | 0.82 | 50.11 |
| | (5.76) | (6.35) | | |
| H.P | | | | |
| 1972-73 to 1995-96 | 217.67 | -2.95** | 0.58 | 30.38 |
| | (17.01) | (-5.21) | | |
| 1972-73 to 1982-83 | 160.73 | 1.08** | 0.48 | 8.43 |
| | (32.28) | (2.90) | | |
| 1983-84 to 1995-96 | 325.73 | -6.59** | 0.73 | 29.74 |
| | (8.51) | (64.99) | | |
| A.P | | | | |
| 1972-73 to 1995-96 | 107.67 | 0.05** | 0.49 | 21.14 |
| | (114.51) | (4.41) | | |
| 1972-73 to 1982-83 | 108.19 | 0.03 | 0.07 | 0.67 |
| | (1.91) | (0.04) | | |
| 1983-84 to 1995-96 | 96.22 | 0.05*** | 0.18 | 2.41 |
| | (38.50) | (1.46) | | |
| U.P | | | | |
| 1972-73 to 1995-96 | 98.73 | 0.16** | 0.38 | 13.48 |
| | (33.79) | (3.53) | | |
| 1972-73 to 1982-83 | 95.81 | 0.26** | 0.54 | 10.59 |
| | (30.79) | (3.24) | | |
| 1983-84 to 1995-96 | 63.18 | 0.59** | 0.57 | 14.58 |
| | (4.61) | (3.43) | | |
| M.P | | | | |
| 1972-73 to 1995-96 | 123.33 | 0.28 | 0.06 | 1.40 |
| | (25.07) | (1.14) | | |
| 1972-73 to 1982-83 | 131.21 | -1.02 | 0.04 | 0.38 |
| | (9.84) | (-0.62) | | |
| 1983-84 to 1995-96 | 140.40 | -0.29 | 0.03 | 0.34 |
| | (9.99) | (-0.56) | | |
| Gujarat | | | | |
| 1972-73 to 1995-96 | 85.63 | 0.26** | 0.41 | 15.29 |
| | (27.66) | (3.73) | | |
| 1972-73 to 1982-83 | 74.99 | 0.66** | 0.45 | 7.43 |
| | (11.01) | (2.73) | | |

*Contd...*

| *States/ Periods* | *Constants* | *Explanatory variable F.C/hectare* | $R^2$ | *F* |
|---|---|---|---|---|
| 1983-84 to 1995-96 | 96.22 | 0.06** | 0.19 | 2.58 |
| | (38.51) | (1.47) | | |
| W. B. | | | | |
| 1972-73 to 1995-96 | 102.74 | 0.15** | 0.74 | 62.61 |
| | (89.27) | (7.60) | | |
| 1972-73 to 1982-83 | 102.71 | 0.16** | 0.54 | 10.83 |
| | (80.32) | (3.29) | | |
| 1983-84 to 1995-96 | 99.99 | 0.19** | 0.43 | 8.30 |
| | (18.51) | (2.65) | | |
| All India | | | | |
| 1972-73 to 1995-96 | 104.51 | -0.007 | 0.08 | 1.91 |
| | (84.89) | (-0.26) | | |
| 1972-73 to 1982-83 | 99.26 | 0.19** | 0.61 | 14.38 |
| | (74.04) | (3.79) | | |
| 1983-84 to 1995-96 | 116.87 | -0.22** | 0.82 | 50.11 |
| | (60.10) | (-6.53) | | |

*Notes:* (i) Figures in the parentheses indicate 't' value.
(ii) * indicates 't' value significant at 1% level.
(iii) ** indicates 't' value significant at 5% level.

*Source:* (i) Banking Statistics, R.B.I.
(ii) Fertilizer Statistics, G.O.I.
(iii) Area and Production of Principal Crops in India, G.O.I.
(iv) Statistical Statements Relating to Co-operative Movements in India, NABARD/R.B.I.

well as the sub-periods. Here the Chow test results reveal that there have been differential influences of FC per hectare on change in CP in all the sample states (except W.B., where F = 0.32 ) and also in India. These states are Punjab, Haryana, Maharashtra, Orissa, Rajasthan, Tamil Nadu, H.P., A.P., Gujarat, U.P., M.P., W.B. and all India. It is to note that the IC has also similar differential impact on CP change in these states. It, therefore, may be concluded that the differential impact of IC may be the prime reason behind the differential influence of FC per hectare on change in CP in these sample states. We have estimated the impact of IC on FC by regressing IC on FC per hectare in the states. The results are given in Table 6. It is observed that the IC per hectare has a significant impact on the FC per hectare in the sample states (except Rajasthan) and

**Table 6**
**Estimates of Simple Regression Equations Concerning Fertiliser Consumption and IC in India by States, Dependent Variable = Fertilizer Consumption/ Hectare, Number of Observation = 24 (1972-73 to 1995-96)**

| *States* | *Constants* | *Explanatory variable I.C/hectare* | $R^2$ | *F* |
|---|---|---|---|---|
| Punjab | 49.71 | 0.17** | 0.71 | 53.86 |
| | (4.44) | (6.98) | | |
| Haryana | 47.46 | 0.11** | 0.45 | 18.00 |
| | (8.18) | (6.66) | | |
| Maharastra | 13.23 | 0.09* | 0.19 | 5.16 |
| | (1.43) | (2.15) | | |
| Orissa | 8.11 | 0.04* | 0.210 | 5.85 |
| | - | (2.10) | | |
| Rajasthan | 11.39 | 0.03 | 0.02 | 0.45 |
| | (2.31) | (0.62) | | |
| Tamil Nadu | 36.43 | 0.10** | 0.53 | 24.81 |
| | (3.41) | (4.78) | | |
| H.P | 17.90 | 0.011* | 0.19 | 5.16 |
| | (8.04) | (2.14) | | |
| A.P | 56.76 | 0.03** | 0.30 | 9.43 |
| | (6.93) | (2.92) | | |
| Gujarat | 11.16 | 0.13** | 0.41 | 15.29 |
| | (1.29) | (3.72) | | |
| U.P | 8.61 | 0.33** | 0.61 | 34.41 |
| | (0.91) | (5.57) | | |
| M.P | 13.14 | 0.03* | 0.22 | 6.21 |
| | (4.85) | (2.36) | | |
| W.B. | 17.44 | 0.17 | 0.27 | 8.14 |
| | (1.39) | (2.69) | | |
| All India | -11.95 | 0.48** | 0.74 | 62.62 |
| | (-1.65) | (7.56) | | |

*Notes:* (i) Figures in the parentheses indicate 't' value.
(ii) * indicates 't' value significant at 1% level.
(iii) ** indicates 't' value significant at 5% level.

*Source:* (i) Banking Statistics, R.B.I.
(ii) Fertilizer Statistics, G.O.I.
(iii) Area and Production of Principal Crops in India, G.O.I
(iv) Statistical Statements Relating to Co-operative Movements in India, NABARD/R.B.I.

**Table 7**

**Estimates of Simple Regression Equations concerning Cropping Pattern and Irrigated Area in India by States , Dependent Variable = Cropping Pattern Index , Number of Observations = 24 (1972-73 to 1995-96)**

| *States/* | *Constants* | *Explanatory variable % of GIA to GCA* | $R^2$ | *F* |
|---|---|---|---|---|
| Punjab | 64.48 | 0.47 | 0.06 | 1.40 |
| | (1.72) | (1.13) | | |
| Haryana | 88.92 | 0.36** | 0.67 | 44.67 |
| | (23.52) | (6.40) | | |
| Maharashtra | 84.33 | -0.48** | 0.34 | 11.33 |
| | (35.76) | (-3.20) | | |
| Orissa | 105.18 | -0.19* | 0.57 | 29.16 |
| | (49.03) | (-1.93) | | |
| Rajasthan | 99.36 | -0.08* | 0.42 | 15.93 |
| | (15.39) | (-3.90) | | |
| Tamil Nadu | 134.67 | -0.47 | 0.03 | 0.68 |
| | (4.52) | (-0.74) | | |
| A.P. | 91.77 | 0.54** | 0.36 | 12.38 |
| | (15.58) | (3.33) | | |
| H.P. | 151.23 | 0.20 | 0.04 | 0.92 |
| | (15.98) | (0.89) | | |
| Gujarat | 74.86 | 0.97** | 0.40 | 14.67 |
| | (12.42) | (3.63) | | |
| U.P. | 80.21 | 0.55** | 0.51 | 22.90 |
| | (13.30) | (4.60) | | |
| M.P. | 114.74 | 1.00*** | 0.10 | 2.44 |
| | (12.42) | (1.50) | | |
| W. B. | 97.71 | 0.50** | 0.39 | 14.07 |
| | (27.31) | (3.64) | | |
| All India | 103.00 | 0.04 | 0.005 | 0.11 |
| | (27.27) | (0.32) | | |

*Notes:* (i) Figures in the parentheses indicate 't' value.
(ii) * indicates 't' value significant at 1% level.
(iii) ** indicates 't' value significant at 5% level.

*Source:* (i) Banking Statistics , R.B.I. ,
(ii) Fertilizer Statistics, G.O.I.
(iii) Area and Production of Principal Crops in India, G.O.I. ,
(iv) Statistical Statements Relating to Co-operative Movements in India, NABARD/R.B.I.

**Table 8**
**Estimates of Simple Regression Equations Concerning Irrigated Area and IC in India by States, Dependent Variable = % of GIA to GCA, Number of Observations = 24 (1972-73 to 1995-96)**

| *States* | *Constants* | *Explanatory variable I.C/hectare* | $R^2$ | *F* |
|---|---|---|---|---|
| Punjab | 78.25 | 0.03** | 0.50 | 22.00 |
| | (28.67) | (4.45) | | |
| Haryana | 62.32 | 0.03** | 0.18 | 4.83 |
| | (21.66) | (2.07) | | |
| Maharashtra | 10.51 | 0.02 | 0.03 | 0.68 |
| | (2.33) | (0.74) | | |
| Orissa | 17.65 | 0.03** | 0.43 | 16.60 |
| | (15.60) | (3.93) | | |
| Rajasthan | 15.02 | 0.07** | 0.67 | 44.67 |
| | (11.85) | (6.43) | | |
| Tamil Nadu | 50.01 | -0.005* | 0.16 | 4.19 |
| | (32.28) | (-1.96) | | |
| H.P | 24.41 | 0.002 | 0.15 | 0.01 |
| | (2.65) | (0.11) | | |
| A.P | 35.62 | 0.001*** | 0.12 | 3.00 |
| | (46.80) | (1.65) | | |
| Gujarat | 15.13 | 0.03** | 0.32 | 10.35 |
| | (6.25) | (3.04) | | |
| U.P | 34.06 | 0.11** | 0.52 | 23.83 |
| | (9.46) | (4.66) | | |
| M.P | 11.71 | 0.01** | 0.30 | 9.43 |
| | (12.73) | (2.91) | | |
| W.B. | 23.72 | 0.006 | 0.73 | 0.14 |
| | (7.14) | (0.37) | | |
| All India | 19.37 | 0.10** | 0.62 | 35.89 |
| | (9.95) | (5.69) | | |

*Notes:* (i) Figures in the parentheses indicate 't' value.
(ii) * indicates 't' value significant at 1% level.
(iii) ** indicates 't' value significant at 5% level.

*Source:* (i) Banking Statistics , R.B.I. ,
(ii) Fertilizer Statistics , G.O.I. ,
(iii) Area and Production of Principal Crops in India, G.O.I. ,
(iv) Statistical Statements Relating to Co-operative Movements in India, NABARD/R.B.I.

**Table 8a**
**Values of Correlation Coefficients Concerning C.P.I., I.C., FC, etc. in India by States**

| *States* | *CPI & IC/ Hectare* | *CPI & % of irrigated area* | *CPI & FC/ hectare* | *% of irrigated area & IC/hectare* |
|---|---|---|---|---|
| Punjab | 0.24 | 0.26 | 0.67* | 0.70* |
| | (1.10) | (1.20) | (4.03) | (4.38) |
| Haryana | 0.82* | 0.80* | 0.80* | 0.79* |
| | (6.40) | (5.96) | (5.96) | (5.76) |
| Maharashtra | 0.20 | -0.22 | 0.13 | 0.11 |
| | (0.91) | (1.00) | (0.59) | (0.49) |
| Orissa | -0.55* | -0.39* | -0.67* | 0.49** |
| | (-2.94) | (-1.89) | (-4.03) | (2.38) |
| Rajasthan | -0.62* | -0.65* | -0.15 | 0.82* |
| | (-3.53) | (-3.82) | (-0.68) | (6.40) |
| Tamil Nadu | 0.67* | -0.16 | 0.85* | 0.40* |
| | (4.03) | (0.72) | (7.21) | (-1.95) |
| H.P. | -0.43* | 0.19 | -0.76* | 0.024 |
| | (-2.13) | (0.86) | (-5.23) | (0.11) |
| A.P. | 0.24 | 0.60* | 0.70* | 0.34 |
| | (1.10) | (3.35) | (4.38) | (1.61) |
| Gujarat | 0.64* | 0.63* | 0.64* | 0.56* |
| | (3.73) | (3.63) | (3.73) | (3.02) |
| U.P. | 0.45** | 0.71* | 0.62* | 0.72* |
| | (2.25) | (4.50) | (3.53) | (8.17) |
| M.P. | 0.18 | 0.31 | 0.24 | 0.54* |
| | (0.82) | (1.46) | (1.10) | (2.87) |
| W. B. | 0.62* | 0.65* | 0.61* | 0.14 |
| | (3.52) | (3.82) | (3.44) | (0.63) |
| All India | 0.16 | 0.07 | -0.06 | 0.78* |
| | (0.72) | (0.31) | (0.26) | (5.57) |

| *States* | *FC/hectare & I.C./hectare* | *C.P.I. & P.C.I or PCSDP* | *Index of agricultural production & IC/hectare* |
|---|---|---|---|
| Punjab | 0.84* | 0.27 | 0.71* |
| | (6.92) | (1.26) | (4.05) |
| Haryana | 0.69* | 0.75* | 0.79* |
| | (4.26) | (5.07) | (5.76) |
| Maharastra | 0.92* | 0.21 | 0.13 |
| | (10.49) | (0.96) | (0.59) |
| Orissa | 0.46** | -0.60 | 0.49** |
| | (2.31) | (-3.35) | (2.38) |

*Contd...*

| *States* | *CPI & IC/ Hectare* | *CPI & % of irrigated* | *CPI & FC/ hectare* | *% of irrigated area & IC/hectare* |
|---|---|---|---|---|
| Rajasthan | 0.14 | -0.60* | | 0.08 |
| | (0.63) | (3.35) | | (0.36) |
| Tamil Nadu | 0.73* | 0.73* | | 0.29 |
| | (4.78) | (4.78) | | (1.36) |
| H.P. | 0.43** | -0.74* | | -0.06 |
| | (2.13) | (-4.92) | | (-0.26) |
| A.P | 0.54* | 0.67* | | 0.21 |
| | (2.87) | (4.03) | | (0.96) |
| Gujarat | 0.64* | 0.50** | | 0.14 |
| | (3.73) | (2.58) | | (0.63) |
| U.P | 0.78* | 0.50** | | 0.77* |
| | (5.57) | (2.58) | | (5.46) |
| M.P. | 0.47** | 0.31 | | 0.14 |
| | (2.31) | (1.46) | | (0.63) |
| W. B. | 0.84* | 0.92* | | 0.86* |
| | (6.12) | (6.78) | | (7.34) |
| All India | 0.86* | 4.71* | | 0.74* |
| | (9.76) | (4.58) | | (4.92) |

*Notes:* (i) Figures in the parentheses indicates 't' value.
(ii) * indicates 't' value significant at 1% level.
(iii) ** indicates 't' value significant at 5% level.

*Source:* (i) Banking Statistics, R.B.I.,
(ii) Fertilizer Statistics, G.O.I.,
(iii) Area and Production of Principal Crops in India, G.O.I.,
(iv) Statistical Statements Relating to Co-operative Movements in India, NABARD/R.B.I.

India. On the other hand, the FC has also made significant impact on CP change in these sample States. So it may be concluded that the IC has made a significant impact on the change in CP through the FC in these sample states and all India.

Let us examine the impact of the percentage of GIA to GCA separately on CP change in the sample states vis-a-vis all India. The results of the estimated simple regression equations are given in Table 7 . In India, the influence of percentage of GIA in GCA on CP change is not significant either in the whole period or the second sub-period, but significant in the first sub-period. On the other hand, it is significant in W.B. for the whole period as well as for the two sub-periods. In the states, namely, Haryana, Orissa, Rajasthan, Gujarat, U.P., M.P. and A.P., the irrigation has influenced the CP change significantly and this fact may have a correspondence with the impact of IC on CP change in these sample states. Let us examine this issue i.e. the impact of IC on CP change through the percentage of GIA in GCA also by regressing GIA as percentage of GCA on the IC. The results of the regression equations are given in Table 8. From this table it is clear that for the states like Punjab, Haryana, Orissa, Rajasthan, Tamil Nadu, A.P., Gujarat, U.P. ,M.P. and India, the IC has made significant impact on change in CP through irrigation.

The relationship between CP change, and different factors influencing it, viz., FC, % of irrigated area, PCI or PCSDP may be explained by a correlation analysis (see Table 8a). It is observed that the value of correlation coefficients (r) between CPI and IC/hectare are significant and positive e.g., 0.82*, 0.67*, 0.64*, 0.45*, 0.62*, respectively for Haryana, Tamil Nadu, Gujarat, U.P. and W.B. For other states these coefficients are statistically not significant even at 10% probability level. The significant and positive correlation between CPI and % irrigated area is observed for the states like Haryana (0.08*) , A.P. (0.60*) , Gujarat (0.63*) , U.P. (0.71*) and W.B. ( 0.65*). On the other hand, the values of r between CPI and the FC per hectare are found to be 0.67*, 0.80*, 0.85*, 0.70*, 0.64*, 0.62* and 0.61* (all significant and positive ) respectively for Punjab, Haryana, Tamil Nadu, A.P., Gujarat, U.P. and W.B. The above results highlight the fact that the larger provision of IC motivates the farmers to change CP through the use of different modern inputs, such as, chemical fertilisers, and artificial irrigation. In other words, the IC may have played a catalytic role behind the use of modern inputs for changing CP in these sample states. It is found that the values for r between the percentage of irrigated area and IC/hectare are 0.70*, 0.79*, 0.49*, 0.82*, 0.56*, 0.72*, 0.54* and 0.78* respectively for Punjab, Haryana, Orissa, Rajasthan, Gujarat, U.P. , M.P. and India as a whole. On the other hand, the value of r between FC per hectare and IC per hectare are found to be 0.84*, 0.69*, 0.92*, 0.46**, 0.73*, 0.43*, 0.54**, 0.64*, 0.78*, 0.47*, 0.84* and 0.86* respectively for Punjab, Haryana, Maharashtra, Orissa, Tamil Nadu, H.P., A.P., Gujarat, U.P. , M.P. , W.B and all India. There are a number of factors explaining the insignificant relationship CP and IC, irrigated area and IC, etc. in the sample states. In practice, the farmers do not get the IC at the time of cultivation and also in required amount. Hence, the credit is not properly utilized for cultivation purposes, but spent for several non-agricultural or unproductive purposes. This type of credit use often does not create significant impact of IC on the agricultural growth through change in CP. The lower volume of available IC is often insufficient to purchase the required amount of agricultural inputs and hence does not create any significant impact on changes in CP. The above correlation analysis also highlight the fact that the IC has played a catalytic role through different modern inputs in changing CP in a number of sample states and in India.

On the other hand, the values of r between CP and PCSDP or PCI are estimated to be 0.75*, 0.73*, 0.67*, 0.50*, 0.50*, 0.92*, 0.71* respectively for Haryana, Tamil Nadu, A.P., Gujarat, U.P., W.B. and India and all of them are statistically significant at 1 to 5 percent probability levels. The r values for other states are not significant. The poor irrigational facilities in these States may explain the low positive or negative value of r between CP and PCI or PCSDP. As a whole, the above correlation analysis shows that the PCI or PCSDP as a demand side factor has contributed significantly to the change in CP in a number of sample states in India.

### Role of IC and other Factors in the Context of Commercialisation of Agriculture in India: State Level Analysis

In the context of commercialisation of Indian agriculture, we may also examine the impact of IC and other related factors, viz., fertiliser consumption, the percentage irrigated area, PCI or PCSDP on the NFA which has also significant influence on agricultural growth.

**Table 8b**

**Coefficient of Variation of the Percentage of Area under Non-foodgrains to GCA(NFA), the I.C. Per Hectare, % of Gross Irrigated Area (GIA) to GCA Across the Sample States of India for Some Sample Years**

| *Year* | *NFA* | *IC/hectare* | *% of GIA to GCA* |
|---|---|---|---|
| 1973-74 | 0.49 | 0.75 | 0.73 |
| 1983-84 | 0.41 | 0.68 | 0.67 |
| 1989-90 | 0.48 | 0.69 | 0.71 |

*Source:* (i) Area and production of Principal Crops in India, Government of India,
(ii) Banking Statistics, R.B.I.

To measure the impact of I.C. and the percentage of irrigated area in the GCA on the CP as measured by the proportion of area under non-foodgrains to the GCA across the sample states of India, we may use the coefficient of variation (CV) analysis for some sample years chosen from the period from 1972-73 to 1995-96. This kind of analysis will give us a rough idea about the relation between C.P. on the one hand and I.C. and percentage of irrigated area in the GCA. on the other. The following Table 8b gives the CV values for each of these variables. It is seen from the value of CV for the value of the variables that the pattern of change in the CVs in the variable IC and percentage irrigated area taking place over the period from 1973-74 to 1989-90 is the same as the pattern of change in CP , measured by NFA. This agreement in the pattern of change in CVs for these variables, of course, does not indicate any cause and effect relationship, but certainly indicates an association between them across the sample states of India. The following regression analysis, however, throws light on the cause and effect relationship between these variables.

Let us first examine the impact of IC per hectare on the NFA for the sample states with all India perspective. The results of the estimated simple regression equations are given in Table 9. The regression results show that the IC has significantly contributed to the commercialisation of agriculture through the change in CP in six sample States,

**Table 9**
**Estimates of Simple Regression Equations Concerning NFA and IC in India by States, Dependent Variable = percentage of Area under Non-foodgrains to GCA (NFA), Number of Observations = 24 (1972-73 to 1995-96)**

| *States* | *Constants* | *Explanatory variable IC/hectare* | $R^2$ | *F* |
|---|---|---|---|---|
| Punjab | 30.74 | -0.02** | 0.36 | 12.38 |
| | (14.10) | (-3.36) | | |
| Haryana | 21.96 | -0.0003 | 0.0003 | 0.007 |
| | (10.31) | (-0.08) | | |
| Maharashtra | 33.33 | -0.02*** | 0.099 | 2.42 |
| | (13.01) | (-1.48) | | |
| Orissa | 239.26 | -1.27 | 0.046 | 1.06 |
| | (1.45) | (-0.98) | | |
| Rajasthan | 27.50 | 0.006 | 0.001 | 0.02 |
| | (21.93) | (0.16) | | |
| Tamil Nadu | -41.02 | 0.55 | 0.024 | 0.54 |
| | (-0.101) | (0.69) | | |
| H.P | 10.24 | -0.003 | 0.08 | 1.91 |
| | (9.51) | (-1.31) | | |
| A.P. | 32.81 | -0.0004 | 0.009 | 0.20 |
| | (16.55) | (-0.14) | | |
| Gujarat | 52.88 | 0.021** | 0.14 | 3.58 |
| | (17.90) | (1.78) | | |
| U.P. | 14.17 | 0.03* | 0.21 | 5.85 |
| | (7.17) | (2.23) | | |
| M.P. | 20.13 | 0.009*** | 0.11 | 2.72 |
| | (15.39) | (1.55) | | |
| W. B. | 13.29 | 0.032** | 0.48 | 20.31 |
| | (9.29) | (4.29) | | |
| All India | 24.86 | 0.02** | 0.26 | 7.73 |
| | (29.45) | (2.63) | | |

*Notes:* (i) Figures in the parentheses indicate 't' value.
(ii) * indicates 't' value significant at 1% level.
(iii) ** indicates 't' value significant at 5% level.

*Source:* (i) Banking Statistics, R.B.I. (ii) Fertilizer Statistics, G.O.I.,
(iii) Area and Production of Principal Crops in India, G.O.I.,
(iv) Statistical Statements Relating to Co-operative Movements in India, NABARD/R.B.I.

**Table 10**
**Estimates of Simple Regression Equations Concerning NFA and Fertiliser Consumption in India by States,, Dependent Variable = Percentage of Area under Non-foodgrains to GCA(NFA), Number of Observations = 24 (1972-73 to 1995-96)**

| *States* | *Constants* | *Explanatory variable Fertilizer consumption/ hectare* | $R^2$ | *F* |
|---|---|---|---|---|
| Punjab | 37.08 | -0.11** | 0.68 | 46.75 |
| | (17.64) | (-6.45) | | |
| Haryana | 0.15 | 16.63 | 0.58 | 30.38 |
| | (9.10) | (5.27) | | |
| Maharashtra | 29.17 | 0.02 | 0.005 | 0.11 |
| | (13.81) | (0.34) | | |
| Orissa | 255.05 | -12.36 | 0.04 | 0.92 |
| | (1.30) | (-0.89) | | |
| Rajasthan | 26.77 | 0.07 | 0.07 | 1.66 |
| | (27.91) | (1.20) | | |
| Tamil Nadu | -501.48 | 8.66** | 0.11 | 17.29 |
| | (-1.04) | (1.57) | | |
| H.P. | 15.66 | -0.301** | 0.44 | 15.93 |
| | (9.14) | (-3.97) | | |
| A.P. | 23.53 | 0.13** | 0.42 | 15.93 |
| | (8.62) | (3.78) | | |
| Gujarat | 55.13 | 0.06 | 0.05 | 1.16 |
| | (19.94) | (1.05) | | |
| U.P. | 14.01 | 0.07** | 0.26 | 7.73 |
| | (7.93) | (2.63) | | |
| M.P. | 14.43 | 0.41** | 0.81 | 93.79 |
| | (16.34) | (9.31) | | |
| W.B. | 16.24 | 0.05 | 0.16 | 4.19 |
| | (10.59) | (1.97) | | |
| All India | 25.31 | 0.04 | 0.36 | 12.38 |
| | (46.29) | (3.37) | | |

*Notes:* (i) Figures in the parentheses indicate 't' value.
(ii) * indicates 't' value significant at 1% level.
(iii) ** indicates 't' value significant at 5% level.

*Source:* (i) Banking Statistics, R.B.I.
(ii) Fertilizer Statistics, G.O.I. ,
(iii) Area and Production of Principal Crops in India , G.O.I. ,
(iv) Statistical Statements Relating to Co-operative Movements in India, NABARD/R.B.I.

**Table 11**
**Estimates of Regression Equations Concerning NFA and Irrigated Area in India by States, , Dependent Variable = Percentage of Area Under Non-foodgrains to GCA (NFA), Number of Observations = 24 (1972-73 to 1995-96)**

| *States* | *Constants* | *Explanatory variable % irrigated area* | $R^2$ | *F* |
|---|---|---|---|---|
| Punjab | 79.42 | -0.62 | 0.77 | 73.65 |
| | | (-8.11) | | |
| Haryana | 31.91 | -0.15 | 0.05 | 1.16 |
| | (3.28) | (-1.05) | | |
| Maharashtra | 33.51 | -0.27* | 0.2 | 5.85 |
| | (18.32) | (-2.32) | | |
| Orissa | 497.48 | -18.65 | 0.03 | 0.68 |
| | (0.92) | (-0.75) | | |
| Rajasthan | 23.90 | 0.17*** | 0.09 | 2.18 |
| | (0.74) | (1.41) | | |
| Tamil Nadu | 2951.07 | -57.04 | 0.06 | (-1.10) |
| H.P. | 9.41 | -0.0005** | 0.000004 | 9.05 |
| | (8.35) | (-9.52) | | |
| A.P. | -21.07 | 1.48** | 0.37 | 12.92 |
| | (-1.32) | (3.40) | | |
| Gujarat | 50.38 | 0.34** | 0.10 | 2.44 |
| | (9.70) | (1.48) | | |
| U.P. | 9.91 | 0.17* | 0.16 | 4.19 |
| | (2.25) | (1.93) | | |
| M.P. | 7.36 | 1.05** | 0.69 | 48.97 |
| | (3.39) | (6.73) | | |
| W.B. | 16.84 | 0.08 | 0.02 | 0.45 |
| | (4.99) | (0.62) | | |
| All India | 20.43 | 0.22** | 0.49 | 21.14 |
| | (13.68) | (4.43) | | |

*Notes:* (i) Figures in the parentheses indicate 't' value.
(ii) * indicates 't' value significant at 1% level.
(iii) ** indicates 't' value significant at 5% level.

*Source:* (i) Banking Statistics, R.B.I.
(ii) Fertilizer Statistics, G.O.I.,
(iii) Area and Production of Principal Crops in India, G.O.I.,
(iv) Statistical Statements Relating to Co-operative Movements in India, NABARD/R.B.I.

namely, Punjab, Maharashtra, Gujarat, U.P., M.P. and W.B. and also in India.

Using the NFA as a measure of CP in the states a cross section analysis of sample states have been made for several years to study the impact of IC on CP. The estimated regression equations are as follows:

| | | | | |
|---|---|---|---|---|
| 1974-75: | NFA = 16.27 | + 0.032** IC; (1.98) | $R^2 = 0.28$ | F = 3.95 |
| 1975-76: | NFA = 10.84 | + 0.0006** IC; (2.18) | $R^2 = 0.32$ | F = 4.74 |
| 1976-77: | NFA = 11.79 | + 0.06** IC; (2.21) | $R^2 = 0.33$ | F = 4.86 |

In the above equations ** indicates coefficients are significant at 5% level and figures in brackets indicate t-ratios. The regression results show that IC has statistically significant impact on the change in CP and the IC can explain 28 to 33% variation in non-foodgrains cultivation across the sample states.

Now let us examine the role of fertiliser consumption to accelerate commercialisation of agriculture in India and sample states. The results of regression equations are given in Table 10. It is found that the fertiliser consumption has made significant impact on commercialisation of agriculture through change in CP in India and in eight sample states e.g., Punjab, Haryana, Tamil Nadu, H.P., A.P., U.P. , M.P. and W.B.

Now we may study the influence of a crucial factor, viz., irrigation on the commercialisation of agriculture in India and the sample states. For this purpose the regression equations have been estimated. The results of these models are given in Table 11. The regression results show that the percentage of GIA in GCA has significantly contributed to the commercialisation of agriculture through the change in C.P. in India and nine sample states viz. , Punjab, Harayana, Maharashtra, Rajasthan, H.P., A.P., Gujarat, U.P. and M.P. .

Thus from the regression results it is established that there is a high degree of relationship between fertiliser consumption and/or the percentage irrigated area on the one hand, and NFA on the other. It has also been established that there exists a strong relationship between IC and NFA. Thus, it may be

**Table 12**
**Estimates of Multiple Regression Equations Concerning NFA and Income in India by States,, Dependent Variable = Percentage Of Area Under Non-foodgrains to GCA(NFA), Number of Observations = 24 (1972-73 to 1995-96)**

| *States* | *Constants* | *Explanatory variable PCI/PCSDP* | $R^2$ | *F* |
|---|---|---|---|---|
| Punjab | 33.33 | -0.005** | 0.91 | 222.44 |
| | (42.11) | (-13.37) | | |
| Maharashtra | 3.39 | -0.000004 | 0.04 | 0.92 |
| | (50.95) | (-0.14) | | |
| Haryana | 19.60 | 0.003 | 0.33 | 10.84 |
| | (8.67) | (3.17) | | |
| Orissa | 206.88 | -0.11 | 0.03 | 0.68 |
| | (0.20) | (-0.71) | | |
| Rajasthan | 26.86 | 0.0008 | 0.02 | 0.45 |
| | (18.95) | (0.64) | | |
| Tamil Nadu | -129.63 | 0.32 | 0.072 | 1.71 |
| | (-0.40) | (0.55) | | |
| H.P. | 14.91 | -0.004** | 0.5002 | 22.02 |
| | (10.77) | (-4.47) | | |
| A.P. | 21.57 | 0.01** | 0.51 | 22.90 |
| | (7.81) | (4.45) | | |
| Gujarat | 53.73 | 0.003* | 0.19 | 5.16 |
| | (25.37) | (2.19) | | |
| U.P. | 14.21 | 0.004** | 0.28 | 8.55 |
| | (8.92) | (4.16) | | |
| M.P. | 0.71 | 0.04** | 0.46 | 18.74 |
| | (0.14) | (4.16) | | |
| W.B. | 10.33 | 0.005* | 0.201 | 9.03 |
| | (2.66) | (2.24) | | |
| All India | 20.46 | 0.004** | 0.44 | 17.29 |
| | (12.24) | (3.94) | | |

*Notes:* (i) Figures in the parentheses indicate 't' value.
(ii) * indicates 't' value significant at 1% level.
(iii) ** indicates 't' value significant at 5% level.

*Source:* (i) Banking Statistics, R.B.I.
(ii) Fertilizer Statistics, G.O.I.,
(iii) Area and Production of Principal Crops in India, G.O.I.,
(iv) Statistical Statements Relating to Co-operative Movements in India, NABARD/R.B.I.

**Table 13**
**Estimates of Simple Regression Equations Concerning HYV Area and IC in India by States, Dependent Variable = % HYV Area, Number of Observations = 24 (1972-73 to 1995-96)**

| *States/ Periods* | *Constants* | *Explanatory variable IC/hectare* | $R^2$ | *F* |
|---|---|---|---|---|
| Punjab | | | | |
| 1972-73 to 1995-96 | -29.92 | 0.41* | 0.14 | 3.58 |
| | (-0.28) | (1.83) | | |
| Haryana | | | | |
| 1972-73 to 1995 -96 | 29.29 | 0.04** | 0.58 | 30.38 |
| | (9.72) | (5.26) | | |
| Maharastra | | | | |
| 1972-73 to 1995 -96 | 13.92 | 0.06* | 0.23 | 6.57 |
| | (2.55) | (2.45) | | |
| Orissa | | | | |
| 1972-73 to 1995-96 | 4.82 | 0.12* | 0.38 | 13.48 |
| | (1.08) | (3.49) | | |
| Rajasthan | | | | |
| 1972-73 to 1995-96 | 7.20 | 0.08*** | 0.19 | 5.16 |
| | (1.11) | (1.52) | | |
| 1972-73 to 1982-83 | 2.54 | 0.101* | 0.92 | 104.38 |
| | (3.82) | (10.22) | | |
| 1983-84 to 1995-96 | 32.53 | -0.06 | 0.03 | 0.34 |
| | (1.64) | (-0.57) | | |
| Tamil Nadu | | | | |
| 1972-73 to 1995-96 | 32.94 | 0.01* | 0.42 | 15.93 |
| | (17.87) | (3.803) | | |
| H.P. | | | | |
| 1972-73 to 1995-96 | 45.51 | 0.01* | 0.17 | 4.51 |
| | (20.18) | (2.02) | | |
| 1972-73 to 1982-83 | 29.32 | 0.10** | 0.83 | 43.52 |
| | (15.56) | (6.61) | | |
| 1983-84 to 1995-96 | 54.68 | 0.002 | 0.02 | 0.22 |
| | (14.97) | (0.41) | | |
| A.P. | | | | |
| 1972-73 to1995-96 | 26.28 | 0.007** | 0.31 | 9.88 |
| | (14.97) | (3.03) | | |
| Gujarat | | | | |
| 1972-73 to 1995-96 | 8.89 | 0.05** | 0.45 | 18.00 |
| | (2.68) | (4.02) | | |
| 1972-73 to 1982-83 | 12.76 | 0.02** | 0.49 | 8.70 |
| | (7.82) | (2.95) | | |

*Contd...*

| *States/ Periods* | *Constants* | *Explanatory variable IC/hectare* | $R^2$ | *F* |
|---|---|---|---|---|
| 1983-84 to 1995-96 | 10.30 | 0.05 | 0.13 | 2.97 |
| | (1.72) | (0.25) | | |
| U.P. | | | | |
| 1972-73 to 1995-96 | 17.11 | 0.15** | 0.41 | 15.29 |
| | (3.11) | (4.44) | | |
| M.P. | | | | |
| 1972-73 to 1995-96 | 16.71 | 0.02*** | 0.12 | 3.00 |
| | (6.55) | (1.65) | | |
| W.B. | | | | |
| 1972-73 to 1995-96 | 13.62 | 0.06* | 0.17 | 4.51 |
| | (2.14) | (2.06) | | |
| All India | | | | |
| 1972-73 to 1995-96 | 6.04 | 0.19** | 0.66 | 42.71 |
| | (1.72) | (6.37) | | |

*Notes:* (i) Figures in the parentheses indicate 't' value.
(ii) * indicates 't' value significant at 1% level.
(iii) ** indicates 't' value significant at 5% level.

*Source:* (i) Banking Statistics, R.B.I.
(ii) Fertilizer Statistics, G.O.I.,
(iii) Area and Production of Principal Crops in India , G.O.I.,
(iv) Statistical Statements Relating to Co-operative Movements in India, NABARD/R.B.I.

safely said that the IC has been instrumental in changing the CP in many of the states in India through the use of chemical fertilisers and irrigation.

The PCI or the PCSDP may have some indirect impact upon the change in CP since the increase in either of them beyond a level enhances the investment capacity of the farmers provided the opportunities for investment are available to them. To examine the impact of PCI and PCSDP on CP, CP is regressed upon PCI and PCSDP. The regression results are presented in Table 12 . It is observed from the regression results that the commercialisation of agriculture, ( reflected by NFA) has made significant progress with the pace of development (indicated by movement of PCI or PCSDP) in India and eight sample states, namely, Punjab, Haryana, H.P., A.P., Gujarat, U.P. , M.P. and W.B .

**Table 14**
**Estimates of Simple Regression Equations Concerning Agricultural Production and IC in India by States, Dependent Variable = Index of Agricultural Production, Number of Observations = 24 (1972-73 to 1995-96)**

| *States* | *Constants* | *Explanatory variable IC/hectare* | $R^2$ | *F* |
|---|---|---|---|---|
| Punjab | | | | |
| 1972-73 to 1995-96 | 120.28 | 0.11* | 0.15 | 3.88 |
| | (4.33) | (1.86) | | |
| 1972-73 to 1982-83 | 108.23 | 0.08 | 0.08 | 0.74 |
| | (5.03) | (0.86) | | |
| 1983-84 to 1995-96 | 430.82 | -0.37** | 0.50 | 11.00 |
| | (5.62) | (-3.11) | | |
| Haryana | | | | |
| 1972-73 to 1995-96 | 113.11 | 0.14** | 0.62 | 35.89 |
| | (11.19) | (5.77) | | |
| 1972-73 to 1982-83 | 102.53 | 0.15* | 0.56 | 11.26 |
| | (10.03) | (3.36) | | |
| 1983-84 to 1995-96 | 206.36 | -0.02 | 0.043 | 0.49 |
| | (11.41) | (-0.63) | | |
| Maharastra | | | | |
| 1972-73 to 1995-96 | 122.51 | 0.15*** | 0.12 | 3.00 |
| | (6.19) | (1.64) | | |
| 1972-73 to 1982-83 | 83.81 | 0.31** | 0.46 | .72 |
| | (4.52) | (2.78) | | |
| 1983-84 to 1995-96 | 209.43 | -0.15 | 0.14 | 1.79 |
| | (6.63) | (-1.20) | | |
| Orissa | | | | |
| 1972-73 to 1995-96 | 94.91 | 0.19** | 0.24 | 6.95 |
| | (9.85) | (2.54) | | |
| 1972-73 to 1982-83 | 90.51 | 0.13 | 0.12 | 1.27 |
| | (8.27) | (1.13) | | |
| 1983-84 to 1995-96 | 131.19 | 0.009 | 0.0009 | 0.009 |
| | (9.29) | (0.092) | | |
| Rajasthan | | | | |
| 1972-73 to 1995-96 | 162.01 | 0.036 | 0.006 | 0.13 |
| | (13.52) | (0.36) | | |
| 1972-73 to 1982-83 | 175.06 | -0.27 | 0.06 | 0.55 |
| | (7.13) | (-0.74) | | |
| 1983-84 to 1995-96 | 191.81 | -0.12 | 0.08 | 0.96 |
| | (8.77) | (-0.88) | | |

*Contd...*

| *States* | *Constants* | *Explanatory variable IC/hectare* | $R^2$ | *F* |
|---|---|---|---|---|
| Tamil Nadu | | | | |
| 1972-73 to 1995-96 | 129.06 | 0.02*** | 0.083 | 1.99 |
| | (20.08) | (1.35) | | |
| 1972-73 to 1982-83 | 139.06 | -0.03 | 0.08 | 0.76 |
| | (12.13) | (-0.87) | | |
| 1983-84 to 1995-96 | 140.66 | 0.005 | 0.02 | 0.22 |
| | (18.23) | (0.42) | | |
| H.P. | | | | |
| 1972-73 to 1995-96 | 114.79 | -0.002 | 0.003 | 0.07 |
| | (34.05) | (-0.26) | | |
| 1972-73 to 1982-83 | 126.03 | -0.08 | 0.05 | 0.46 |
| | (8.49) | (-0.68) | | |
| 1983-84 to 1995-96 | 111.25 | 0.002 | 0.024 | 0.27 |
| | (53.21) | (0.47) | | |
| A.P. | | | | |
| 1972-73 to 1995-96 | 145.00 | 0.006 | 0.05 | 1.16 |
| | (33.84) | (0.99) | | |
| 1972-73 to 1982-83 | 110.78 | 0.16** | 0.70 | 21.403 |
| | (16.52) | (4.63) | | |
| 1983-84 to 1995-96 | 157.09 | -0.001 | 0.012 | 0.13 |
| | (39.40) | (-0.33) | | |
| Gujarat | | | | |
| 1972-73 to 1995-96 | 101.92 | 0.02 | 0.02 | 0.45 |
| | (17.72) | (0.66) | | |
| 1972-73 to 1982-83 | 97.69 | 0.04 | 0.05 | 0.47 |
| | (8.29) | (0.71) | | |
| 1983-84 to 1995-96 | 103.06 | -0.008 | 0.06 | 0.70 |
| | (33.47) | (0.78) | | |
| U.P. | | | | |
| 1972-73 to 1995-96 | 86.73 | 0.25** | 0.61 | 34.41 |
| | (12.14) | (5.45) | | |
| 1972-73 to 1982-83 | 88.63 | 0.18** | 0.45 | 7.36 |
| | (10.53) | (2.73) | | |
| 1983-84 to 1995-96 | 140.80 | -0.02 | 0.07 | 0.83 |
| | (31.71) | (-0.82) | | |
| M.P. | | | | |
| 1972-73 to 1995-96 | 132.71 | 0.01 | 0.02 | 0.45 |
| | (25.61) | (0.64) | | |
| 1972-73 to 1982-83 | 124.63 | 0.07 | 0.007 | 0.06 |
| | (6.23) | (0.26) | | |
| 1983-84 to 1995-96 | 139.66 | 0.002 | 0.005 | 0.05 |
| | (37.86) | (0.21) | | |

*Contd...*

| *States* | *Constants* | *Explanatory variable IC/hectare* | $R^2$ | *F* |
|---|---|---|---|---|
| W.B. | | | | |
| 1972-73 to 1995-96 | 103.26 | 0.20* | 0.21 | 5.85 |
| | (6.13) | (2.29) | | |
| 1972-73 to 1982-83 | 94.67 | 0.010*** | 0.22 | 2.54 |
| | (10.69) | (1.60) | | |
| 1983-84 to 1995-96 | 177.11 | -0.04 | 0.02 | 0.22 |
| | (8.22) | (-0.41) | | |
| All India | | | | |
| 1972-73 to 1995-96 | 34.84 | 0.70** | 0.55 | 26.89 |
| | (2.18) | (4.18) | | |
| 1972-73 to 1982-83 | 31.15 | 0.70* | 0.33 | 4.43 |
| | (1.09) | (2.09) | | |
| 1983-84 to 1995-96 | 124.14 | 0.06 | 0.006 | 0.07 |
| | (3.17) | (0.23) | | |

*Notes:* (i) Figures in the parentheses indicate 't' value.
(ii) * indicates 't' value significant at 1% level.
(iii) ** indicates 't' value significant at 5% level.

*Source:* (i) Banking Statistics, R.B.I.
(ii) Fertilizer Statistics, G.O.I.,
(iii) Area and Production of Principal Crops in India, G.O.I.,
(iv) Statistical Statements Relating to Co-operative Movements in India, NABARD/R.B.I.

With the emergence of HYVs in Indian agriculture the change in CP has got boost and this change has taken place from traditional crops, say paddy, and other low yielding inferior crops, such as, maize, bajra and even pulses to the HYV crops like HYV paddy, HYV wheat. etc. It is observed from Table 13 that the availability of higher amount of IC per hectare has significantly contributed to the increases in the area under HYV crops, thereby changing the CP in all the states under study and India.

From the above analysis it may be concluded that the IC has made a significant impact on change in CP which,in turn, influenced agricultural growth in India and the sample states. This may also be examined from the estimated simple regression equations. The specifications of these models are given in Table 14 for the whole period and two sub-periods. The above regression analysis shows that the IC has influenced significantly the agricultural growth in most of the sample states and India as a whole. The chow test reveals that there exists differential influences of IC, on the agricultural production in India and 9 sample states. For the states like Gujarat, H.P. and M.P. such type of differential impact has not been observed. (where F = 1.27, 0.71 and 1.58 respectively for H.P. , Gujarat and M.P.). So it proves the fact that IC has a significant role to accelerate agricultural growth through change in CP in India and sample states.

The values of r between index of agricultural production and IC per hectare be 0.71*, 0.79*, 0.49*, 0.77*, 0.86* and 0.74* respectively for Punjab, Haryana, Orissa, U.P., W. B. and all India. [*, ** indicate significant respectively at 1% and 5% level]. It is insignificant for other sample states. The non-availability of IC in right-time, the lower volume of available IC, etc. are the factors explaining the insignificant correlation between the availability of IC and agricultural production. As a whole, it can be said that the larger provision of IC can influence the agricultural growth in sample states of India.

**Relation Between Credit and Cropping Pattern in India by States: A Cropwise Study**

A change in the percentage of area under a particular crop is also a measure of change in the CP. Therefore, finding the relation between the change in the percentage of area under a crop and the IC will unravel the relation between the CP and IC. Here we have made an attempt to find out this relation cropwise over the period 1980-94 for the some states in India.

Considering the change in the area under paddy as a percentage of total area and its relation with the credit provided by PACS during 1980-81 to 1993-94 measured by the correlation coefficient it is found that the correlation coefficients are 0.38, 0.98*, 0.94*, 0.90* for A.P. Gujarat, Haryana and Punjab respectively. In the case of wheat cultivation the correlation coefficients are found to be 0.96*, 0.82*, 0.33, 0.99* and 0.96* respectively for Haryana, H.P., Punjab, Rajasthan and U.P. Again the correlation coefficients for cotton is 0.87*, 0.85*, 0.99*, 0.46, 0.13 respectively for A.P., Haryana , Rajasthan, U.P. and Tamil Nadu. For sugarcane the correlation coeffcients have been found to be 0.20, 0.53*, 0.99*, 0.27*, 0.78*, 0.66*, 0.62* respectively for Gujarat ,

Haryana , M.P., Orissa, Punjab, Rajasthan and U.P. (* and ** indicate that the value is significant at 1% and 5% level respectively). It again proves that the credit plays a catalytic role in changing CP across the sample states.

## Credit and Cropping Pattern by Size Classes of Land Holdings: India and W.B.

A study relating to credit and CP according to size classes of land holding also highlight the positive correlation between credit availability and change in CP for different size classes of landholding in both India and W.B. (see Tables 14a and b). For example, it is observed for India that the percentage of area under foodgrains has increased over years for the size classes of land holdings (LHs) say, up to 2.0 and 2.0-4.0 hectares on the one hand and the percentage share of those LHs in total PACS credit also increased over years on the other. This positive association implies that these larger availability of credit may have a catalytic role in changing CP for the farmers belonging to LHs upto 4 hectares in India. The same is true for W.B. where the CP has changed in favour of non-foodgrains.

**Table 14a**

**Credit and Cropping Pattern of Size Classes of Land Holding in India**

| *Size of* | *% of area under foodgrains* | | *PACS credit (%)* | | |
|---|---|---|---|---|---|
| *L.H. (in hectare)* | *1980-81* | *1985-86* | *1976-77* | *1983-84* | *1994-95* |
| Up to 2.0 | 85.15 | 86.90 | 16.68 | 13.52 | 19.58 |
| 2.0-4.0 | 81.20 | 82.00 | 22.10 | 31.23 | 23.43 |
| Above 4.0 | 74.90 | 75.35 | 22.27 | 20.86 | 18.70 |

*Source:* (i) Census of Agricultural Holding–All India, (ii) Statistical Statements Relating to Co-operative Movement in India, R.B.I / NABARD.

**Table 14b**

**Credit and Cropping Pattern of Size Classes of Land Holdings in W.B.**

| *Size of* | *% of area under foodgrains* | | *PACS credit (%)* | | |
|---|---|---|---|---|---|
| *L.H. (in hectare)* | *1985-86* | *1990-91* | *1980-81* | *1985-86* | *1987-88* |
| Up to 2.0 | 14.03 | 16.74 | 34.18 | 37.91 | 38.18 |
| 2.0-4.0 | 12.58 | 15.43 | 17.08 | 19.29 | 18.55 |
| Above 4.0 | 22.17 | 53.51 | 7.27 | 2.43 | 2.54 |

*Source:* (i) Agricultural Census, Government of West Bengal, (ii) Statistical Statements Relating to Co-operative Movement in India, R.B.I./ NABARD.

But this type of positive association between credit and CP is not seen for the cultivators having land holdings above 4.0 hectares in both India and W.B. Thus it may be concluded that the role of credit on change in CP is probably more significant for the cultivators belonging to smaller size classes of LHs.

## Impact of IC and Other Factors On Change in CP in W.B.: District Level Analysis

For examining the impact of area under irrigation as percentage of the GCA, I.C. and fertilizer consumption per hectare on the CP we have estimated the CAGR of the factors. It is seen from Tables 15, 15a, 15b that there has been significant growth of all these factors during the period from 1972-73 to 1995-96 in all the districts of W.B. It is observed that almost all the growth rates are significant at either 1% or 5% probability level. It may be presumed that the high growth rate of these factors might have contributed to the change in the CP in the districts of W.B. during the period, 1972-73 to 1995-96 (Tables 15, 15a, 15b). However, to identify the factors contributing to the change in CP we regress CP upon IC , the percentage irrigated area, the fertilizer consumption. The results are given in the following tables.

In this section we make a district level analysis of the impact of IC on the CP. At the state level analysis the impact of IC and as such, of any factor or factors on the CP is assessed in aggregative terms where the variation in its impact on CP among the districts are not captured. There may be some districts where IC may have significant effect on CP and in some other districts it may not have any impact. It is, therefore, necessary to identify the districts experiences significant effect of IC on CP and isolate them from those having no such experience at all. Once this is done, it will be easier for us to identify the factors contributing to this differential effects of IC on CP among the districts in the states.This exercise of identifying the factors at the district level will be more meaningful and useful than an exercise at the state level where aggregative data are very much likely to conceal the strength

**Table 15**
**Compound Annual Growth Rates (CAGRs) of Some Selected Factors Influencing Cropping Pattern in W.B. by Districts during 1972-73 to 1995-96**

| *State/ Districts* | *Commercial Bank Credit/ hectare* | *Fertilizer/ hectare* | *Irrigation (% GIA to GCA)* |
|---|---|---|---|
| W.B. | 13.88* | 9.85* | 3.05* |
| | (17.70) | (17.70) | (5.93) |
| Burdwan | 19.72* | 9.53* | 2.29* |
| | (18.85) | (13.72) | (3.67) |
| Birbhum | 23.37* | 9.31* | 2.02* |
| | (29.09) | (22.55) | (5.57) |
| Bankura | 25.86* | 7.36* | 1.92* |
| | (14.75) | (14.80) | (2.82) |
| Coochbehar | 22.14* | 19.72* | 6.18 |
| | (10.01) | (17.58) | (1.22) |
| Hoogly | 16.18* | 8.65* | 3.66* |
| | (12.15) | (9.04) | (2.82) |
| Howrah | 19.72* | 12.64* | 6.18* |
| | (18.01) | (16.60) | (2.80) |
| Jalpaiguri | 5.55* | 19.84* | 6.18** |
| | (3.13) | (10.78) | (2.06) |
| Malda | 25.86* | 13.77* | 6.39* |
| | (11.80) | (15.99) | (3.21) |
| Midnapore | 28.40* | 11.29* | 1.21*** |
| | (14.50) | (16.18) | (1.62) |
| Mursidabad | 22.14* | 10.41* | 2.33** |
| | (11.22) | (13.73) | (2.46) |
| Nadia | 16.18* | 10.74 | 3.04* |
| | (10.65) | (7.00) | (5.57) |
| 24-Parganas | 19.72* | 7.47* | 4.08** |
| | (18.27) | (2.67) | (1.95) |
| Purulia | 19.72* | 15.83* | 6.18* |
| | (20.91) | (23.09) | (3.34) |
| West Dinajpur | 29.96* | 17.35* | 8.33** |
| | (7.31) | (12.54) | (2.24) |
| Darjeeling | 6.72* | 18.41* | 1.10 |
| | (5.79) | (11.48) | (0.92) |

*Notes:* (i) Figures in the parentheses indicate 't' values.
(ii) * indicates 't' value significant at 1% level.
(iii) ** indicates 't' value significant at 5% level.

*Source:* (i) Banking Statistics, R.B.I.
(ii) Statistical Abstract, Government of West Bengal.
(iii) Economic Review, Government of West Bengal.

**Table 15a**
**Consumption of Fertilizer Per Unit of Gross Cropped Area in West Bengal by District**

*(Kg/hectare)*

| | *1972-1973* | *1977-1978* | *1982-1983* | *1987-1988* | *1993-1994* | *1995-1996* | *2001-2002* |
|---|---|---|---|---|---|---|---|
| Bankura | 19.51 | 28.09 | 26.42 | 63.95 | 74.53 | 76.46 | 115.23 |
| Birbhum | 25.89 | 33.24 | 44.73 | 77.13 | 119.65 | 120.58 | 149.56 |
| Burdwan | 23.03 | 30.67 | 59.96 | 94.77 | 118.03 | 138.46 | 157.54 |
| Coochbehar | 3.08 | 6.05 | 21.75 | 54.57 | 73.51 | 87.79 | 127.17 |
| Darjeeling | 9.58 | 11.19 | 44.63 | 75.70 | 139.78 | 170.32 | 184.95 |
| Hoogly | 28.94 | 48.74 | 99.43 | 164.33 | 176.36 | 181.46 | 270.45 |
| Howrah | 22.17 | 49.83 | 138.91 | 216.20 | 271.24 | 317.84 | 400.61 |
| Jalpaiguri | 4.86 | 3.88 | 16.82 | 41.18 | 59.96 | 82.66 | 103.92 |
| Malda | 10.32 | 18.36 | 48.14 | 83.99 | 98.52 | 95.01 | 106.74 |
| Midnapore | 9.51 | 15.13 | 24.38 | 57.94 | 70.29 | 89.19 | 118.14 |
| Nadia | 16.09 | 34.32 | 43.82 | 70.19 | 86.63 | 95.87 | 81.72 |
| 24-Parganas | 12.02 | 21.35 | 32.23 | 73.89 | 186.07 | 119.17 | 135.49 |
| Purulia | 5.93 | 10.07 | 22.86 | 56.26 | 84.65 | 99.43 | 152.62 |
| West Dinajpur | 2.84 | 12.64 | 24.66 | 50.61 | 56.31 | 61.99 | 88 33 |
| Murshidabad | 11.30 | 19.94 | 41.95 | 63.85 | 79.06 | 82.36 | 77.88 |
| West Bengal | 13.01 | 21.60 | 37.35 | 66.74 | 93.54 | 105.88 | 129.23 |

*Source:* Statistical Abstract, Govt. of West Bengal.

**Table 15b**
**Percentage of Gross Irrigated Area to Gross Cropped Area in West Bengal by District**

| | *1972-1973* | *1977-1978* | *1982-1983* | *1988-1989* | *1993-1994* | *1995-1996* | *2001-2002* |
|---|---|---|---|---|---|---|---|
| Bankura | 22.73 | 32.81 | 16.20 | 35.34 | 40.10 | 24.39 | 73.59 |
| Birbhum | 38.15 | 38.59 | 33.38 | 50.82 | 52.10 | 62.81 | 67.93 |
| Burdwan | 44.69 | 45.71 | 43.96 | 81.80 | 87.10 | 68.35 | 67.30 |
| Coochbehar | 9.08 | 15.50 | 25.11 | 32.80 | 33.15 | 55.57 | 67.00 |
| Darjeeling | 2.71 | 5.57 | 2.16 | 5.12 | 6.03 | 7.19 | 10.00 |
| Hoogly | 25.80 | 27.34 | 29.16 | 64.12 | 65.00 | 69.96 | 45.89 |
| Howrah | 22.88 | 22.57 | 24.38 | 35.24 | 38.90 | 27.56 | 11.88 |
| Jalpaiguri | 2.71 | 2.00 | 1.72 | 6.86 | 8.20 | 17.10 | 20.64 |
| Malda | 12.30 | 23.78 | 29.18 | 22.07 | 33.01 | 33.11 | 35.45 |
| Midnapore | 8.70 | 24.95 | 21.68 | 21.25 | 24.15 | 37.11 | 23.13 |
| Nadia | 13.73 | 18.07 | 28.22 | 35.23 | 40.03 | 65.13 | 69.45 |
| 24-Parganas | 12.85 | 11.89 | 17.80 | 29.28 | 30.98 | 20.40 | 26.12 |
| Purulia | 1.18 | 14.05 | 15.27 | 25.00 | 26.25 | 24.05 | 14.58 |
| West Dinajpur | 0.06 | 17.10 | 19.19 | 25.75 | 19.70 | 14.63 | 15.50 |
| Murshidabad | 7.37 | 26.59 | 26.20 | 28.86 | 27.09 | 15.49 | 12.34 |

*Source:* 1. Statistical Abstract, Govt. of West Bengal.
2. Ministry of Rural Development, Govt. of West Bengal.

**Table 16**
**Estimates of Simple Regression Equations Concerning Cropping Pattern and IC in W.B. by Districts, Dependent Variable = Cropping Pattern Index; Number of Observations = 24 (1972-73 to 1995-96)**

| *Districts/ Period* | *Constants* | *Explanatory variable IC/hectare* | $R^2$ | *F* |
|---|---|---|---|---|
| Burdwan | | | | |
| 1972-73 to 1995-96 | 107.59 | 0.12* | 0.55 | 26.89 |
| | (45.02) | (4.96) | | |
| 1972-73 to 1982-83 | 112.87 | 0.03 | 0.01 | 0.13 |
| | (21.38) | (0.36) | | |
| 1983-84 to 1995-96 | 105.07 | 0.15** | 0.55 | 13.44 |
| | (15.10) | (3.32) | | |
| Nadia | | | | |
| 1972-73 to 1995-96 | 104.54 | 0.24* | 0.39 | 14.07 |
| | (16.80) | (3.58) | | |
| 1972-73 to 1982-83 | 108.40 | 0.11 | 0.11 | 1.15 |
| | (14.85) | (1.07) | | |
| 1983-84 to 1995-96 | 117.78 | 0.16*** | 0.22 | 3.10 |
| | (10.68) | (1.59) | | |
| Murshidabad | | | | |
| 1972-73 to 1995-96 | 109.50 | 0.35** | 0.63 | 37.46 |
| | (26.29) | (5.83) | | |
| 1972-73 to 1982-83 | 115.83 | 0.08*** | 0.19 | 2.13 |
| | (42.32) | (1.46) | | |
| 1983-84 to 1995-96 | 107.05 | 0.43** | 0.63 | 8.73 |
| | (11.57) | (3.94) | | |
| Midnapore | | | | |
| 1972-73 to 1995-96 | 113.11 | 0.04* | 0.15 | 3.88 |
| | (41.45) | (1.91) | | |
| 1972-73 to 1982-83 | 103.98 | 0.27*** | 0.19 | 2.13 |
| | (23.26) | (1.46) | | |
| 1983-84 to 1995-96 | 129.41 | -0.04*** | 0.21 | 2.92 |
| | (28.57) | (-1.49) | | |
| Jalpaguri | | | | |
| 1972-73 to 1995-96 | 97.12 | 0.007 | 0.008 | 0.18 |
| | (27.16) | (0.41) | | |
| 1972-73 to 1982-83 | 103.98 | -0.012*** | 0.16 | 1.75 |
| | (44.28) | (-1.32) | | |
| 1983-84 to 1995-96 | 92.31 | 0.06*** | 0.23 | 3.29 |
| | (16.91) | (1.64) | | |

*Contd...*

| *Districts/ Period* | *Constants* | *Explanatory variable IC/hectare* | $R^2$ | *F* |
|---|---|---|---|---|
| Howrah | | | | |
| 1972-73 to 1995-96 | 87.75 | 0.09** | 0.49 | 21.14 |
| | (33.37) | (4.41) | | |
| 1972-73 to 1982-83 | 92.06 | 0.06 | 0.07 | 0.67 |
| | (18.43) | (0.82) | | |
| 1983-84 to 1995-96 | 86.04 | 0.104* | 0.49 | 10.57 |
| | (13.31) | (2.96) | | |
| Birbhum | | | | |
| 1972-73 to 1995-96 | 101.41 | 0.04 | 0.04 | 0.92 |
| | (25.52) | (0.91) | | |
| 1972-73 to 1982-83 | 100.47 | 0.11 | 0.06 | 0.54 |
| | (16.01) | (0.74) | | |
| 1983-84 to 1995-96 | 86.53 | 0.17* | 0.25 | 3.67 |
| | (8.02) | (1.72) | | |
| Hooghly | | | | |
| 1972-73 to 1995-96 | 122.39 | 0.04** | 0.19 | 5.16 |
| | (34.07) | (2.15) | | |
| 1972-73 to 1982-83 | 122.17 | 0.05*** | 0.23 | 2.76 |
| | (27.86) | (1.66) | | |
| 1983-84 to 1995-96 | 114.68 | 0.07 | 0.13 | 1.64 |
| | (8.12) | (1.14) | | |
| Bankura | | | | |
| 1972-73 to 1995-96 | 97.26 | 0.26** | 0.66 | 42.71 |
| | (39.88) | (6.27) | | |
| 1972-73 to 1982-83 | 93.79 | 0.402** | 0.55 | 10.95 |
| | (27.41) | (3.31) | | |
| 1983-84 to 1995-96 | 114.49 | 0.04 | 0.22 | 3.10 |
| | (17.17) | (0.45) | | |
| Malda | | | | |
| 1972-73 to 1995-96 | 114.31 | 0.17** | 0.55 | 26.89 |
| | (36.48) | (4.93) | | |
| 1972-73 to 1982-83 | 114.35 | 0.19** | 0.43 | 6.92 |
| | (33.16) | (2.63) | | |
| 1983-84 to 1995-96 | 107.27 | 0.22* | 0.39 | 7.03 |
| | (9.63) | (2.41) | | |
| Darjeeling | | | | |
| 1972-73 to 1995-96 | 106.12 | 0.009 | 0.005 | 0.11 |
| | (8.25) | (0.31) | | |
| 1972-73 to 1982-83 | 108.91 | -0.07*** | 0.18 | 1.96 |
| | (29.41) | (-1.41) | | |

*Contd...*

| *Districts/ Period* | *Constants* | *Explanatory variable IC/hectare* | $R^2$ | *F* |
|---|---|---|---|---|
| 1983-84 to 1995-96 | 84.33 | 0.09*** | 0.202 | 2.78 |
| | (3.83) | (1.07) | | |
| Coochbehar | | | | |
| 1972-73 to 1995-96 | 55.87 | 0.02 | 0.014 | 0.31 |
| | (29.98) | (0.54) | | |
| 1972-73 to 1982-83 | 56.31 | 0.01 | 0.004 | 0.033 |
| | (25.44) | (0.18) | | |
| 1983-84 to 1995-96 | 49.15 | 0.14 | 0.101 | 1.24 |
| | (6.30) | (1.004) | | |
| W. Dinajur | | | | |
| 1972-73 to 1995-96 | 96.303 | -0.01*** | 0.11 | 2.72 |
| | (98.51) | (-1.59) | | |
| 1972-73 to 1982-83 | 95.96 | -0.025 | 0.006 | 0.054 |
| | (36.11) | (-0.23) | | |
| 1983-84 to 1995-96 | 97.25 | -0.01* | 0.34 | 5.67 |
| | (77.99) | (-2.15) | | |
| 24 Parganas | | | | |
| 1972-73 to 1995-96 | 108.35 | 0.01 | 0.045 | 1.04 |
| | (44.84) | (0.97) | | |
| 1972-73 to 1982-83 | 95.83 | 0.13 | 0.56 | 11.65 |
| | (28.96) | (3.14)** | | |
| 1983-84 to 1995-96 | 114.14 | -0.002 | 0.003 | 0.03 |
| | (26.96) | (-0.17) | | |
| Purulia | | | | |
| 1972-73 to 1995-96 | 106.69 | -0.02 | 0.002 | 0.04 |
| | (25.51) | (-0.19) | | |
| 1972-73 to 1982-83 | 113.11 | -0.14 | 0.09 | 0.87 |
| | (35.11) | (-0.93) | | |
| 1983-84 to 1995-96 | 79.83 | 0.405 | 0.41 | 7.64 |
| | (8.67) | (2.51)** | | |

*Notes:* (i) Figures in the parentheses indicate 't' values.
(ii) *indicates 't' value significant at 1% level.
(iii) **indicates 't' value significant at 5% level.

*Source:* (i) Banking Statistics, R.B.I.
(ii) Statistical Abstract, Government of West Bengal.
(iii) Economic Review, Government of West Bengal.

and weaknesses of a district's agriculture. Keeping this in mind we have carried out the same exercise of estimating the effect of IC on CP at the district level. For this purpose we have regressed CP on IC for each district. The results are presented in Table 16. From the results of the estimated equations it is observed that the IC has influenced significantly the change in CP either in the sub-periods or in the whole period for all the districts except Coochbehar. The Chow test results reveal that there exist differential influences of IC on change in CP between two sub-periods in most of the districts of W.B. The Chow F values are found to be 3.13*, 10.24**, 33.19**, 22.22**, 26.08**, 0.59, 1.84, 0.60, 6.40**, 0.86, 5.84**, 0.47, 0.79, 20.46** and 9.02** respectively for Burdwan, Nadia, Murshidabad, Midnapore, Jalpaiguri, Howrah, Birbhum, Hooghly, Bankura, Malda, Darjeeling, Coochbehar, West Dinajpur, Purulia and 24-Parganas. (*, ** indicate significant at 5% and 1% level respectively). These results indicate that in as many as nine districts the impact of I.C. on C.P. has been different between two sub-periods and in six districts no such differences between two sub-periods are noticed. Of the nine districts in three districts ,viz., Midnapore, Bankura, 24-Parganas the effect of IC on CP in the sub-period I was higher than that in the sub-period II and in six districts ,viz., Burdwan, Nadia, Murshidabad, Jalpaiguri, Darjeeling, and Purulia the effect of IC on CP in the second sub-period was higher than that in the first sub-period.

In this paragraph we estimate the effects of the factors like fertiliser consumption per hectare (FC) and the percentage of GIA in the GCA on the CP in the districts of W.B. by regressing CP on FC and GIA as a percentage of GCA for the periods under study. The regression results are presented in Table 17. The results show that in Burdwan the GIA as a percentage of GCA has made significant contribution to the change in CP during the period from 1972-73 to 1995-96 and also during the two sub-periods from 1972-73 to 1982-83 and from 1983-84 to 1995-96. In Nadia, both fertiliser consumption per hectare and irrigation have made significant impact on CP change only in the second sub-period. In the districts of Purulia, West Dinajpur and Birbhum none of these factors have any significant influence on the change in CP. In Murshidabad, the factors have made significant influence on CP change only in the whole period, but no such significant influence is noticed during any of the two sub-periods. In Midnapore, only the fertiliser consumption has significantly influenced the CP change both in the whole period and the first sub-period. In Jalpaiguri, the irrigation has significant influence on CP change in all the periods. The irrigation has significant influence in CP change only in the second sub-period in Howrah and only in the first sub-period in Hooghly. In the case of Bankura, while the percentage of GIA in GCA and FC have made significant impact on the CP change in the whole period, only GIA/GCA is seen to have significant effect in the first sub-period. In Malda, both the factors have significantly influenced the change in CP in the sub-periods. In Coochbehar and 24-Parganas, the percentage of GIA in GCA has made significant contribution to CP change in the first sub-period and the whole period respectively. In Darjeeling, only the fertiliser consumption has significantly influenced the CP change in the whole period. The Chow test results show that there exist differential influences of fertiliser consumption per hectare and the percentage of GIA to GCA on the change in CP in most of the districts of W.B., namely, Burdwan, Nadia, Murshidabad, Midnapore, Jalpaiguri, Howrah, Birbhum, Hoogly, Bankura, Malda, Darjeeling, Coochbehar, West Dinajpur, 24-Parganas and Purulia.

In the above analysis we have obtained certain apparently unacceptable results. For example, when none of the two factors are seen to have any significant effect on the CP during the two sub-periods, both of them are found to have significant effect upon the CP during the entire period that consists of the two sub-periods. This paradox may be resolved by keeping in mind that during a short period of say ten to twelve years (as our sub-periods are) the effect of the variables may not be significant (t values being small) due to (i) high degree of multicollinearity between the variables and/or (ii) small variance of one or both of the explanatory variables. When the period is long, the degree of multicollinearity will be low and the variances of the explanatory variables will be, in all probability, reasonably high. As a result, the effects of these variables may be expected to be significant.

**Table 17**
**Estimates of Multiple Regression Equations Concerning Cropping Pattern, Fertiliser Consumption and Irrigated Area in W.B. by Districts, Dependent Variable = Cropping Pattern Index, Number of Observations = 24 (1972-73 to 1995-96)**

| | | *Explanatory variables* | | | |
|---|---|---|---|---|---|
| *Districts/ Periods* | *Constant* | *FC/ Hectare* | *% of GIA to GCA* | $R^2$ | *F* |
| Burdwan | | | | | |
| 1972-73 to 1995-96 | 97.32 | 0.06 | 0.30* | 0.24 | 3.32 |
| | (19.93) | (1.21) | (4.22) | | |
| 1972-73 to 1982-83 | 97.02 | 0.16 | 0.21*** | 0.24 | 1.27 |
| | (8.64) | (1.20) | (1.44) | | |
| 1983-84 to 1995-96 | 80.35 | -0.01 | 0.59* | 0.55 | 6.11 |
| | (5.22) | (-0.25) | (3.12) | | |
| Nadia | | | | | |
| 1972-73 to 1995-96 | 86.93 | 0.004 | 1.76 | 0.68 | 22.31 |
| | (21.54) | (0.10) | (0.12) | | |
| 1972-73 to 1982-83 | 44.97 | -1.09** | 6.48** | 0.83 | 19.19 |
| | (3.80) | (-3.17) | (4.99) | | |
| 1983-84 to 1995-96 | 122.33 | -0.06 | 0.63 | 0.18 | 1.10 |
| | (6.19) | (-0.67) | (0.99) | | |
| Murshidabad | | | | | |
| 1972-73 to 1995-96 | 105.55 | 0.36 | 0.51*** | 0.57 | 13.92 |
| | (17.92) | (3.38) | (1.89) | | |
| 1972-73 to 1982-83 | 113.96 | 0.08 | 0.19 | 0.11 | 0.48 |
| | (18.21) | (0.42) | (0.99) | | |
| 1983-84 to 1995-96 | 129.74 | 0.12 | 0.24 | 0.10 | 0.56 |
| | (8.40) | (0.77) | (0.48) | | |
| Midnapore | | | | | |
| 1972-73 to 1995-96 | 109.82 | 0.25* | -0.30 | 0.33 | 5.17 |
| | (19.78) | (3.06) | (-0.12) | | |
| 1972-73 to 1982-83 | 105.74 | 0.34 | -0.12 | 0.20 | 0.99 |
| | (28.78) | (1.41) | (-0.52) | | |
| 1983-84 to 1995-96 | 119.49 | 0.12 | -0.09 | 0.18 | 1.10 |
| | (12.44) | (1.31) | (-0.008) | | |
| Jalpaguri | | | | | |
| 1972-73 to 1995-96 | 91.66 | 0.008 | 1.61* | 0.44 | 8.25 |
| | (42.97) | (0.07) | (1.97) | | |
| 1972-73 to 1982-83 | 96.16 | -0.05 | 0.39 | 0.08 | 0.34 |
| | (60.34) | (-0.32) | (0.77) | | |
| 1983-84 to 1995-96 | 82.99 | 0.10 | 2.21*** | 0.60 | 7.50 |
| | (15.40) | (0.49) | (1.58) | | |

*Contd...*

| | Explanatory variables | | | | |
|---|---|---|---|---|---|
| *Districts/ Periods* | *Constant* | *FC/ Hectare* | *% of GIA to GCA* | $R^2$ | *F* |
| Howrah | | | | | |
| 1972-73 to 1995-96 | 91.29 | 0.06* | -0.06 | 0.42 | 7.60 |
| | (32.37) | (2.58) | (-0.21) | | |
| 1972-73 to 1982-83 | 72.79 | -0.03 | 8.18 | 0.28 | 0.05 |
| | (4.73) | (-0.25) | (1.27) | | |
| 1983-84 to 1995-96 | 87.44 | 0.07*** | 0.002 | 0.28 | 1.94 |
| | (7.21) | (1.75) | (0.004) | | |
| Birbhum | | | | | |
| 1972-73 to 1995-96 | 102.03 | 0.033 | 0.006 | 0.18 | 2.30 |
| | (7.43) | (0.42) | (0.01) | | |
| 1972-73 to 1982-83 | 80.19 | 0.31 | 0.33 | 0.09 | 0.39 |
| | (2.40) | (0.79) | (0.45) | | |
| 1983-84 to 1995-96 | 96.69 | 0.07 | 0.03 | 0.07 | 0.38 |
| | (4.39) | (0.70) | (0.06) | | |
| Hooghly | | | | | |
| 1972-73 to 1995-96 | 124.43 | 0.015 | 0.08 | 0.11 | 1.78 |
| | (34.20) | (0.38) | (0.64) | | |
| 1972-73 to 1982-83 | 107.96 | 0.08 | 0.54*** | 0.36 | 2.26 |
| | (10.68) | (1.1) | (1.67) | | |
| 1983-84 to 1995-96 | 117.75 | 0.04 | 0.11 | 0.14 | 0.81 |
| | (10.14) | (0.58) | (0.74) | | |
| Bankura | | | | | |
| 1972-73 to 1995-96 | 84.79 | 1.14** | 0.55** | 0.49 | 10.09 |
| | (10.68) | (1.71) | (2.53) | | |
| 1972-73 to 1982-83 | 67.46 | 0.11 | 0.53** | 0.59 | 5.66 |
| | (6.15) | (0.23) | (2.33) | | |
| 1983-84 to 1995-96 | 104.55 | 0.16*** | 0.07 | 0.31 | 2.25 |
| | (11.57) | (1.87) | (0.48) | | |
| Malda | | | | | |
| 1972-73 to 1995-96 | 118.67 | 0.19* | -0.25 | 0.39 | 6.71 |
| | (25.33) | (3.30) | (-0.72) | | |
| 1972-73 to 1982-83 | 109.93 | 1.02** | -0.43 | 0.71 | 9.69 |
| | (33.19) | (4.13) | | | |
| 1983-84 to 1995-96 | 193.84 | -0.55** | -2.65* | 0.56 | 6.70 |
| | (6.92) | (-1.93) | (-3.01) | | |
| Darjeeling | | | | | |
| 1972-73 to 1995-96 | 99.58 | 0.19** | 0.60 | 0.23 | 3.14 |
| | (17.10) | (2.36) | (0.08) | | |

*Contd...*

| | | *Explanatory variables* | | | |
|---|---|---|---|---|---|
| *Districts/ Periods* | *Constant* | *FC/ Hectare* | *% of GIA to GCA* | $R^2$ | *F* |
| 1972-73 to 1982-83 | 103.01 | 0.05 | -0.0001 | 0.05 | 0.19 |
| | (56.20) | (0.61) | (-0.01) | | |
| 1983-84 to 1995-96 | 98.94 | 0.27 | -0.30 | 0.16 | 0.95 |
| | (3.34) | (1.24) | (-0.28) | | |
| Coochbehar | | | | | |
| 1972-73 to 1995-96 | 55.49 | 0.06*** | -0.009 | 0.13 | 1.57 |
| | (49.92) | (1.69) | (-0.26) | | |
| 1972-73 to 1982-83 | 60.57 | -0.02 | -1.21 | 0.57 | 5.39 |
| | (29.49) | (-0.25) | (-3.01) | | |
| 1983-84 to 1995-96 | 54.86 | 0.07 | -0.0009 | 0.201 | 1.26 |
| | (28.50) | (1.34) | (0.02) | | |
| West Dinajpur | | | | | |
| 1972-73 to 1995-96 | 96.39 | 0.03 | -0.08 | 0.06 | 0.67 |
| | (56.54) | (0.53) | (-1.10) | | |
| 1972-73 to 1982-83 | 96.31 | -0.06 | -0.02 | 0.01 | 0.05 |
| | (26.42) | (-0.28) | (-0.08) | | |
| 1983-84 to 1995-96 | 102.89 | -0.05 | -0.12*** | 0.29 | 2.04 |
| | (17.87) | (-0.61) | (-1.82) | | |
| 24-Parganas | | | | | |
| 1972-73 to 1995-96 | 98.81 | 0.25 | 0.59* | 0.22 | 2.96 |
| | (12.41) | (0.35) | (2.31) | | |
| 1972-73 to 1982-83 | 90.81 | 0.76 | -0.37 | 0.91 | 2.64 |
| | (9.30) | (1.24) | (-0.49) | | |
| 1983-84 to 1995-96 | 49.08 | 0.13 | 2.87 | 0.47 | 4.43 |
| | (0.90) | (2.22)* | (1.00) | | |
| Purulia | | | | | |
| 1972-73 to 1995-96 | 104.42 | -0.14*** | 0.41 | 0.07 | 0.79 |
| | (25.68) | (-1.99) | (1.20) | | |
| 1972-73 to 1982-83 | 111.72 | -0.19 | 0.11 | 0.13 | 0.58 |
| | (36.23) | (-1.03) | (0.23) | | |
| 1983-84 to 1995-96 | 77.75 | 0.22 | 0.45 | 0.51 | 5.20 |
| | (8.85) | (1.14) | (0.37) | | |

*Notes:* (i) Figures in the parentheses indicate 't' values. ii) * indicates 't' value significant at 1% level.
(iii) **indicates 't' value significant at 5% level.

*Source:* (i) Banking Statistics, R.B.I.
(ii) Statistical Abstract, Government of West Bengal. (iii) Economic Review, Government of West Bengal

However, when this problem is resolved by invoking the econometric reasons, the problem of significant effect of the variables in one or both the sub-periods, but insignificant effect of either or both of the variables in the whole period, i.e. , the problem of apparent inconsistency between the results of the whole period and the sub-periods can also be addressed by referring to (i) the presence of multicollinearity between the explanatory variables and (ii) the variance of the explanatory variables.

To overcome this problem we use only one explanatory variable at a time in the regression analysis which will take care of multicollinearity to a large extent. However, the small variance of the explanatory variables may lead to the acceptance of the null hypothesis that there is no significant effect of the explanatory variable on the dependent variable.

We first consider the fertiliser consumption per hectare as the only explanatory variable and regress the CP on the fertiliser consumption in each of the districts in W.B. The regression results are given in Table 18. The fertiliser consumption per hectare has significantly influenced the CP change in thirteen out of fifteen districts of W.B. (except Birbhum and West Dinajpur ) in either whole period or in one of the sub-periods. The Chow test results show differential influences of fertiliser consumption on the CP change in seven districts between the two sub-periods [ where F = 1.05, 19.13**, 17.21**, 13.12**, 6.67**, 0.20, 1.06, 1.06, 5.97**, 1.28, 0.41, 0.48, 0.17, 2.74*, 24.97** respectively for Burdwan, Nadia, Murshidabad, Midnapore, Jalpaiguri, Howrah, Birbhum, Hoogly, Bankura, Malda, Darjeeling, Coochbehar, West Dinajpur, 24-Parganas and Purulia.] (Where *, ** indicate significant at 5% and 1% respectively).

Further, it is observed that the IC per hectare has a significant impact on the fertiliser consumption per hectare in the districts of WB. The regression results on the impact of IC on FC are presented in Table 19. Therefore, it may be concluded that the IC has made significant influence on the change in CP through the fertiliser consumption in these districts of W.B.

**Table 18**
**Estimates of Simple Regression Equations concerning Cropping Pattern and Fertiliser Consumption in W.B. by Districts, Dependent Variable = Cropping Pattern Index; Number of Observations =24 (1972-73 to 1995-96)**

| *Districts/ Periods* | *Constants* | *Explanatory variable FC/hectare* | $R^2$ | *F* |
|---|---|---|---|---|
| **Burdwan** | | | | |
| 1972-73 to 1995-96 | 113.62 | 0.13* | 0.18 | 4.83 |
| | (28.02) | (2.09) | | |
| 1972-73 to 1982-83 | 111.02 | 0.08 | 0.04 | 0.41 |
| | (18.60) | (0.64) | | |
| 1983-84 to 1995-96 | 126.94 | 0.01 | 0.003 | 0.03 |
| | (24.05) | (0.15) | | |
| **Nadia** | | | | |
| 1972-73 to 1995-96 | 111.33 | 0.27** | 0.29 | 8.98 |
| | (20.29) | (2.84) | | |
| 1972-73 to 1982-83 | 99.49 | 0.49* | 0.29 | 3.68 |
| | (11.31) | (1.92) | | |
| 1983-84 to 1995-96 | 140.60 | -0.08 | 0.08 | 0.96 |
| | (20.37) | (-0.86) | | |
| **Murshidabad** | | | | |
| 1972-73 to 1995-96 | 112.70 | 0.44** | 0.48 | 20.31 |
| | (23.43) | (4.32) | | |
| 1972-73 to 1982-83 | 117.00 | 0.08 | 0.02 | 0.18 |
| | (22.64) | (0.43) | | |
| 1983-84 to 1995-96 | 135.44 | 0.13 | 0.07 | 0.83 |
| | (14.54) | (0.85) | | |
| **Midnapore** | | | | |
| 1972-73 to 1995-96 | 109.23 | 2.25** | 0.33 | 10.84 |
| | (37.09) | (3.15) | | |
| 1972-73 to 1982-83 | 104.19 | 0.29*** | 0.17 | 1.87 |
| | (22.74) | (1.37) | | |
| 1983-84 to 1995-96 | 118.76 | 0.12*** | 0.18 | 2.41 |
| | (29.60) | (1.4) | | |
| **Jalpaiguri** | | | | |
| 1972-73 to 1995-96 | 94.15 | 0.118** | 0.33 | 10.84 |
| | (50.97) | (3.12) | | |
| 1972-73 to 1982-83 | 97.02 | -0.04 | 0.009 | 0.08 |
| | (86.65) | (-0.29) | | |
| 1983-84 to 1995-96 | 85.11 | 0.37** | 0.48 | 10.15 |
| | (15.11) | (2.89) | | |

*Contd...*

| *Districts/ Periods* | *Constants* | *Explanatory variable FC/hectare* | $R^2$ | *F* |
|---|---|---|---|---|
| Howrah | | | | |
| 1972-73 to 1995-96 | 91.27 | 0.06** | 0.41 | 15.29 |
| | (33.19) | (3.74) | | |
| 1972-73 to 1982-83 | 91.66 | 0.06 | 0.14 | 1.42 |
| | (23.09) | (1.19) | | |
| 1983-84 to 1995-96 | 87.46 | 0.07** | 0.28 | 4.28 |
| | (9.36) | (1.86) | | |
| Birbhum | | | | |
| 1972-73 to 1995-96 | 102.26 | 0.03 | 0.02 | 0.45 |
| | (24.38) | (0.61) | | |
| 1972-73 to 1982-83 | 93.81 | 0.30 | 0.07 | 0.64 |
| | (6.91) | (0.80) | | |
| 1983-84 to 1995-96 | 97.94 | 0.07 | 0.07 | 0.83 |
| | (11.33) | (0.81) | | |
| Hooghly | | | | |
| 1972-73 to 1995-96 | 125.62 | 0.04*** | 0.09 | 2.17 |
| | (40.49) | (1.43) | | |
| 1972-73 to 1982-83 | 122.66 | 0.09 | 0.14 | 1.45 |
| | (22.66) | (1.20) | | |
| 1983-84 to 1995-96 | 121.51 | 0.06 | 0.08 | 0.96 |
| | (11.94) | (0.91) | | |
| Bankura | | | | |
| 1972-73 to 1995-96 | 93.70 | 0.38** | 0.58 | 30.38 |
| | (27.16) | (5.27) | | |
| 1972-73 to 1982-83 | 80.73 | 0.84* | 0.31 | 3.96 |
| | (7.05) | (1.99) | | |
| 1983-84 to 1995-96 | 108.09 | 0.15* | 0.29 | 4.49 |
| | (21.38) | (1.91) | | |
| Malda | | | | |
| 1972-73 to 1995-96 | 116.59 | 0.21** | 0.38 | 13.48 |
| | (31.92) | (3.51) | | |
| 1972-73 to 1982-83 | 111.89 | 0.44** | 0.43 | 6.83 |
| | (26.30) | (2.61) | | |
| 1983-84 to 1995-96 | 119.52 | 0.17 | 0.07 | 0.83 |
| | (6.67) | (0.77) | | |
| Darjeeling | | | | |
| 1972-73 to 1995-96 | 99.72 | 0.19* | 0.23 | 6.57 |
| | (18.39) | (2.46) | | |
| 1972-73 to 1982-83 | 102.99 | 0.05 | 0.05 | 0.43 |
| | (61.93) | (0.66) | | |

*Contd...*

| Districts/ Periods | Constants | Explanatory variable FC/hectare | $R^2$ | F |
|---|---|---|---|---|
| 1983-84 to 1995-96 | 93.46 | 0.26 | 0.15 | 1.94 |
| | (4.93) | (1.28) | | |
| Coochbehar | | | | |
| 1972-73 to 1995-96 | 55.48 | 0.06** | 0.13 | 3.29 |
| | (51.23) | (1.76) | | |
| 1972-73 to 1982-83 | 55.53 | 0.11 | 0.11 | 0.95 |
| | (34.21) | (0.98) | | |
| 1983-84 to 1995-96 | 54.86 | 0.07*** | 0.201 | 2.77 |
| | (30.47) | (1.51) | | |
| W.Dinajpur | | | | |
| 1972-73 to 1995-96 | 95.56 | -0.002 | 0.001 | 0.02 |
| | (62.10) | (-0.05) | | |
| 1972-73 to 1982-83 | 96.11 | -0.06 | 0.01 | 0.098 |
| | (37.24) | (-0.31) | | |
| 1983-84 to 1995-96 | 95.78 | -0.004 | 0.002 | 0.02 |
| | (20.18) | (-0.05) | | |
| 24-Parganas | | | | |
| 1972-73 to 1995-96 | 103.96 | 0.08 | 0.45 | 18.00 |
| | (51.61) | (4.03)** | | |
| 1972-73 to 1982-83 | 94.83 | 0.48 | 0.38 | 5.51 |
| | (18.62) | (2.35)* | | |
| 1983-84 to 1995-96 | 3.51 | 0.08 | 0.401 | 7.36 |
| | (22.13) | (2.46)* | | |
| Purulia | | | | |
| 1972-73 to 1995-96 | 107.07 | -0.03 | 0.008 | 0.18 |
| | (30.93) | (-0.39) | | |
| 1972-73 to 1982-83 | 112.52 | -0.17 | 0.102 | 1.03 |
| | (45.46) | (-1.01) | | |
| 1983-84 to 1995-96 | 79.24 | 0.36 | 0.46 | 9.37 |
| | (9.20) | (2.76)** | | |

*Notes:* (i) Figures in the parentheses indicate 't' values.
(ii) *indicates 't' value significant at 1% level.
(iii) **indicates 't' value significant at 5% level.

*Source:* (i) Banking Statistics, R.B.I.
(ii) Statistical Abstract, Government of West Bengal. (3) Economic Review, Government of West Bengal.

Let us now examine the impact of the percentage of GIA to GCA on CP change in the districts of W. B. The results of estimated regression equations are given in Table 20. The irrigation has made significant impact on change in CP in several districts, namely, Burdwan, Nadia, Murshidabad, Jalpaiguri, Howrah, Hoogly, Bankura, and 24-Parganas. But the impact is not significant for other districts, viz., Midnapore, Birbhum, Malda, Darjeeling, Coochbehar, West Dinajpur and Purulia during the whole period. But only in case of West Dinajpur, the irrigation has made significant influence on the CP change in the second sub-period. It is further observed that the IC per hectare has made significant impact on the irrigation in the districts like Burdwan, Nadia, Murshidabad, Howrah, Hoogly, Bankura, Coochbehar, 24-Parganas, Purulia and Birbhum (Table 21) . Therefore, it may be concluded that the IC has made significant contribution to CP change through irrigation in these districts.

Besides estimating regression coefficients of various explanatory variables in order to measure the impact of these variables upon the CP in the districts of the states, we have estimated correlation coefficients between CPI on the one hand, and IC per hectare, GIA/GCA , and fertiliser consumption (FC) per hectare on the other for the sample districts in the states to measure the type and degree of association between these two variables. It is observed that the values of correlation coefficient (r) between CPI and IC/hectare (Table 21a) are 0.79*, 0.82*, 0.74*, 0.70*, 0.43*, 0.81* and 0.74* respectively for Murshidabad, Nadia, Burdwan, Howrah, Hoogly, Bankura and Malda, but that is not significant for other sample districts of W.B. The values of r between, CPI and percentage of irrigated area in GCA have been found to be significant at 1% probability level (0.55*, 0.82*, 0.73*, 0.66*, 0.65*, 0.46*) for Murshidabad, Nadia, Burdwan, Jalpaiguri, Bankura, and 24-Parganas respectively) and at 5% probability level (0.46**) for Howrah. But that is not significant for other sample districts. On the other hand, the values of r between CPI and the fertiliser consumption per hectare are significant at 1 percentage probability level

Table 19
**Estimates of Simple Regression Equation concerning Fertiliser Consumption and IC in W.B. by Districts, Dependent Variable = Fertilizer Consumption (F.C.)/hectare, Number of Observations = 24 (1972-73 to 1995-96)**

| *Districts* | *Constants* | *Explanatory variable IC/hectare* | $R^2$ | F |
|---|---|---|---|---|
| Burdwan | 28.19 | 0.29** | 0.26 | 7.73 |
| | (2.13) | (2.63) | | |
| Nadia | 10.79 | 0.47** | 0.38 | 13.48 |
| | (0.87) | (3.49) | | |
| Murshidabad | 15.30 | 0.43** | 0.38 | 13.48 |
| | (1.80) | (3.47) | | |
| Midnapore | 26.82 | 0.03 | 0.02 | 0.45 |
| | (4.009) | (0.59) | | |
| Jalpaiguri | 33.03 | -0.05 | 0.04 | 0.92 |
| | (3.01) | (-0.96) | | |
| Howrah | -1.01 | 1.23** | 0.88 | 161.33 |
| | (-0.07) | (12.12) | | |
| Birbhum | 8.29 | 0.78** | 0.85 | 124.67 |
| | (1.36) | (10.80) | | |
| Hooghly | -0.11 | 0.61** | 0.62 | 35.89 |
| | (-5.10) | (5.73) | | |
| Bankura | 14.66 | 0.58** | 0.82 | 100.22 |
| | (4.05) | (9.45) | | |
| Malda | 7.90 | 0.57* | 0.71 | 53.86 |
| | (1.05) | (6.96) | | |
| Darjeeling | 75.60 | -0.06 | 0.03 | 0.68 |
| | (2.41) | (-0.80) | | |
| Coochbehar | -1.07 | 0.55** | 0.24 | 6.95 |
| | (-1.11) | (2.54) | | |
| W.Dinajpur | 22.45 | 0.1006* | 0.31 | 9.88 |
| | (4.98) | (3.03) | | |
| 24-Parganas | 53.41 | 0.21*** | 0.10 | 2.44 |
| | (2.80) | (1.48) | | |
| Purulia | -8.80 | 1.26* | 0.88 | 161.33 |
| | (-1.94) | (12.02) | | |

*Notes:* (i) Figures in the parentheses indicate 't' values.
(ii) *indicates 't' value significant at 1% level.
(iii) **indicates 't' value significant at 5% level.

*Source:* (i) Banking Statistics, R.B.I.
(ii) Statistical Abstract, Government of West Bengal.
(iii) Economic Review, Government of West Bengal.

**Table 20**
**Estimates of Simple Regression Equations concerning Cropping Pattern and Irrigated Area in W.B. by Districts, Dependent Variable = Cropping Pattern Index, Number of Observations = 24 (1972-73 to 1995-96)**

| *Districts* | *Constants* | *Explanatory variable % of GIA to GCA* | $R^2$ | *F* |
|---|---|---|---|---|
| Burdwan | 98.69 (20.54) | 0.34** (4.89) | 0.54 | 25.83 |
| Nadia | 86.89 (14.37) | 1.77** (6.59) | 0.68 | 46.75 |
| Murshidabad | 112.43 (16.47) | 0.90** (2.95) | 0.30 | 9.43 |
| Midnapore | 116.23 (18.99) | 0.02 (0.07) | 0.0003 | 0.007 |
| Jalpaiguri | 91.64 (40.81) | 1.67** (3.98) | 0.44 | 17.29 |
| Howrah | 95.16 (35.12) | 0.50* (2.29) | 0.21 | 5.85 |
| Birbhum | 99.45 (8.25) | 0.11 (0.42) | 0.009 | 0.20 |
| Hooghly | 124.57 (35.18) | 0.12*** (1.53) | 0.11 | 2.72 |
| Bankura | 80.68 (10.19) | 0.78** (3.82) | 0.42 | 15.93 |
| Malda | 130.29 (34.33) | -0.44 (-1.02) | 0.05 | 1.16 |
| Darjeeling | 110.20 (26.73) | -0.009 (-0.34) | 0.007 | 0.15 |
| Coochbehar | 56.64 (61.02) | 0.01 (0.35) | 6.34 | 0.13 |
| W.Dinajpur | 96.75 (62.91) | -0.07 (-0.98) | 0.05 | 1.16 |
| 24-Parganas | 102.81 (29.70) | 0.59* (2.33) | 0.21 | 5.85 |
| Purulia | 104.56 (25.46) | 0.09 (0.41) | 0.008 | 0.18 |

*Notes:* (i) Figures in the parentheses indicate 't' values.
(ii) *indicates 't' value significant at 1% level.
(iii) **indicates 't' value significant at 5% level.

*Source:* (i) Banking Statistics, R.B.I.
(ii) Statistical Abstract, Government of West Bengal.
(iii) Economic Review, Government of West Bengal.

**Table 21**
**Estimates of Simple Regression Equations Concerning Irrigated Area and IC in W.B. by Districts, Dependent Variable = Percentage of Irrigated Area, Number of Observations =24 (1972-73 to 1995-96)**

| *Districts* | *Constants* | *Explanatory variable IC/hectare* | $R^2$ | *F* |
|---|---|---|---|---|
| Burdwan | 40.99 | 0.23** | 0.40 | 14.67 |
| | (3.31) | (3.66) | | |
| Nadia | 11.02 | 0.12** | 0.47 | 19.51 |
| | (4.07) | (4.22) | | |
| Murshidabad | 13.18 | 0.12* | 0.20 | 5.50 |
| | (3.52) | (2.25) | | |
| Midnapore | 19.32 | 0.007 | 0.009 | 0.20 |
| | (8.40) | (0.44) | | |
| Jalpaiguri | 5.05 | -0.005 | 0.03 | 0.57 |
| | (3.58) | (-0.78) | | |
| Howrah | 0.55 | 0.08** | 0.47 | 19.51 |
| | (0.22) | (4.25) | | |
| Birbhum | 37.46 | 0.11** | 0.37 | 12.92 |
| | (13.67) | (3.45) | | |
| Hooghly | 17.57 | 0.14** | 0.30 | 9.43 |
| | (1.91) | (2.94) | | |
| Bankura | 29.24 | 0.17** | 0.43 | 16.60 |
| | (11.05) | (3.88) | | |
| Malda | 5.70 | 0.01 | 0.01 | 0.22 |
| | (2.44) | (0.52) | | |
| Darjeeling | -28.85 | 0.16 | 0.02 | 0.45 |
| | (-0.46) | (0.64) | | |
| Coochbehar | -10.99 | 0.53* | 0.99 | 4.88 |
| | (-1.01) | (2.21) | | |
| W.Dinajpur | 16.62 | 0.03 | 0.06 | 1.40 |
| | (5.19) | (1.17) | | |
| 24-Parganas | 9.42 | 0.02* | 0.22 | 6.20 |
| | (5.46) | (2.38) | | |
| Purulia | 2.03 | 0.39* | 0.74 | 62.61 |
| | (0.91) | (7.52) | | |

*Notes:* (i) Figures in the parentheses indicate 't' values.
(ii) *indicates 't' value significant at 1% level.
(iii) **indicates 't' value significant at 5% level.

*Source:* (i) Banking Statistics, R.B.I.
(ii) Statistical Abstract, Government of West Bengal.
(iii) Economic Review, Government of West Bengal.

**Table 21a**
**Values of Correlation Coefficients Concerning C.P.I., I.C., FC, etc. in the Districts of W.B.**

| *Districts* | *CPI & IC/ hectare* | *CPI & % of irrigated area* | *CPI & FC/ hectare* |
|---|---|---|---|
| Darjeeling | 0.06 | -0.07 | 0.48** |
| | (0.26) | (0.31) | (2.45) |
| Murshidabad | 0.79* | 0.55* | 0.69* |
| | (5.76) | (2.59) | (4.26) |
| Nadia | 0.62* | 0.82* | 0.53* |
| | (3.53) | (6.40) | (2.79) |
| Burdwan | 0.74* | 0.73* | 0.42** |
| | (4.92) | (4.78) | (2.10) |
| Coochbehar | 0.12 | 0.08 | 0.36 |
| | (0.54) | (0.36) | (1.72) |
| W. Dinajpur | -0.33 | -0.21 | -0.01 |
| | (-1.56) | (0.96) | (0.04) |
| Jalpaiguri | 0.09 | 0.66 * | 0.57* |
| | (0.40) | (3.93) | (3.10) |
| Howrah | 0.70* | 0.46** | 0.64* |
| | (4.38) | (2.31) | (3.73) |
| Birbhum | 0.20 | 0.09 | 0.13 |
| | (0.91) | (0.40) | (0.59) |
| Hooghly | 0.43* | 0.32 | 0.30 |
| | (2.13) | (1.51) | (1.40) |
| Bankura | 0.81* | 0.65* | 0.76* |
| | (6.18) | (3.82) | (5.23) |
| 24-Parganas | 0.21 | 0.46* | 0.67* |
| | (0.96) | (2.31) | (4.03) |
| Purulia | -0.04 | 0.09 | -0.08 |
| | (0.18) | (0.40) | (-0.36) |
| Malda | 0.74* | -0.22 | 0.61* |
| | (4.92) | (1.00) | (3.47) |

| *Districts* | *% of irrigated area & IC/ hectare* | *FC/hectare & IC/hectare* | *Index of agricultural production & IC/hectare* |
|---|---|---|---|
| Darjeeling | 0.14 | -0.17 | -0.01 |
| | (0.63) | (0.76) | (0.04) |
| Murshidabad | 0.45* | 0.61* | 0.50* |
| | (2.52) | (3.44) | (2.58) |
| Nadia | 0.68* | 0.61* | 0.56* |
| | (4.15) | (3.44) | |
| Burdwan | 0.63* | 0.50** | 0.86* |
| | (3.63) | (2.58) | (7.34) |

*Contd...*

| *Districts* | *% of irrigated area & IC/ hectare* | *FC/hectare & IC/hectare* | *Index of agricultural production & IC/hectare* |
|---|---|---|---|
| Coochbehar | 0.44** | 0.49** | 0.73* |
| | (2.16) | (2.38) | (4.78) |
| W. Dinajpur | 0.25 | 0.56* | 0.41** |
| | (1.16) | (3.02) | (2.01) |
| Jalpaiguri | -0.16 | -0.20 | 0.13 |
| | (-0.72) | (0.91) | (0.591) |
| Howrah | 0.69* | 0.93 | 0.86* |
| | (4.26) | (11.31) | (7.34) |
| Birbhum | 0.61 | 0.92* | 0.59* |
| | (3.44) | (10.49) | (3.27) |
| Hooghly | 0.55* | 0.79* | 0.62* |
| | (2.95) | (5.76) | (3.53) |
| Bankura | 0.65* | 0.90* | 0.73* |
| | (3.82) | (9.23) | (4.78) |
| 24-Parganas | 0.47** | 0.31 | 0.47* |
| | (2.31) | (1.45) | (2.31) |
| Purulia | 0.86* | 0.93* | 0.61* |
| | (3.27) | (11.31) | (3.47) |
| Malda | 0.11 | 0.84* | 0.59* |
| | (0.49) | (6.92) | (3.27) |

*Notes:* (i) Figures in the parentheses indicate 't' values.
(ii) *indicates 't' value significant at 1% level.
(iii) **indicates 't' value significant at 5% level.

*Source:* (i) Banking Statistics, R.B.I.
(ii) Statistical Abstract, Government of West Bengal.
(iii) Economic Review, Government of West Bengal.

for Darjeeling, Burdwan, Nadia, Jalpaiguri, Howrah, Bankura, 24-Parganas and Malda and significant at 5% probability level for Murshidabad, but significant for other six sample districts. The correlation between IC on the one hand, and the factors such as use of chemical fertiliser per hectare and proportion (%) of irrigated area to the GCA have also been found to measure the nature and degree of association between them. It is found (Table 21a ) that correlation coefficient for IC and FC are statistically significant at 1% level for the districts of Murshidabad, Nadia, Burdwan, Howrah, Hoogly, Bankura, Coochbehar, and Purulia and at 5% probability level for 24-Parganas, but not significant for the other seven sample districts in the state. The correlation coefficients between IC and percentage of irrigated area to the GCA are also statistically significant at 1% level for the districts, viz., Murshidabad, Nadia, Burdwan, Howrah, Hoogly, Bankura, Coochbehar, and Purulia, but not significant for other districts.

We may mention a number of factors explaining the insignificant association between CP and IC, irrigation and IC and fertiliser consumption and IC in many districts of W.B. It is observed in reality that the farmers do not get the IC at the time when they require it and hence the credit is not properly utilised for the cultivation purpose and generally used either for various unproductive purposes or for the purchase of consumer goods. This type of credit use naturally does not create significant impact of IC on agricultural growth through change in CP. Again, the available amount of IC is not sufficient to purchase the required amount of agricultural inputs. In such case IC can hardly be expected to have made significant contribution to the agricultural growth by effecting change in CP.

## Role of IC and Other Factors in the Context of Commercialisation of Agriculture in W.B.: District Level Analysis

In the context of commercialisation of agriculture, we may also examine the impact of IC and other related factors, viz., fertiliser consumption, the percentage of GIA to GCA on the NFA which has also a significant influence on agricultural growth in the districts of W.B.

To measure the impact of IC and the percentage of irrigated area in the GCA on the CP as measured by the NFA across the district of W.B., we may use the CV analysis for some sample years chosen from the period from 1972-73 to 1995-96. This kind of analysis will give us a rough idea about the relation between CP on the one hand and IC and the percentage of irrigated area in GCA on the other. The Table 21b gives the CV values for each of these variables. It is seen from the values of CV for the variables that the pattern of change in the CVs in the variables IC and the percentage irrigated area taking place over the period from 1972-73 to

**Table 21b**
**Coefficient of Variation of Percentage of Area under Non-Foodgrains to GCA (NFA), the IC Per Hectare, % of GIA to GCA Across the Districts of W.B. for Sample Years**

| *Year* | *NFA* | *IC/hectare* | *% of GIA to GCA* |
|---|---|---|---|
| 1972-73 | 0.62 | 2.04 | 0.93 |
| 1984-85 | 0.47 | 0.61 | 0.80 |
| 1992-93 | 0.51 | 0.81 | 0.85 |

*Source:* Statistical Abstract, Government of West Bengal.

1992-93 is the same as the pattern of change in CP, measured by NFA. This agreement in the pattern of changes in CVs for these variables, of course, does not indicate any cause and effect relationship, but certainly indicates an association between them across the districts of W.B. The following regression analysis, however, throws light on the cause and effect relationship between these variables in the districts of W.B.

Let us first examine the impact of IC per hectare on the NFA by regressing NFA on IC for the districts of W.B. The results of estimated regression equations are given in Table 22. From the estimates of the regression coefficients it can be concluded that the IC per hectare has made significant influence on commercialisation of agriculture through change in CP in the districts of Murshidabad, Nadia, Burdwan, Midnapore, Howrah, Bankura, Coochbehar, and West Dinajpur.

A regression analysis has also been made by using cross section data for some years on the variables of our concern to verify the above results. The estimates of the regression coefficients and their levels of significance show that IC has made significant contribution to the agricultural growth in quite a good number of districts in the state through its significant impact upon the growth of non-foodgrains production in the districts of the state. The regression results are given below.

| | | | |
|---|---|---|---|
| 1973-74 : | NFA = 12.41 | + 0.04** IC; | $R^2 = 0.54$ ; F = 15.36 |
| | | (3.92) | |
| 1976-77 : | NFA = 12.87 | + 0.03** IC; | $R^2 = 0.69$ ; F = 29.57 |
| | | (5.44) | |

*Contd...*

| | | | |
|---|---|---|---|
| 1979-80 : | NFA = 7.38 | + 0.04** IC;<br>( 3.79 ) | $R^2$ = 0.52 ; F = 14.36 |
| 1980-81 : | NFA = 11.83 | + 0.02** IC;<br>(2.51) | $R^2$ = 0.33 ; F = 6.29 |
| 1986-87 : | NFA = 7.06 | + 0.026* IC;<br>(2.25) | $R^2$ = 0.28 ; F = 5.06 |

In the above equations *, ** indicate coefficient significant at 5% and % level respectively and figures in brackets indicate t ratios.

A study on the impact of fertiliser consumption per hectare and irrigation using regression analysis reveals that both of them have significantly contributed to the growth of non-foodgrains production (Tables 23, 24). It is also observed from the regression equations given in Table 25 that the IC has significantly helped to expand the HYVs cultivation in most of the districts of West Bengal.

The above analysis highlights the important role of IC on agricultural growth through change in CP in the districts of W.B. So there is a need for direct analysis to examine the impact of IC on the agricultural production in the districts of West Bengal. The specifications of estimated regression equations are given in Table 26. The regression analysis shows that the IC has influenced significantly the agricultural growth in most of the districts of W.B. The Chow test reveals the differential influences of IC on agricultural growth in two sub-periods, for all the districts (except Birbhum), [where F = 2.70**, 8.19**, 17.08*, 12.43**, 0.21, 21.13**, 1.16, 36.25**, 11.83**, 12.40**, 13.85**, 11.97**, 34.33**, 15.35** and 21.65** respectively for Burdwan, Nadia, Murshidabad, Midnapore, Jalpaiguri, Howrah, Birbhum, Hoogly, Bankura, Malda, Darjeeling, Coochbehar, West Dinajpur, 24-Parganas, and Purulia (** indicate significant at 1% level).

Therefore it can be concluded that the IC plays an important role to accelerate agricultural growth through change in CP in the districts of W.B. This may be again proved by the correlation analysis. It is found that the value of r between index of agricultural production and IC per hectare

**Table 22**
**Estimates of Simple Regression Equations Concerning NFA and IC in W.B. by Districts, Dependent Variable = Percentage of Area Under Non-foodgrains to GCA(NFA), Number of Observations =24 (1972-73 to 1995-96)**

| *Districts* | *Constants* | *Explanatory variable IC/hectare* | $R^2$ | *F* |
|---|---|---|---|---|
| Burdwan | 8.50 | 0.05** | 0.57 | 29.16 |
| | (7.20) | (5.15) | | |
| Nadia | 22.08 | 0.10** | 0.26 | 7.73 |
| | (6.34) | (2.67) | | |
| Murshidabad | 17.46 | 0.09** | 0.39 | 14.07 |
| | (10.30) | (3.58) | | |
| Midnapore | 6.62 | 0.01* | 0.21 | 5.86 |
| | (9.96) | (2.27) | | |
| Jalpaiguri | 32.18 | -0.01 | 0.07 | 1.91 |
| | (9.27) | (-1.25) | | |
| Howrah | 13.27 | -0.01*** | 0.08 | 5.85 |
| | (8.31) | (-1.32) | | |
| Birbhum | -2.45 | 0.37 | 0.06 | 1.40 |
| | (-0.09) | (1.17) | | |
| Hooghly | 38.46 | -0.03 | 0.01 | 0.22 |
| | (2.99) | (-0.54) | | |
| Bankura | 7.29 | 0.05** | 0.47 | 19.51 |
| | (9.32) | (4.19) | | |
| Malda | 14.61 | 0.008 | 0.01 | 0.22 |
| | (9.86) | (0.52) | | |
| Darjeeling | 34.87 | -0.01 | 0.02 | 0.45 |
| | (4.84) | (-0.70) | | |
| Coochbehar | 8.05 | 0.31** | 0.50 | 22.00 |
| | (3.50) | (5.20) | | |
| W.Dinajpur | 22.03 | -0.03** | 0.26 | 7.73 |
| | (18.52) | (-2.63) | | |
| 24-Parganas | 15.06 | -0.002 | 4.81 | 0.09 |
| | (12.31) | (-0.31) | | |
| Purulia | 6.74 | -0.03 | 0.02 | 0.45 |
| | (3.67) | (-0.79) | | |

*Notes:* (i) Figures in the parentheses indicate 't' values.
(ii) *indicates 't' value significant at 1% level.
(iii) **indicates 't' value significant at 5% level.

*Source:* (i) Banking Statistics, R.B.I.
(ii) Statistical Abstract, Government of West Bengal.
(iii) Economic Review, Government of West Bengal.

**Table 23**
**Estimates of Simple Regression Equations concerning NFA and Fertiliser Consumption in W.B. by Districts, Dependent Variable = Percentage of Area Under Non-food grains to GCA(NFA), Number of Observations =24 (1972-73 to 1995-96)**

| *Districts* | *Constants* | *Explanatory variable FC/hectare* | $R^2$ | *F* |
|---|---|---|---|---|
| Burdwan | 10.82 | 0.05* | 0.19 | 5.16 |
| | (6.88) | (2.21) | | |
| Nadia | 24.78 | 0.12* | 0.20 | 5.50 |
| | (8.40) | (2.27) | | |
| Murshidabad | 17.57 | 0.13** | 0.90 | 19.80 |
| | (10.56) | (3.63) | | |
| Midnapore | 5.82 | 0.06** | 0.032 | 0.73 |
| | (7.80) | (3.07) | | |
| Jalpaiguri | 30.56 | -0.10*** | 0.09 | 2.18 |
| | (14.21) | (-1.44) | | |
| Howrah | 13.04 | -0.01 | 0.07 | 1.66 |
| | (8.56) | (-1.23) | | |
| Birbhum | 14.21 | 0.58** | 0.12 | 3.00 |
| | (-0.53) | (1.64) | | |
| Hooghly | 33.13 | -9.74 | 0.0007 | 0.02 |
| | (3.14) | (-0.12) | | |
| Bankura | 0.28 | 0.08** | 0.47 | 19.51 |
| | (6.33) | (4.21) | | |
| Malda | 14.26 | 0.02 | 0.03 | 0.68 |
| | (9.78) | (0.81) | | |
| Darjeeling | 37.54 | -0.14** | 0.39 | 14.06 |
| | (13.79) | (-3.61) | | |
| Coochbehar | 21.91 | 0.04 | 0.03 | 0.68 |
| | (14.77) | (0.89) | | |
| W.Dinajpur | 22.39 | -0.07 | 0.07 | 1.66 |
| | (11.34) | (-1.23) | | |
| 24-Parganas | 13.83 | 0.01 | 0.05 | 1.16 |
| | (10.54) | (1.02) | | |
| Purulia | 6.39 | -0.02 | 0.02 | 0.45 |
| | (4.18) | (-0.63) | | |

*Notes:* (i) Figures in the parentheses indicate 't' values.
(ii) *indicates 't' value significant at 1% level.
(iii) **indicates 't' value significant at 5% level.

*Source:* (i) Banking Statistics, R.B.I.
(ii) Statistical Abstract, Government of West Bengal.
(iii) Economic Review, Government of West Bengal.

**Table 24**
**Estimates of Simple Regression Equations Concerning NFA and Irrigated Area in W.B. by Districts, Dependent Variable = NFA, Number of Observations =24(1972-73 to 1995-96)**

| *Districts* | *Constants* | *Explanatory variable % of GIA to GCA* | $R^2$ | *F* |
|---|---|---|---|---|
| Burdwan | 9.39 | 0.07* | 0.14 | 3.58 |
| | (3.63) | (1.82) | | |
| Nadia | 19.50 | 0.52* | 0.23 | 6.57 |
| | (4.05) | (2.42) | | |
| Murshidabad | 18.43 | 0.21** | 0.17 | 3.29 |
| | (7.49) | (2.02) | | |
| Midnapore | 6.69 | 0.05 | 0.02 | 0.45 |
| | (4.40) | (0.69) | | |
| Jalpaiguri | 31.68 | -0.84*** | 0.11 | 2.72 |
| | (12.24) | (-1.58) | | |
| Howrah | 12.39 | -1.00 | 0.04 | 0.92 |
| | (9.43) | (-0.91) | | |
| Birbhum | -102.32 | 2.77** | 0.12 | 3.00 |
| | (-1.34) | (1.68) | | |
| Hooghly | 30.60 | 0.03 | 0.008 | 0.18 |
| | (2.52) | (0.12) | | |
| Bankura | 6.04 | 0.10*** | 0.12 | 3.00 |
| | (2.42) | (1.65) | | |
| Malda | 14.36 | 0.13 | 0.04 | 0.92 |
| | (11.80) | (0.98) | | |
| Darjeeling | 29.59 | -0.01 | 0.03 | 0.68 |
| | (12.88) | (0.85) | | |
| Coochbehar | 23.09 | -0.03 | 0.02 | 0.45 |
| | (19.33) | (-0.68) | | |
| W. Dinajpur | 19.59 | 0.04 | 0.01 | 0.22 |
| | (9.41) | (0.48) | | |
| 24-Parganas | 12.71 | 0.17 | 0.08 | 1.91 |
| | (6.83) | (1.28) | | |
| Purulia | 6.53 | -0.05 | 0.01 | 0.22 |
| | (3.59) | (-0.57) | | |

*Notes:* (i) Figures in the parentheses indicate 't' values.
(ii) *indicates 't' value significant at 1% level.
(iii) **indicates 't' value significant at 5% level.

*Source:* (i) Banking Statistics, R.B.I.
(ii) Statistical Abstract, Government of West Bengal.
(iii) Economic Review, Government of West Bengal.

**Table 25**
**Estimates of Simple Regression Equations Concerning HYV Area and IC in W.B. by Districts, Dependent Variable = % HYV Area, Number of Observation = 24 (1972-73 to 1995-96)**

| *Districts/ Periods* | *Constants* | *Explanatory variable IC/hectare* | $R^2$ | *F* |
|---|---|---|---|---|
| Burdwan | | | | |
| 1972-73 to 1995-96 | 28.42 | 0.18** | 0.82 | 100.22 |
| | (12.09) | (8.69) | | |
| Nadia | | | | |
| 1972-73 to 1995-96 | 10.67 | 0.14** | 0.27 | 8.14 |
| | (2.06) | (2.53) | | |
| 1972-73 to 1982-83 | 15.78 | 0.02* | 0.33 | 4.44 |
| | (25.95) | (2.11) | | |
| 1983-84 to 1995-96 | 9.06 | 0.20 | 0.14 | 1.79 |
| | (0.41) | (0.98) | | |
| Mursidabad | | | | |
| 1972-73 to 1995-96 | 25.26 | 0.12* | 0.76 | 69.67 |
| | (22.24) | (7.29) | | |
| Midnapur | | | | |
| 1972-73 to 1995-96 | 21.60 | 0.07** | 0.52 | 23.83 |
| | (10.57) | (4.27) | | |
| 1972-73 to 1982-83 | 19.83 | 0.03*** | 0.08 | 0.78 |
| | (5.23) | (1.39) | | |
| Jalpaiguri | | | | |
| 1972-73 to 1995-96 | 21.55 | -0.01 | 0.02 | 0.45 |
| | (5.49) | (-0.62) | | |
| 1972-73 to 1982-83 | 12.75 | 0.01* | 0.24 | 2.93 |
| | (6.94) | (1.71) | | |
| 1983-84 to 1995-96 | 10.19 | 0.14** | 0.61 | 17.21 |
| | (2.02) | (3.10) | | |
| Howrah | | | | |
| 1972-73 to 1995-96 | 16.87 | 0.13** | 0.88 | 161.33 |
| | (10.90) | (11.03) | | |
| Birbhum | | | | |
| 1972-73 to 1995-96 | 25.92 | 0.21** | 0.80 | 88.00 |
| | (13.89) | (8.23) | | |
| Hooghly | | | | |
| 1972-73 to 1995-96 | 38.83 | 0.01 | 0.002 | 0.04 |
| | (3.58) | (0.19) | | |

*Contd...*

| *Districts/ Periods* | *Constants* | *Explanatory variable IC/hectare* | $R^2$ | *F* |
|---|---|---|---|---|
| 1972-73 to 1982-83 | 41.89 | 0.02* | 0.24 | 2.83 |
| | (24.45) | (1.68) | | |
| 1983-84 to 1995-96 | 38.21 | 0.07 | 0.12 | 1.50 |
| | (2.23) | (0.91) | | |
| Bankura | | | | |
| 1972-73 to 1995-96 | 18.72 | 0.12*** | 0.05 | 1.16 |
| | (3.51) | (1.34) | | |
| 1972-73 to 1982-83 | 15.01 | 0.32** | 0.66 | 17.18 |
| | (6.89) | (4.15) | | |
| 1983-84 to 1995-96 | 21.06 | 0.22** | 0.56 | 14.00 |
| | (3.51) | (2.79) | | |
| Malda | | | | |
| 1972-73 to 1995-96 | 9.46 | 0.06 | 0.06 | 1.40 |
| | (4.01) | (1.11) | | |
| 1972-73 to 1982-83 | 18.73 | 0.09** | 0.78 | 32.74 |
| | (25.73) | (5.72) | | |
| 1983-84 to 1995-96 | 41.49 | -0.05 | 0.11 | 1.36 |
| | (5.77) | (-0.82) | | |
| Darjeeling | | | | |
| 1972-73 to 1995-96 | 15.32 | 0.003 | 0.02 | 0.45 |
| | (5.90) | (0.58) | | |
| Coochbehar | | | | |
| 1972-73 to 1995-96 | 17.32 | 0.32 | 0.02 | 0.45 |
| | (0.79) | (0.60) | | |
| W. Dinajpur | | | | |
| 1972-73 to 1995-96 | 15.98 | 0.06* | 0.32 | 10.35 |
| | (8.57) | (3.09) | | |
| 24-Parganas | | | | |
| 1972-73 to 1995-96 | 20.88 | 0.02*** | 0.09 | 2.18 |
| | (6.13) | (1.43) | | |
| Purulia | | | | |
| 1972-73 to 1995-96 | 14.05 | 0.11 | 0.04 | 0.92 |
| | (3.16) | (1.04) | | |

*Notes:* (i) Figures in the parentheses indicate 't' values.
(ii) * indicates 't' value significant at 1% level.
(iii) ** indicates 't' value significant at 5% level.

*Source:* (i) Banking Statistics, R.B.I.
(ii) Statistical Abstract, Government of West Bengal.
(iii) Economic Review, Government of West Bengal

**Table 26**
**Estimates of Simple Regression Equations Concerning Agricultural Production and IC in W.B. by Districts, Dependent Variable = Index of Agricultural Production, Number of Observations = 24 (1972-73 to 1995-96)**

| *Districts/ Periods* | *Constants* | *Explanatory variable IC/hectare* | $R^2$ | *F* |
|---|---|---|---|---|
| Burdwan | | | | |
| 1972-73 to 1995-96 | 72.63 | 0.77** | 0.74 | 62.62 |
| | (5.96) | (7.64) | | |
| 1972-73 to 1982-83 | 79.79 | 0.58* | 0.27 | 3.29 |
| | (3.58) | (1.81) | | |
| 1983-84 to 1995-96 | 104.26 | 0.58** | 0.58 | 15.19 |
| | (3.99) | (3.53) | | |
| Nadia | | | | |
| 1972-73 to 1995-96 | 87.99 | 0.95** | 0.32 | 10.35 |
| | (3.06) | (3.05) | | |
| 1972-73 to 1982-83 | 95.86 | 0.31** | 0.55 | 10.96 |
| | (13.67) | (3.31) | | |
| 1983-84 to 1995-96 | 261.79 | 0.38 | 0.05 | 0.58 |
| | (4.14) | (-0.69) | | |
| Murshidabad | | | | |
| 1972-73 to 1995-96 | 106.51 | 0.51** | 0.26 | 7.73 |
| | (7.97) | (2.65) | | |
| 1972-73 to 1982-83 | 115.49 | -0.07 | 0.02 | 0.23 |
| | (17.28) | (-0.48) | | |
| 1983-84 to 1995-96 | 160.41 | 0.03 | 0.0004 | 0.004 |
| | (3.79) | (0.06) | | |
| Midnapore | | | | |
| 1972-73 to 1995-96 | 105.57 | 0.36** | 0.52 | 23.83 |
| | (10.16) | (4.68) | | |
| 1972-73 to 1982-83 | 106.86 | -0.32 | 0.058 | 0.56 |
| | (10.46) | (-0.75) | | |
| 1983-84 to 1995-96 | 157.95 | 0.13 | 0.11 | 1.36 |
| | (6.47) | (0.99) | | |
| Jalpaiguri | | | | |
| 1972-73 to 1995-96 | 111.19 | 0.28 | 0.02 | 0.45 |
| | (1.17) | (0.61) | | |
| 1972-73 to 1982-83 | 143.57 | 0.203 | 0.003 | 0.02 |
| | (0.45) | (0.16) | | |
| 1983-84 to 1995-96 | 123.91 | 0.08** | 0.67 | 22.33 |
| | (42.86) | (4.28) | | |

*Contd...*

| *Districts/ Periods* | *Constants* | *Explanatory variable IC/hectare* | $R^2$ | *F* |
|---|---|---|---|---|
| Howrah | | | | |
| 1972-73 to 1995-96 | 70.92 | 0.54** | 0.75 | 66.00 |
| | (7.16) | (7.70) | | |
| 1972-73 to 1982-83 | 93.18 | 0.07 | 0.03 | 0.28 |
| | (10.46) | (-0.75) | | |
| 1983-84 to 1995-96 | 116.22 | 0.34** | 0.53 | 12.40 |
| | (5.86) | (3.17) | | |
| Birbhum | | | | |
| 1972-73 to 1995-96 | 81.39 | 0.51** | 0.36 | 12.37 |
| | (6.35) | (3.35) | | |
| 1972-73 to 1982-83 | 100.99 | -0.19 | 0.07 | 0.67 |
| | (9.94) | (-0.82) | | |
| 1983-84 to 1995-96 | 88.83 | 0.49 | 0.102 | 1.25 |
| | (1.65) | (1.01) | | |
| Hooghly | | | | |
| 1972-73 to 1995-96 | 82.59 | 0.48 | 0.39 | 14.06 |
| | (3.08) | (3.57) | | |
| 1972-73 to 1982-83 | 99.25 | 0.18** | 0.603 | 13.67 |
| | (12.33) | (3.71) | | |
| 1983-84 to 1995-96 | 287.10 | -0.31 | 0.06 | 0.70 |
| | (3.37) | (-0.73) | | |
| Bankura | | | | |
| 1972-73 to 1995-96 | 74.96 | 1.01** | 0.54 | 25.83 |
| | (6.15) | (4.87) | | |
| 1972-73 to 1982-83 | 97.76 | -0.33 | 0.09 | 0.86 |
| | (9.80) | (-0.92) | | |
| 1983-84 to 1995-96 | 105.03 | 0.72 | 0.13 | 1.64 |
| | (2.21) | (1.19) | | |
| Malda | | | | |
| 1972-73 to 1995-96 | 109.28 | 0.44** | 0.36 | 12.37 |
| | (9.05) | (3.32) | | |
| 1972-73 to 1982-83 | 110.31 | 0.14*** | 0.19 | 2.12 |
| | (23.95) | (1.45) | | |
| 1983-84 to 1995-96 | 191.37 | -0.18 | 0.03 | 0.34 |
| | (4.55) | (0.51) | | |
| Darjeeling | | | | |
| 1972-73 to 1995-96 | 142.95 | -0.004 | 0.0002 | 0.004 |
| | (4.68) | (-0.06) | | |

*Contd...*

| *Districts/ Periods* | *Constants* | *Explanatory variable IC/hectare* | $R^2$ | *F* |
|---|---|---|---|---|
| 1972-73 to 1982-83 | 144.18 | -0.05* | 0.35 | 4.91 |
| | (13.76) | (-2.21) | | |
| 1983-84 to 1995-96 | 84.38 | 0.22* | 0.24 | 3.47 |
| | (1.78) | (1.71) | | |
| Coochbehar | | | | |
| 1972-73 to 1995-96 | 81.96 | 0.79** | 0.54 | 25.83 |
| | (11.26) | (4.84) | | |
| 1972-73 to 1982-83 | 88.50 | 0.29*** | 0.23 | 2.64 |
| | (17.69) | (1.62) | | |
| 1983-84 to 1995-96 | 120.86 | 0.19 | 0.013 | 0.14 |
| | (3.92) | (0.34) | | |
| W.Dinajpur | | | | |
| 1972-73 to 1995-96 | 115.00 | 0.093* | 0.17 | 4.51 |
| | (18.47) | (2.03) | | |
| 1972-73 to 1982-83 | 88.11 | 0.59** | 0.43 | 6.88 |
| | (15.69 | (2.62) | | |
| 1983-84 to 1995-96 | 140.71 | 0.02 | 0.02 | 0.22 |
| | (17.95) | (0.38) | | |
| 24-Parganas | | | | |
| 1972-73 to 1995-96 | 126.91 | 0.11* | 0.22 | 6.20 |
| | (12.94) | (2.39) | | |
| 1972-73 to 1982-83 | 112.29 | 0.04 | 0.005 | 0.05 |
| | (7.47) | (0.22) | | |
| 1983-84 to 1995-96 | 107.85 | 0.04 | 0.07 | 0.83 |
| | (10.92) | (0.82) | | |
| Purulia | | | | |
| 1972-73 to 1995-96 | 86.70 | 0.85** | 0.38 | 13.48 |
| | (8.38) | (3.52) | | |
| 1972-73 to 1982-83 | 129.01 | -1.70* | 0.29 | 3.66 |
| | (6.70) | (-1.91) | | |
| 1983-84 to 1995-96 | 118.73 | 0.41** | 0.30 | 4.71 |
| | (10.09) | (1.99) | | |

*Notes:* (i) Figures in the parentheses indicate 't' values.
(ii) *indicates 't' value significant at 1% level.
(iii) ** indicates 't' value significant at 5% level.

*Source:* (i) Banking Statistics, R.B.I.
(ii) Statistical Abstract, Government of West Bengal.
(iii) Economic Review, Government of West Bengal.

be 0.50**, 0.56*, 0.86*, 0.73*, 0.41*, 0.86*, 0.59*, 0.62*, 0.73*, 0.47*, 0.61* and 0.59* respectively for Murshidabad, Nadia, Burdwan, Coochbehar, West Dinajpur, Jalpaiguri, Howrah, Birbhum, Hoogly, Bankura, 24-Parganas, Purulia and Malda (*, ** indicate significant at 1% and 5% level respectively). It is evident from above correlation analysis that the IC can play a significant role in changing CP in the districts of W.B.

## Role of IC and Other Factors in Change in CP in Midnapore: Block Level Analysis

Let us now examine the trend/growth of factors like fertiliser consumption per hectare and % of GIA to GCA during 1993 to 2003. For the purpose of analysis , the triennia average has been calculated for the periods 1993-96, 1997-99, and 2000-03. It is observed that the fertiliser consumption per hectare has shown a rising trend for all blocks and the % of GIA to GCA has also shown an increasing trend for 53 blocks (Table 27). In case of only Midnapore block the percentage irrigated area has declined from 33.68 in 1993-96 to 25.42 in 2000-03 (Table 27). The CAGRs of both the irrigated area and fertiliser consumption have been significant for all blocks.

In this section, an attempt has been made to explore the impact of IC and other factors like fertiliser consumption per hectare and % of GIA to GCA on the change in CP influencing agricultural growth across different blocks of Midnapore district. Since the data on credit is not available for the whole period under study ,so we have made a cross section study across the blocks for the years 1999-2000, 2000-2001, 2001-2002 & 2002-2003 to examine the impact of IC availability on change in CP. To examine the relationship between the availability of IC and change in CP as measured by the percentage of area under foodgrains to GCA (FA) across the blocks we may use the CV analysis for the sample years 1999-2000, 2000-01, 2001-02, 2002-03. This kind of analysis gives a rough idea about the relation between CP change on the one hand and availability of IC on the other. The values of CV are given below:

**Tables 27**

**Trends/CAGROF Percentage of Irrigated Area to GCA and Fertilizer Consumption/Hectare in Blocks of Midnapore**

| Block | % irrigated area | | F. C per hectare | | (Kg/Ha) | | CAGR of % Irrigated Area | CAGR of F.c./Ha |
|---|---|---|---|---|---|---|---|---|
| | 1993-1996 | 1997-1999 | 2000-2003 | 1993-1996 | 1997-1999 | 2000-2003 | | |
| Jhargram | 21.29 | 57.43 | 70.78 | 166.34 | 146.09 | 171.1 | 16.18* | 14.31* |
| Binpur-1 | 11.6 | 31.86 | 88.94 | 25.58 | 78.18 | 122.67 | 18.54* | 15.32* |
| Binpur-ii | 7.8 | 29.06 | 40.75 | 35.47 | 101.53 | 113.59 | 7.57* | 16.35* |
| Jamboni | 10.47 | 29.64 | 35.31 | 50.96 | 180.09 | 181.23 | 17.35* | 14.61* |
| Nayagram | 7.74 | 13.91 | 25.62 | 28.51 | 100.68 | 130.63 | 24.61* | 15.21* |
| Sankrail | 13.95 | 40.91 | 47.79 | 44.68 | 141.23 | 135.21 | 16.18* | 16.34* |
| Gopiballabpur-i | 24.59 | 37.97 | 45.61 | 50.42 | 170.68 | 204.41 | 20.92* | 21.64* |
| Gopiballabpur-ii | 10.19 | 23.77 | 34.46 | 38.88 | 147.41 | 141.98 | 14.32* | 22.06* |
| Debra | 11.11 | 34.42 | 31.94 | 49.02 | 49.02 | 124.03 | 25.32* | 12.35* |
| Pingla | 21.8 | 37.13 | 46.29 | 44.56 | 121.72 | 149.79 | 31.65* | 24.12* |
| Keshiary | 8.55 | 29.24 | 29.59 | 32.8 | 124.94 | 128.98 | 12.36* | 19.64* |
| Dantan-i | 21.79 | 22.74 | 35.41 | 35.59 | 125.78 | 149.04 | 26.65* | 18.65* |
| Dantan-ii | 23.32 | 33.18 | 25.78 | 40.71 | 129.86 | 176.48 | 24.94* | 17.36* |
| Narayangarh | 20.61 | 24.71 | 22.45 | 29.39 | 88.22 | 98.83 | 14.52* | 13.65* |
| Mohanpur | 15.24 | 32.76 | 56.18 | 61.23 | 169.29 | 221.66 | 21.55* | 25.41* |
| Salboni | 12.71 | 31.42 | 33.72 | 41.41 | 135.09 | 144.29 | 32.12* | 24.56* |
| Keshpur | 12.75 | 28.61 | 36.54 | 25.16 | 75.95 | 83.29 | 24.62* | 21.54* |
| Garbeta-i | 29.77 | 38.67 | 46.91 | 51.69 | 160.56 | 243.57 | 15.31* | 15.98* |
| Garbeta-ii | 18.67 | 39.52 | 42.24 | 53.49 | 164.91 | 160.34 | 18.65* | 18.57* |

Contd...

| Block | % irrigated area 1993-1996 | 1997-1999 | 2000-2003 | F. C per hectare (Kg/Ha) 1993-1996 | 1997-1999 | 2000-2003 | CAGR of % Irrigated Area | CAGR of F.c./Ha |
|---|---|---|---|---|---|---|---|---|
| Garbeta-iii | 29.08 | 33.91 | 38.5 | 53.54 | 161.79 | 165.05 | 16.37* | 18.69* |
| Midnapur | 33.68 | 21.16 | 25.42 | 52.83 | 147.85 | 136.33 | 19.32* | 14.35* |
| Sabang | 28.96 | 35.5 | 36.51 | 29.35 | 82.6 | 98.86 | 25.31* | 17.35* |
| Kharagpur-i | 8.68 | 15.07 | 26.11 | 41.67 | 189.63 | 227.36 | 23.12* | 27.35* |
| Kharagpur-ii | 15.8 | 30.8 | 48.02 | 31.31 | 85.15 | 115.3 | 15.36* | 13.65* |
| Chandrakona-i | 17.23 | 27.98 | 37.59 | 32.05 | 93.77 | 149.01 | 14.23* | 14.23* |
| Chandrakona-ii | 17.54 | 26.8 | 25.51 | 30.88 | 102.01 | 172.91 | 19.31* | 15.32* |
| Ghatal | 8.09 | 42.42 | 51.86 | 37.35 | 131.66 | 162.44 | 17.32* | 16.32* |
| Daspur-i | 15.65 | 50.74 | 50.07 | 51.44 | 192.11 | 199.92 | 15.23* | 17.35* |
| Daspur-ii | 16.62 | 41.94 | 49.86 | 43.86 | 131.06 | 190.71 | 17.41* | 21.35* |
| Tamluk-i (Tamluk) | 3.45 | 44.91 | 48.08 | 40.59. | 155.03 | 164.7 | 21.55* | 26.32* |
| Tamluk-ii (Sahid Matangini) | 11.17 | 56.37 | 59.34 | 45.16 | 133.47 | 143.05 | 23.66* | 22.32* |
| Panskura-i | 4.52 | 50.72 | 53.87 | 99.26 | 173.44 | 203.98 | 24.77* | 14.32* |
| Panskura-ii | 12.89 | 38.72 | 5.29 | 41.21 | 125.12 | 161.4 | 26.34* | 19.32* |
| Moyna | 13.11 | 42.43 | 27.38 | 30.35 | 82.78 | 98.85 | 14.52* | 17.23* |
| Mahisadal-i (Nandakumar) | 1.42 | 33.69 | 38.89 | 28.91 | 84.72 | 93.65 | 12.36* | 18.32* |
| Mahisadal-ii (Mahisadal) | 35.81 | 46.83 | 53.51 | 28.18 | 78.53 | 103.98 | 18.71* | 19.23* |
| Nandigram-i | 1.31 | 13.76 | 13.63 | 39.03 | 128.84 | 142.39 | 16.32* | 15.23* |
| Nandigram-ii | 1.74 | 18.8 | 19.25 | 61.07 | 188.75 | 201.63 | .32* | 18.23* |
| Nandigram-iii (Chandipur) | 10.81 | 18.08 | 23.97 | 43.07 | 116.79 | 133.76 | 15.21* | 22.13* |
| Sutahata-i (Sutahata) | 1.58 | 28.94 | 25 | 50.44 | 170.57 | 174.46 | 28.31* | 25.41* |

Contd...

| Block | % irrigated area 1993-1996 | 1997-1999 | 2000-2003 | F. C per hectare (Kg/Ha) 1993-1996 | 1997-1999 | 2000-2003 | CAGR of % Irrigated Area | CAGR of F.c./Ha |
|---|---|---|---|---|---|---|---|---|
| Sutahata-ii (Haldia) | 7.31 | 26.91 | 20.6 | 56.67 | 110.28 | 141.93 | 26.31* | 21.31* |
| Potashpur-i | 3.03 | 46.64 | 38.06 | 43.69 | 84.94 | 105.54 | 29.12* | 25.31* |
| Potashpur-ii | 2.51 | 37.82 | 37.83 | 40 | 117.18 | 110.44 | 25.94* | 24.32* |
| Khejuri-i | 0.26 | 13.21 | 12.44 | 6.52 | 45.08 | 23.68 | 12.31* | 22.15* |
| Khejuri-ii | 0.33 | 13.37 | 11.97 | 5.99 | 39.54 | 42.38 | 12.54* | 17.23* |
| Bhagwanpur-i | 4.73 | 29.57 | 31.79 | 14.44 | 74.25 | 68.71 | 10.24* | 15.23* |
| Bhagwanpur-ii | 3.08 | 29.84 | 30.15 | 21.91 | 75.28 | 71.6 | 16.31* | 19.36* |
| Ramnagar-i | 4.19 | 50 | 40.12 | 61.75 | 159.95 | 110.71 | 21.31* | 17.23* |
| Ramnagar-ii | 7.84 | 29.32 | 65.05 | 33.31 | 101.5 | 146.09 | 23.21* | 16.32* |
| Egra-i | 2.59 | 42.45 | 45.47 | 40.82 | 162.63 | 117.31 | 23.15* | 19.23* |
| Egra-ii | 4.43 | 39.04 | 39.26 | 40.73 | 107.51 | 102.64 | 21.35* | 17.56* |
| Contai-i | 0.14 | 16.12 | 14.63 | 30.39 | 87.89 | 89.97 | 20.16* | 21.35* |
| Contai-ii | 3.68 | 14.97 | 14.23 | 33.33 | 66.33 | 83.37 | 15.34* | 16.32* |
| Contai-iii | 3.32 | 21.01 | 18.98 | 30.68 | 88.32 | 76.02 | 16.35* | 14.53* |

*Notes:* CAGR = Compound Annual Growth Rate, F.C. = Fertiliser Consumption.

*Source:* Annual Plan Of Purba and Paschim Midnapur District and District Statistical Handbook (Various Issues)

| *Year* | *FA* | *IC/Ha* |
|---|---|---|
| 1999-2000 | 0.39 | 0.76 |
| 2000-01 | 0.21 | 0.66 |
| 2001-02 | 0.25 | 0.76 |
| 2002-03 | 0.28 | 0.76 |

It is seen from the values of CV for the variables that the pattern of change in CV of IC/Ha is almost same as that of FA during 1999-2000 to2002-03. This agreement in the pattern of changes in CV for these variables, of course, does not indicate any cause and effect relationship, but certainly indicates an association between them across the blocks of Midnapore. The following regression analysis across the blocks ,however, throws light on the cause and effect relationship between these variables. The regression results are given below:

| | | | |
|---|---|---|---|
| 1999-2000: | FA=35.39<br>(9.88) | +0.01*IC/Ha,<br>(4.52) | $R^2$=0.28, F=20.42 |
| 2000-01: | FA=62.93+<br>(19.59) | 0.013*IC/Ha.,<br>(6.41) | $R^2$=0.44, F=40.92 |
| 2001-02 | FA=69.41<br>(16.68) | +0.0047*IC/Ha,<br>(2.29) | $R^2$=0.09, F=5.28 |
| 2002-03: | FA=56.19+ | 0.009*IC/Ha,<br>(15.64) | $R^2$=0.40, F=34<br>(5.89) |

*indicates that the coefficients are significant at 5% level of significance. The above regression results show that the availability of IC has significantly influenced the change in CP influencing agricultural growth and the availability of IC can explain 9 to 44% variation in CP in the blocks of Midnapur district.

Since IC plays a catalytic role through the use of fertiliser, irrigation, etc. to change CP, so an analysis can be made to examine the role of IC in fertiliser consumption and irrigation and the impact of those factors on change in CP. To examine the impact of IC on fertiliser consumption we have estimated following regression equations across the blocks:

| | | | |
|---|---|---|---|
| 1999-2000: | FC = 110.73 | +0.14 IC/Ha (1.29) | $R^2$ = 0.03, F= 1.44 |
| 2000-01: | FC = 107.39 | +0.04*IC/Ha (3.98) | $R^2$ =0.23, F= 15.82 |

where FC= fertiliser consumption per hectare. * indicates that coefficients are significant at 5% level. The regression analysis shows that the provision of IC has made significant impact on fertiliser consumption. Let us now examine the impact of fertiliser consumption on change in CP. The estimated regression equations are as follows:

| | | | |
|---|---|---|---|
| 1997-98: | FA = 81.47 | +0.26*FC (3.53) | $R^2$= 0.19, F=12.45 |
| 1999-2000: | FA= 97.16 | +0.09*FC (1.86) | $R^2$= 0.06, F=3.44 |
| 2000-01: | FA= 73.73+ | 0.31*FC (4.54) | $R^2$=0.28, F=20.64 |

* indicates that coefficients are significant at 5% level. These regression equations show the significant impact of fertiliser consumption on change in CP. Thus the IC has made significant impact through the use of fertiliser on the change in CP across the blocks in Midnapur.

Let us now examine the impact of availability of IC on irrigation across the blocks. The estimated regression equations are as follows:

| | | | |
|---|---|---|---|
| 1999-2000: | Ir = 71.79 | +0.02*IC (3.74) | $R^2$=0.21, F= 13.99 |
| 2000-01: | Ir = 101.28 | + 0.04*IC (3.24) | $R^2$=0.17, F= 10.47 |

where Ir = GIA to GCA, * indicates that coefficients are significant at 5% level. The regression results show that the availability of IC has made significant impact on irrigation across the blocks. Let us now examine the impact of irrigation on change in CP. The estimated regression equations are as follows:

| | | | |
|---|---|---|---|
| 1996-97: | FA = 94.11 | +0.19*Ir , (2.81) | $R^2$ =0.13, F= 7.87 |

| | | | |
|---|---|---|---|
| 1997-98: | FA = 84.86 | +0.28 *Ir , (4.81) | $R^2$=0.31, F= 23.18 |
| 1999-2000: | FA = 87.06 | +0.26*Ir, (4.81) | $R^2$= 0.31, F= 23.15 |
| 2000-01: | FA = 94.03 | +0.19*Ir, (2.87) | $R^2$= 0.14, F=8.26 |

* indicates that coefficients are significant at 5% level. The regression results show that the irrigation has made significant impact on change in CP across the blocks. Thus the availability of IC has made significant impact through the spread of irrigation on change in CP across the blocks of Midnapore.

From the above regression analysis it may be concluded that the IC through the use of technological inputs has made significant impact on change in CP influencing agricultural growth across the blocks in Midnapore. Thus, the availability of agricultural inputs has played an important role in accelerating agricultural growth in the blocks of Midnapore district.

## A Summing-Up

It is observed that the bank loans (after the nationalisation of commercial banks) have created a significant impact on the change in CP in India and most of the sample states. On the other hand, the larger provision of credit, particularly, IC, larger irrigational facilities, massive use of fertilizer, etc. on the supply side have contributed to change in CP in India, the sample states, the districts of W.B. and the blocks of Midnapore district. Side by side, the significant growth of PCI or PCSDP on the demand side has also made significant contribution to change in C.P. in India and most of the sample states. Finally, the IC has played a catalytic role in accelerating agricultural growth in India, sample states, the districts of W.B. as well as in the blocks of Midnapore district. The role of IC on change in CP is probably more significant for the cultivators belonging to smaller size classes of land holdings in India and W.B.

# Chapter–10

## ECONOMICS OF CHANGE IN CROPPING PATTERN IN RELATION TO CREDIT: A MICRO LEVEL STUDY IN WEST BENGAL

The previous chapter has established the positive and significant impact of credit on change in CP at the macro level. But the economics of change in CP in relation to credit has not been so far discussed. The present chapter attempts to analyse the economics of change in CP in relation to credit at the FH level in Midnapore district of W.B.

To focus on the economics of CP of the sample households in relation to credit we need to examine the following issues:

(i) Whether there exists any significant difference of the change in CP between the different groups of cultivators classified on the basis of use of credit or no credit, viz., FHs with IC, FHs with NIC, FHs with both IC and NIC and FHs without credit i.e., with OF only.

(ii) Whether there exists any significant relationship between CP on the one hand, and profitability and employment per unit of cultivated land, etc. among these groups of cultivators.

(iii) Whether there exists any significant differences in CP between sample households belonging to different size classes of land holdings.

(iv) Whether there exists any significant differences in the availability of credit among the sample households with different size of land holdings.

To examine the above relations we formulate the following hypotheses.

*Hypothesis–I:* The larger the availability of credit, larger is the percentage of area under non-foodgrains (NF), viz, betelvine, flowers, potato, jute, etc. as well as boro (HYV) paddy. This hypothesis is formulated on the supposition that the NF cultivation requires use of greater amount of inputs like water for irrigation, fertilizers, insecticides and pesticides, etc., which, in turn, require more fund to purchase these inputs as and when they are required. Credit from different sources provides necessary fund for these purposes.

*Hypothesis–II:* The larger the percentage of area under NF, larger is the amount of gross profit (GP) per acre.This hypothesis has its basis on the supposition that the profit margin for most of the NF crops is much higher than that for foodgrains.

*Hypothesis–III:* The smaller the size of land holdings, larger is the percentage of area under NF for those households who receive loans. The basis of this hypothesis is that the marginal and small farmers have look for earning maximum income from agriculture, as they do not normally have any other significant sources of income. As a result, they will go for cultivation of those crops that guarantee higher profit and it is the NF that fetch higher profit as compared to the foodgrains.

*Hypothesis–IV:* The larger the percentage of area under NF, larger is the employment of labour per acre. Cultivation of NF, in addition to foodgrains, keeps the land engaged throughout the year and, as a result, employment is created throughout the year.

In the following paragraphs we shall test these hypothesis and draw the inference on the relation between the change in CP and credit and between total credit and different types of credit on the one hand and income of the FHs, employment generated, etc. on the other.

In order to establish the relationship between credit and CP for the sample FHs we construct a frequency distribution of the percentage of area under NF for the households before and after they received loans either from institutional or non-institutional sources or from both of them. It is seen from Table 1 that while 10 households of the total 90 sample households (i.e. 11.11%) cultivated NF on 36% to 65% of their total cultivated area in the pre-loan period, i.e. the year 1992, as many as 28 the households (i.e. about 31%) cultivated NF on the same percentage of their total cultivated area after they received farm loans from institutional and/or non-institutional sources (columns 2, 3 & 4 ). Again, while 4.4% of the sample FHs, i.e., 4 out of 90, cultivated NF on about 60% of their cultivable land before they received loans, about 16% of them cultivated NF on the same percentage of cultivable land after they received loans (columns 2, 3 & 4). Another interesting thing to note is that while as many as 27% of the sample FHs, (24 out of 90) did not grow any of the NF before they received loans, perhaps because of the fund constraint, all the FHs cultivated NF on their land, though in different proportions, after they received loans from institutional and/or non-institutional sources.The variations in the proportions of area under NF in the total area under cultivation among the sample households may be accounted for by the amount of loans borrowed as well as their own funds available for cultivation purposes.

## Institutional Credit and Cropping Pattern

So far as the IC is concerned, it is seen from Table 1 that 30% of the sample FHs receiving only IC did not grow any of the NF before they received loans, but after they received loans all the sample FHs opted for the cultivation of some NF. Though it is seen from the table that 70% of the sample FHs cultivated NF even when they did not receive loans, both the number of households and the proportion of area under NF increased significantly when they received loans. For example, while 17% of the households receiving IC had about 10% of their total cultivated area under NF before they received loans,

**Table 1**
**Frequency Distribution of Sample Farm Households (FHs) by Percentage of Area under Non-foodgrains**

| | | *FHs with credit* | | *FHs with IC* | | *FH with Non-IC,* | | *FH with both I.C. & Non-I.C.,* | | *F.H. without Credit* | |
|---|---|---|---|---|---|---|---|---|---|---|---|
| *Percentage of Area under Non-foodgrains* | *Class mark* | *Pre-loan period* | *Post-loan period* | *Pre-loan period* | *Post-loan period* | *Pre-loan period* | *Post-loan period* | *Pre-loan period* | *Post-loan period* | *1992* | *1997* |
| *1* | *2* | *3* | *4* | *5* | *6* | *7* | *8* | *9* | *10* | *11* | *12* |
| 0 | | 24 (26.67) | 0 (0) | 9 (30) | 0 (0) | 7 (23.33) | 0 (0) | 8 (26.67) | 0 (0) | 8 (11.43) | 6 (8.57) |
| 1-5 | 3 | 22 (24.44) | 15 (16.67) | 8 (26.67) | 3 (10) | 8 (26.67) | 5 (16.67) | 6 (20) | 7 (23.33) | 12 (17.14) | 14 (20) |
| 6-15 | 10.5 | 19 (21.11) | 23 (25.56) | 5 (16.67) | 6 (20) | 7 (23.33) | 9 (30) | 7 (23.33) | 8 (23.67) | 20 (28.57) | 28 (40) |
| 16-35 | 25.5 | 15 (24.44) | 24 (26.67) | 5 (16.67) | 10 (33.33) | 5(16.67) | 7 (23.33) | 5 (16.67) | 7 (23.33) | 15 (21.43) | 14 (20) |
| 36-55 | 45.5 | 6 (6.67) | 14 (15.56) | 2 (6.67) | 6(20) | 2(6.67) | 4 (13.33) | 2 (6.67) | 4 (13.33) | 10 (14.28) | 7 (10) |
| 56-65 | 60.5 | 4 (4.44) | 14 (15.56) | 1 (3.33) | 5 (16.67) | 1(3.33) | 5 (16.67) | 2 (6.67) | 4 (13.33) | 5 (7.14) | 1 (1.43) |
| Total | | 90 (100) | 90 (100) | 30 (100) | 30 (100) | 30 (100) | 30 (100) | 30 (100) | 30 (100) | 70 (100) | 70 (100) |

*Source:* Field Level Survey.
*Note:* Figure in brackets indicate percentage to total.

20% of them brought the same percentage of area under NF as they received loans. While about 7% of the sample households cultivated NF on 45% of the total cultivated area before they received loans from institutional sources, 20% of them cultivated NF on the same percent of their total cultivated land, implying thereby an increase by about 200% in the number of households (from 2 households to 6 households) opting for cultivation of NF on as much as 45% of their total cultivated area. Further, we see that there has been 40% increase in the number of households (from 1 household to 5 households) opting for cultivation of NF on as much as 60% of their total cultivated area, when they borrowed from various sources. The same pattern is also observed for the households receiving credit from non-institutional sources or from both.

Side by side, we have examined the pattern of change over time from 1992 to 1997 in the number of households receiving no credit and growing NF on some fixed percent of total cropped area. It is seen from Table 1 that the number of sample FHs receiving no credit from any sources and cultivating NF on a given percentage of area has, in general ,declined over these two time points. For example, while 10 sample FHs (column 11) cultivated NF on about 45% of their total cultivated land (column 2) in 1992, it was only 7 in 1997. Again, it is seen from the table that 5 FHs cultivated NF (column 11) on about 60% of their total cultivable land (column 2) in 1992. This number declined to only 1 in 1997.

Thus, it is seen from the above table that the availability of credit to the farmers has motivated them to cultivate NF on more and more areas out of their fixed land.

## Cropping Pattern and Profitability

Now the question that automatically arises is why the farmers go for cultivating more area under NF than the area under foodgrains. Profit per acre from NF *vis-à-vis* the profit per acre from foodgrains may be one of the motivating factors. To test this hypothesis a comparative analysis of profit per acre of foodgrains and NF for different sizes of land holdings is made here. Table 2 provides us with some data on this issue. It is seen from Table 2 that the profit per acre from NF for the sample households receiving no credit is estimated to be Rs. 57,375.00, while that from the foodgrains is only Rs. 4,625.00 only. That is, the profit per acre from NF is about 13 times the profit per acre from foodgrains. In the case of the sample FHs receiving credit from various sources, the profit per acre from the foodgrains is Rs. 5493.75 and the profit per acre from the NF is Rs. 67,625.00 (Table 3) which is also almost 13 times that from foodgrains. From the table we observe two important things.

**Table 2**
**Average Profit Per Acre Per Size Classes of Land Holdings for Farm Households without Credit**

| *Size of Land Holdings (acres)* | *No. of Households* | *Average profit per Acre from Non-Foodgrains (Rs.)* | *Average profit per Acre from Foodgrains (Rs.)* |
|---|---|---|---|
| Less than 1.0 | 8 | 42500.00 | 4000.00 |
| 1.0-Less than 2.0 | 20 | 60000.00 | 4500.00 |
| 2.0-Less than 3.0 | 20 | 62000.00 | 5000.00 |
| 3.0 & above | 22 | 65000.00 | 5000.00 |
| Total | 70 | 57375.00 (Av.) | 4625.00 (Av.) |

*Source:* Field Level Survey.

**Table 3**
**Average Profit Per Acre Per Size Class of Land Holdings for Farm Households with Credit**

| *Size of Land Holdings (acres)* | *No. of Households* | *Average profit per Acre from Non-Foodgrains (Rs.)* | *Average profit per Acre from Foodgrains (Rs.)* |
|---|---|---|---|
| Less than 1.0 | 29 | 73500.00 | 5900.00 |
| 1.0-Less than 2.0 | 43 | 70000.00 | 5675.00 |
| 2.0-Less than 3.0 | 11 | 65000.00 | 5200.00 |
| 3.0 & above | 7 | 62000.00 | 5200.00 |
| Total | 90 | 67625.00 (Av.) | 5493.75 (Av.) |

*Source:* Field Level Survey.

1. The profit per acre from NF for the 90 sample households with credit is observed to be higher than that from the NF by Rs. 10,250.00 for the 70 sample households without credit.
2. The profit per acre from foodgrains for the households with credit is higher than that from the foodgrains for the households without credit by Rs. 868.75 only.

Thus, several things emerge from these results.

1. The profit per acre from NF is significantly higher than that from foodgrains for all the sample households whether they use credit or no credit for cultivation of the crops.
2. The profit per acre from NF for the sample households using credit for cultivation is much higher than that

for the sample households using no credit for cultivation of these crops.

3. The profit per acre from foodgrains for the sample households using credit for cultivation is higher than that for the sample households using no credit for cultivation of these crops. This implies that the sample households receiving loans use the credit to buy and use more of modern inputs such as fertilizers, pesticides and insecticides, irrigation water, etc. on time and at the required amount, which increase the land productivity and the profit per acre. But the households who do not receive loans are not in a position to buy and use these inputs either on times or in required amount. As a result, the land productivity and, therefore, profit per acre have been lower for them in relation to those sample households receiving loans and using loans for increasing land productivity through the use of modern inputs.

Now we consider the differences in the profit earnings per acre from foodgrains and NF of the marginal, small, medium sample FHs. Let us first consider the profit earnings of the different categories of farmers who do not avail themselves of any credit facilities. It is seen from Table 2 that the profit per acre, whether from foodgrains or from NF, increases with the increase in the size of land holdings. For example, the profit per acre from NF increases steadily from Rs. 42,500.00 to Rs.65,000.00 as the land holding size increases from less than 1 acre to 3 acres or above. In the case of foodgrains the profit increases from Rs. 4000.00 per acre to Rs. 5000.00 per acre with the increases in land holding size from less than 1 acre to 3 acres or more. But the situations are completely reversed for the farmers who have used credit for cultivation of foodgrains and NF. There has been steady fall in the profit per acre for both NF and foodgrains. For example, in the case of NF, the marginal farmer's (less than one acre of land holdings) profit earning was the highest, i.e. , Rs. 73,500.00 per acre. Compared to this the profit earnings per acre for the farmers with land holdings of 2 acres and less than 3 acres and 3 acres and above were Rs. 65,000.00 and Rs. 62,000.00 respectively. In the case of foodgrains, the profit earnings of the marginal farmer was Rs. 5,900.00, and that of the small farmers was Rs. 5,675.00. Compared to this the profit earnings of the medium and large farmers were Rs. 52,000.00 only. Several things are worth mentioning from the above findings.

1. The profit earnings per acre from NF are several time higher than those from foodgrains.
2. The profit earnings per acre from NF for the FHs with credit are significantly higher, (Rs.68,000.00 approximately) than those for the FHs without credit (Rs. 57,000.00 approximately).
3. The availability of credit facility increases the financial strength of the marginal and small farmers more significantly than that of the FHs with 3 acres or more of land holdings, whether they produce foodgrains or NF.

This difference in the per acre profit between the small and marginal farmers, on the one hand, and medium and large on the other, whether with credit or without credit and whether from foodgrains or NF, may be explained in the following way.

1. The imputed costs, namely, costs of family labour, family bullocks, cost of OF, i.e., interest on OF, cost of family farm yard manures and cost of own supervision, etc., which constitute a significant part of the cost in the case of marginal and small farmers–are not included in the estimation of total production cost.
2. The marginal and small farmers do the supervision of their agricultural activities on the farm land more efficiently than the middle and large farmers.
3. Moreover, a large percentage of the FHs receiving credit (90) from different sources belong to marginal farmers (32.22) and small farmers(43.00). In other words, about 75% of the FHs receiving credit belong to marginal and small FHs groups (Table 4). As a

result, the average land productivity and profit per acre per FH are higher for the marginal and small farmers than for the medium farmers.

4. It is further seen that out of the total 37 marginal sample FHs with less than one acre of land holding, as many as 29 households, i.e. 78%, and of the total 63 small sample FHs with landholdings of 1 acre to less than 2 acres, as many as 43, i.e. 68% have received credit from various sources. In other words, out of total 100 marginal and small farmers as many as 72 households, i.e., 72% obtained credit. On the other hand, the percentage of medium FHs with operational land holdings of 2 acres or more receiving credit is comparatively much less. For example, of the total 60 FHs with land holdings of 3 acres and more, only eighteen, i.e., only 30% have received credit.

**Table 4**
**Distribution of Farm Households with and without Credit by the size of Land Holdings**

| | *Farm Households* | | *FH with Credit* | | | *FH without Credit* | | |
|---|---|---|---|---|---|---|---|---|
| *Size of Land Holdings (in acres)* | *No.* | *%* | *No.* | *% of the FHs with credit belonging to each class* | *% of the FHs belonging to each class* | *No.* | *% of the FHs without credit to each class* | *% of the FHs belonging to each class* |
| Less than 0.5 | 17 | 10.62 | 12 | 13.33 | 71.00 | 5 | 7.14 | 29.00 |
| 0.5- < 1.0 | 20 | 12.50 | 17 | 18.89 | 85.00 | 3 | 4.20 | 15.00 |
| 1.0 -< 1.5 | 27 | 16.87 | 23 | 25.56 | 85.00 | 4 | 5.71 | 15.00 |
| 1.5 -< 2.0 | 36 | 22.50 | 20 | 22.22 | 56.00 | 16 | 22.86 | 44.00 |
| 2.0 -< 3.0 | 31 | 19.37 | 11 | 12.22 | 36.00 | 20 | 28.57 | 64.00 |
| 3.0 &above | 29 | 18.12 | 7 | 7.78 | 24.00 | 22 | 31.43 | 76.00 |
| Total | 160 | 100 | 90 | 100 | 56 | 70 | 100 | 44.00 |

*Source:* Field Level Survey.

Thus, the availability of credit to most of the marginal and small FHs as against a negligible proportion of the FHs with larger sizes of land holdings and the use of family inputs in large quantities by the small and marginal farmers are some of the factors that explain, to a large extent, why profit per acre is relatively much higher for those farmers having smaller land holdings.

In this context, one question arise that if the profit per acre of non- food grains is larger than that of foodgrains, then why there is not a wholesale shift for the cultivation of non-food grains. It can be said that higher cost of cultivation for non-foodgrains (Table 4a), limited availability of land, capital etc., food security, fluctuation of price of foodgrains, risk aversion, non -availability of credit at the right time and in the right quantity, etc. are the factors inhibiting marginal and small

**Table 4(a)**

**Cost of Cultivation and Net Income of Different Crops in Midnapore**

| *Crops* | *Cost of Cultivation Per Acre(in Rs.)* | *Net Income per Acre (in Rs.)* |
|---|---|---|
| *Food grains paddy* | | |
| (a)Aman | 1811.64 | 3580.43 |
| (b) Boro(HYV) paddy | 3197.61 | 4200.23 |
| (c) Wheat | 1653.80 | 1200.55 |
| *Non food grains* | | |
| Jute | 1847.22 | 2011.36 |
| Potato | 7527.49 | 6229.79 |
| Mustard | 1187.60 | 990.29 |
| Sugar Cane | 2534.02 | 3326.38 |
| Gram | 381.16 | 985.29 |
| Lentil | 588.12 | 509.43 |
| *Betelvine* | | |
| (a) Mitha | 110000.00 | 150000.00 |
| (b) Bangla | 66000.00 | 70000.00 |
| Ground Nut | 2350.00 | 4290.00 |
| Til | 917.19 | 2258.53 |
| *Flower* | | |
| (a) Rose | 7000.00 | 30250.00 |
| (b) Tuberose | 9000.00 | 16280.00 |
| (c) Bela | 12000.00 | 25500.00 |
| (d)Gladiolus | 16000.00 | 26840.00 |
| (e) Marigold | 8000.00 | 20000.00 |
| (f) Jui | 17000.00 | 17500.00 |

*Source:* (1) Field level Study.
(2) Farm Management Survey, Govt. of West Bengal.

farmers to make a wholesale shift towards the non-foodgrains production. Besides, in case of medium and large farmers, the higher risk of production, agro-ecological suitability, storing and marketing problems, non-availability of credit at the right time and in the right quantity, etc. are also the factors inhibiting the wholesale shift towards the non-foodgrains production. In case of sample FHs one of the most important reasons of not wholesale shifting from foodgrain to non-foodgrains is mixed crop-livestock farming system which is widely prevalent across the country. To meet the fodder requirement of animals foodgrains are sine qua non in India as most of our livestock thrive on crop residues and crop by products.

## Changes in Cropping Pattern in Relation to Credit According to Size of Land Holdings

There may be some association between the total amount of credit from institutional source or the credit from non-institutional source or from both the sources and the OF, on the one hand and the ratio of area under NF to the area under foodgrains (NFA/FA) or the ratio of area under HYV paddy to the total area under foodgrains (HYVA/FA) on the other. It may be assumed that the greater the amount of credit per hectare, the greater will be capability of the FHs to buy more of modern agricultural inputs, such as, fertilizers, pesticides and insecticides etc., to create more irrigation facilities and to spend more on the hired labour in the form of wages. As a result, the FHs would cultivate more of those crops which require greater use of these inputs, but at the same time, would fetch more profit and/or ensure employment and/or food security. We test this hypothesis from the data collected from the sample FHs, which are presented below. However, the variables, mentioned above, have been presented according to the size of land holding (LH) since it is the land holding that is one of the most important factors determining the availability of credit per acre for most of the FHs. We have used the correlation analysis to test this hypothesis.

The correlation has been calculated between IC/(IC + OF) and NFA/FA and also between IC/(IC+OF) and HYVA/FA to verify the nature of association between these variables (Table 5). The correlation between IC/(IC+OF) and NFA/FA has been calculated to be 0.83 for those FHs who received IC and used them in cultivation. This correlation has been found to be statistically significant at 1% probability level (t = 5.96). But the correlation coefficient between IC/(IC+OF) and HYV paddy/total foodgrains i.e., HYVA/FA, has been found to be -0.43, which has been found to be statistically insignificant at even 10% probability level (t = - 0.95) (see Correlation Matrix- I). This implies that the IC has not been any significant motivation for the farmers to go for cultivation of more HYV paddy in relation to the total area under total foodgrains.

**Table 5**
**Changes in Cropping Pattern in Relation to Institutional Credit Availability According to Size of Land Holdings**

| *Size of Land Holdings (L.H.)(acres)* | *Mean (LH) (acres)* | *No. of House-holds* | *N.F.A./ F.A.* | *I.C./acre (Rs.)* | *I.C./ (IC+OF)* | *HYV paddy/ F.A.* |
|---|---|---|---|---|---|---|
| Less than 0.5 | 0.4 | 5 | 2.00 | 2500 | 0.55 | 0.50 |
| 0.5 -< 1.0 | 0.6 | 6 | 1.67 | 2100 | 0.34 | 0.67 |
| 1.0- < 1.5 | 1.0 | 8 | 1.50 | 3000 | 0.43 | 0.75 |
| 1.5 -< 2.0 | 1.4 | 5 | 1.33 | 3500 | 0.30 | 0.67 |
| 2.0- < 3.0 | 1.9 | 4 | 0.66 | 4100 | 0.29 | 0.50 |
| 3.0 & above | 2.3 | 2 | 0.66 | 4100 | 0.25 | 0.67 |
| Total | 1.27 | 30 | 1.32 | 3216.67 | 0.13 | 0.68 |

*Source:* Field Level Survey.

It is also observed that there exist negative correlations between size of the land holding and IC/(IC + OF) and between land holding size and NFA/FA and positive correlation between NFA/FA and IC/(IC+OF) for the FHs with IC only.

To check the nature of association between NIC and CP we have constructed another table (Table 6) showing the amount of NIC per acre, NFA/FA and HYVA/FA. The correlation between NIC/(NIC+OF) and NFA/FA is found to be 0.69 which is statistically significant at 5% probability level (t = 1.86). Similarly, the associations between the NIC/(NIC+OF) on the one hand and HYVA/FA, on the other, has been ascertained by means of correlation coefficient. The observation on the ratio of area under NF to the area under foodgrains corresponding to the ratio of NIC to (NIC+OF) for different land holding sizes are presented in Table 6. It is observed that the medium and large farm households get relatively large share of NIC and it may be an important factor behind the change in CP in favour of NF cultivation for the relatively large FHs. It is found from the correlation matrix-II that tnere exist positive correlations between LH and NIC/(NIC+OF) and between LH and NFA/FA and between NIC/(NIC+OF) and NFA/FA. It is to note that the values of all these correlation coefficients are statistically significant. So it may be argued that the non-IC has made relatively larger contribution to change in CP toward commercial crops cultivations for the relatively large cultivators. On the other hand, the correlation between (NIC)/(NIC+OF) and (HYVA/FA) is found to be negative (–0.22) for the FHs with NIC which is, however, not statistically significant (see Correlation Matrix-II).

**Table 6**

**Changes in Cropping Pattern in Relation to Non-IC Across the Different Size Classes of Land Holdings**

| *Size of Land Holdings (L.H.)(acres)* | *Mean (LH) (acres)* | *No. of House-holds* | *N.F.A./ F.A.* | *Non-I.C./ acre (Rs.)* | *Non-I.C./ (Non-IC+OF)* | *HYV paddy/ F.A.* |
|---|---|---|---|---|---|---|
| Less than 0.5 | 0.4 | 5 | 3.00 | 4000 | 0.44 | 0.50 |
| 0.5 -< 1.0 | 0.7 | 7 | 3.67 | 6500 | 0.45 | 0.87 |
| 1.0 -< 1.5 | 1.2 | 7 | 4.17 | 7200 | 0.42 | 0.71 |
| 1.5 -< 2.0 | 1.6 | 6 | 4.29 | 8500 | 0.77 | 0.67 |
| 2.0 -< 3.0 | 1.9 | 3 | 4.56 | 10000 | 0.74 | 0.50 |
| 3.0 & above | 2.5 | 2 | 5.67 | 11050 | 0.69 | 0.75 |
| Total | 1.59 | 30 | 1.96 | 47250 | 0.43 | 0.71 |

*Source:* Field Level Survey.

**Correlation Matrix-I**

**Correlation Results for Sample Households with IC and OF**

| | *L.H.* | *IC/IC+OF* | *NFA/FA* | *HYVA/FA* |
|---|---|---|---|---|
| LH | 1 | -0.81 | -0.97 | – |
| | – | (-2.78) | (-7.92) | – |
| IC/IC+OF | -0.81 | 1 | 0.83 | -0.43 |
| | (-2.78) | – | (5.96) | (-0.95) |
| NFA/FA | -0.97 | 0.83 | 1 | – |
| | (-7.92) | (5.96) | – | – |
| HYVA/FA | – | -0.43 | – | 1 |
| | – | (-0.95) | – | – |

(Figure in the parentheses indicate t values).

**Correlation Matrix-II**

**Correlation Results for Sample Households with NIC and OF**

| | *LH* | *NIC/NIC+OF* | *NFA/FA* | *HYVA/FA* |
|---|---|---|---|---|
| LH | 1 | 0.78 | 0.98 | – |
| | – | (2.49) | (9.80) | – |
| NIC/(NIC+OF) | 0.78 | 1 | 0.69 | -0.22 |
| | (2.49) | – | (1.86) | (-0.44) |
| NFA/FA | 0.98 | 0.69 | 1 | – |
| | (9.80) | (1.86) | – | – |
| HYVA/FA | – | -0.22 | – | 1 |
| | – | (-0.44) | – | – |

(Figure in the parentheses indicate t values).

**Correlation Matrix-III**

**Correlation Results for Sample Households with Institutional and Non-Institutional Credit and OF**

| | *LH* | *NIC+IC/ NIC+IC+OF* | *NFA/FA* | *HYVA/FA* |
|---|---|---|---|---|
| LH | 1 | -0.89 | -0.79 | – |
| | – | (-3.88) | (-2.56) | – |
| NIC+IC/ | -0.89 | 1 | 0.96 | 0.63 |
| (NIC+IC+OF) | (-3.88) | – | (6.96) | (1.63) |
| NFA/FA | -0.79 | 0.96 | 1 | – |
| | (-2.56) | (6.79) | – | – |
| HYVA/FA | – | 0.63 | – | 1 |
| | – | (1.63) | – | – |

(Figure in the parentheses indicate t values).

**Correlation Matrix-IV**
**Correlation Results for Sample Households without any Credit**

| | *LH* | *OF/acre* | *NFA/FA* | *HYVA/FA* |
|---|---|---|---|---|
| LH | 1 | 0.99 | 0.96 | – |
| | – | (14.00) | (6.79) | – |
| OF/acre | 0.99 | 1 | 0.94 | 0.56 |
| | (14.00) | – | (5.43) | (1.35) |
| NFA/FA | 0.96 | 0.94 | 1 | – |
| | (6.79) | (5.43) | – | – |
| HYVA/FA | – | 0.56 | – | 1 |
| | – | (1.35) | – | – |

(Figure in the parentheses indicate t values).

We have tried to find out the relation between the credit from institutional and non-institutional sources as a proportion of the total fund used for cultivation by the sample FHs and the CP as measured by (1) ratio of NFA to FA and (2) the ratio of HYVA to FA for the sample FHs classified according to land holding size. The data on these variables are presented in Table-7. The associations between (1) (IC + NIC)/(IC + NIC + OF) and NFA / FA and (2) ( IC + NIC)/(IC + NIC + OF) & HYVA / FA have been measured by means of correlation coefficient. It is seen that the correlation coefficient for (IC + NIC)/(IC + NIC + OF) and NFA/FA is 0.96 which is statistically significantly different from zero at 1% probability level. The correlation coefficient for (IC + NIC)/(IC + NIC + OF) and HYVA/FA is found to be 0.63. The correlation matrix-III also shows that there exist negative correlation between LH and (IC + NIC)/( IC + NIC + OF ) and between LH and NFA/FA, while there is positive correlation between (IC + NIC)/(IC + NIC + OF) and NFA/FA.

It is found from the values of r that there exist an inverse relationship between size of LHs and the availability of IC, but a positive relationship between size of LHs and availability of NIC. Let us explain the reasons behind these relationships. Due to the factors like larger participation of family labour, closer supervision, etc., the cost of cultivation of smaller FHs is lower as compared to the larger FHs. On the other hand, due to the priority sector lending policy and/or pro active bank policy the share of small holders in total institutional loans be higher. The factors like inability to provide tangible collaterals, exorbitant rate of interest, larger provision of IC, lower cost of cultivation, higher risk of repayment, etc. have influenced smaller FHs to take smaller amount of loans from non-institutional sources. But the factors like ability to provide tangible collaterals, higher cost of cultivation, non-availability at the right time and of the required amount of IC, higher transaction cost of IC, etc. have influenced larger FHs to receive larger amount of loans from non-institutional sources. Under these circumstances, there exist an inverse relationship between size of LHs and the availability of IC and a positive relationship between size of LHs and availability of NIC. On the other hand, in case of FHs with either IC or IC+NIC, an inverse relationship between size of LHs and change in CP in favour of non-foodgrains is observed. Thus, the credit has influenced change in CP significantly in case of lower size of LHs. But that relationship is the opposite for FHs with only NIC. Let us explain the reasons behind these relationships. In case of FHs with either IC or IC+NIC the factors like larger participation of family labour, closer supervision, more relaxation of financial constraint through larger availability of IC and subsidised interest rate for IC, lower cost of cultivation, etc. have helped smaller FHs to make larger change in CP in favour of non-foodgrains. On the other hand, in case of FHs with NIC only the factors like ability to provide tangible collaterals, availability at the right time and of the required amount of NIC, more relaxation of financial constraints through larger availability of NIC, lesser risk of repayment from the point of view of lenders, etc. have helped larger FHs to make larger change in CP in favour of non-foodgrains. Again there exists a positive relationship between credit (IC or NIC or both) availability and change in CP in favour of non-foodgrain cultivation. But the credit availability has made significant contribution on change in CP in favour of HYV cultivation only in case of FHs getting credit from both sources. The joint influence of IC and NIC has helped the sample farmers to change CP in favour of HYV cultivation.

It is observed from the field level study that the financial constrant in case of FHs getting IC or NIC only is the factor inhibiting the farmers to change CP in favour of cultivation of HYV crops. Thus, the credit availability has changed the CP of farmers , particularly, small and marginal ones.

**Table 7**

**Changes in Cropping Pattern in Relation to IC and Non-IC Across the Different Size Classes of Land Holdings**

| *Size of Land Holdings (L.H.)(acres)* | *Mean (LH) (acres)* | *No. of House-holds* | *N.F.A./ F.A.* | *I.C.+Non-I.C/acre (Rs.)* | *I.C.+N.I.C./ (IC+N.I.C.+ OF)* | *HYV paddy/ F.A.* |
|---|---|---|---|---|---|---|
| Less than 0.5 | 0.42 | 2 | 3.00 | 6500 | 0.76 | 0.75 |
| 0.5 -< 1.0 | 0.7 | 4 | 2.00 | 7500 | 0.65 | 0.83 |
| 1.0 -< 1.5 | 1.3 | 8 | 2.00 | 9700 | 0.66 | 0.70 |
| 1.5- < 2.0 | 1.5 | 9 | 1.64 | 10800 | 0.56 | 0.65 |
| 2.0- < 3.0 | 2.0 | 4 | 1.50 | 10700 | 0.55 | 0.75 |
| 3.0 & above | 4.0 | 3 | 1.25 | 12500 | 0.47 | 0.60 |
| Total | 1.6 | 30 | 1.60 | 57700 | 0.57 | 0.67 |

*Source:* Field Level Survey.

## Changes in Cropping Pattern in relation to Own Fund According to Size of Land Holdings

In case of FHs without credit it is observed that there exists a positive correlation between LH and OF/acre; LH and NFA/ FA, and between OF/acre and NFA/FA (Table 8 and Correlation Matrix-IV). The values of these correlation coefficients are statistically significant at different levels of significance.

The values of the correlation coefficients and the data presented in Table 8 clearly show that as the land holding size increases the amount of OF used per acre also increases, which has contributed to the increase in area under NF as a proportion of the area under foodgrains. In other words, it means that the FHs other than marginal and small FHs cultivate NF on greater percentage of total cultivated area with their OF. But, as we have seen earlier, the profit earnings per acre of these relatively large FHs are not as high as those of the marginal and small FHs. One explanation for this relatively lower profit per acre for the relatively large FHs may be the non-availability of sufficient fund for use of modern inputs on their relatively larger land holding. Their greater interest perhaps lying elsewhere, for example, in the more profitable non-agricultural activities, they do not borrow or a sample percentage of them borrow for cultivation purposes and remain satisfied with the amount of profit that they obtain from the use of their own fund in the cultivation of NF.

**Table 8**
**Changes in Cropping Pattern in Relation to OF Across the Different Size classes of Land Holdings**

| *Size of Land Holdings (L.H.)(acres)* | *Mean (LH) (acres)* | *No. of House-holds* | *N.F.A./ F.A.* | *O.F./acre (Rs.)* | *HYV paddy /F.A.* |
|---|---|---|---|---|---|
| Less than 0.5 | 0.4 | 5 | 1.50 | 8500 | 0.75 |
| 0.5- less than1.0 | 0.7 | 3 | 2.00 | 11500 | 0.80 |
| 1.0- < 1.5 | 1.2 | 4 | 2.50 | 14700 | 0.82 |
| 1.5 -< 2.0 | 1.5 | 16 | 2.73 | 19300 | 0.80 |
| 2.0 -< 3.0 | 2.0 | 20 | 2.93 | 22200 | 0.82 |
| 3.0 & above | 2.2 | 22 | 3.00 | 26500 | 0.80 |
| Total | 1.59 | 70 | 2.78 | 102700 | 0.81 |

*Source:* Field Level Survey.

However, our above findings show that the credit, whether from institutional or non-institutional sources, has played a big role in improving the financial conditions of all categories of FHs, by inspiring them to cultivate those crops that ensure maximum income (profit) as in the case of NF, and/or the crops that guarantee food security, as in the case of HYV paddy in relation to foodgrains. However, among all categories of FHs, it is the marginal and small FHs a large percenatge of whom avail themselves of credit facilities and who have earned the highest profit (income) per acre by making least use of credit facilities.

Thus, the credit facility available to the FHs does guarantee the maximum profit earnings and food security to the small and marginal farmers and gives a boost to the realization of the economic objective of equal distribution of income among the FHs.

**Different Types of Credit and Cropping Pattern-A Regression Analysis with Dummy Variables**

In the following paragraphs the statistical tests of the hypotheses regarding the relation between total credit and CP as well as between different types of credit, taken separately, and CP are carried out.

To estimate the relation between credit as a whole, i.e., the total credit from all sources and CP, measured by the ratio of area under NF to the area under foodgrains, we formulate the following model:

$$Y_i = \beta_1 + \beta_2 X_{i2} + \xi_i \qquad (i = 1, \ldots, n)$$

$Y_i$ = the percentage of area under NF to the area under all crops for the households.

$X_{i2}$ = 1 if $Y_i$ corresponds to the households with credit.

= 0 otherwise, i.e., if $Y_i$ corresponds to the households without credit.

$\xi_i$ = disturbance term satisfying all the assumptions of classical normal regression equations.

$\beta_2$ measures the difference in the percentage of area under NF for the FHs with credit and the FHs without credit. Hence, the testing of the hypothesis $H_0 : \beta_2 = 0$ against the alternative hypothesis $H_1 : \beta_2 \neq 0$ is to be carried out to test if the difference is statistically significantly different from zero. To test the hypothesis we estimate the above regression equation. The estimated equation is:

$$\underset{(49.56)}{Y_i = 37.43} + \underset{(19.41)}{19.54}\, X_{i2} + \xi_i \qquad (i = 1, \ldots, n \text{ households})$$

$R^2 = 0.71$, $\bar{R}^2 = 0.70$, N = 160, F = 376.85. (Figure in the parentheses are t values).

The t value corresponding to $\beta_2$ is seen to be greater than the tabled value of t at 1% probability level. Hence, the alternative hypothesis that $\beta_2 \neq 0$ is accepted at 1% probability level. This implies that the effect of credit on the percent of area under NF to the area under all crops is statistically significant at 1% probability level.

We now make an attempt to statistically measure and test the impact of different types of credit on the CP in the study area. The FHs have three sources of credit, viz., OF , IC and NIC. However, the FHs may avail themselves of any combination of these three sources. Considering four possible sources of loans, namely, (1) OF, (2) OF and IC, (3)OF and NIC and (4) OF, IC and NIC, we develop a regression model to measure the effect of different sources of credit on the change in CP for sample FHs and to test the statistical significance of these effects. The model is

$$Y_i = \alpha_1 + \alpha_2 D_{i2} + \alpha_3 D_{i3} + \alpha_4 D_{i4} + \xi_i$$

Where

$D_{i2} = \{$ 1 if $Y_i$ corresponds to FHs with IC and OF.
0 otherwise.

$D_{i3} = \{$ 1 if $Y_i$ corresponds to FHs with NIC + OF.
0 otherwise.

$D_{i4} = \{$ 1 if $Y_i$ corresponds to FHs with IC + NIC + OF.
0 otherwise.

From the above model we find

$E(Y_i \mid D_{i2} = 0, D_{i3} = 0, D_{i4} = 0) = \alpha_1$

$E(Y_i \mid D_{i2} = 1, D_{i3} = 0, D_{i4} = 0) = \alpha_1 + \alpha_2$

$E(Y_i \mid D_{i2} = 0, D_{i3} = 1, D_{i4} = 0) = \alpha_1 + \alpha_3$

$E(Y_i \mid D_{i2} = 0, D_{i3} = 0, D_{i4} = 1) = \alpha_1 + \alpha_4$

$\alpha_1 \rightarrow$ mean percentage of area under NF to the area under all crops when the households use own fund for cultivation.

$\alpha_1 + \alpha_2 \rightarrow$ mean percentage of area under NF to the area under all crops when the households use OF and IC (Di2 = 1, others zero).

$\alpha_1 + \alpha_3 \rightarrow$ mean percentage of area under NF to total area under all crops when the households use OF and NIC, (Di3 = 1, others zero) and

$\alpha_1 + \alpha_4 \rightarrow$ mean percentage of area under NF to the area under all crops when the households use OF, IC as well as NIC for cultivation of crops (Di4 = 1, others zero).

Therefore, $\alpha_1$ measures the effect of OF on Yi ,$\alpha_2$ measures the difference in the effect of IC+OF over OF on Yi. Similarly, $\alpha_3$ measures the difference in the effect of NIC+ OF over OF on Yi and $\alpha_4$ measures the difference in the effect of IC+NIC+OF over OF on Yi.

To measure the statistical significance of $\alpha_1$ , $\alpha_2$ , $\alpha_3$ and $\alpha_4$ the following hypotheses are formulated.

$$H_0 : \alpha_1 = 0$$
$$H_1 : \alpha_1 \neq 0 \qquad (1)$$

$$H_0 : \alpha_2 = 0$$
$$H_1 : \alpha_2 \neq 0 \qquad (2)$$

$$H_0 : \alpha_3 = 0$$
$$H_1 : \alpha_3 \neq 0 \qquad (3)$$

$$H_0 : \alpha_4 = 0$$
$$H_1 : \alpha_4 \neq 0 \qquad (4)$$

Further, (i) to test whether the effect of (OF+IC) on $Y_i$ is equal to the effect of (OF+NIC) on $Y_i$ we formulate the hypotheses:

$$H_0 : \alpha_2 = \alpha_3$$
$$H_1 : \alpha_2 \neq \alpha_3 \qquad (5)$$

(ii) To test whether the effect of (OF + IC + NIC) taken together on $Y_i$ is equal to that of (OF+IC), we formulate the hypotheses.

$$H_0 : \alpha_2 = \alpha_4$$
$$H_1 : \alpha_2 \neq \alpha_4 \qquad (6)$$

(iii) To test whether the effect of (OF + IC + NIC) taken together on $Y_i$ is equal to that of (OF+NIC), the following hypothesis can be tested.

$$H_0 : \alpha_3 = \alpha_4$$
$$H_1 : \alpha_3 \neq \alpha_4 \qquad (7)$$

All the tests (1) to (7) can be carried out by t test. While $t = \alpha_i / s\alpha_i$ (i = 1, 2, 3 and 4) for the hypotheses (1) to (4) for the hypotheses (5) to (7)

$$t = \frac{\alpha j - \alpha k}{s\,(\alpha j - \alpha k)}$$

where $E(\alpha_j - \alpha_k) = \alpha_j - \alpha_k$ and

$$var\ (\alpha_j - \alpha_k) = var\ (\alpha_j) + var\ (\alpha_k) - 2cov\ (\alpha_i\ \alpha_j)$$

$$j, k, = 2,3 \text{ and } 4 : j > k.$$

The estimated regression equation is given by

$$Y_i = 48.98 + 3.28D_{i2} + 5.34D_{i3} + 13.81D_{i4}$$

$$(103.87)\ (3.81)\ (6.21)\ (16.05)$$

$$R^2 = 0.63,\ R^2 = 0.62,\ F = 87.24,\ N = 160.$$

(Figures in the parentheses are t values).

The t values for the hypotheses (5) , (6) and (7) are calculated to be $t_5 = 2.04$, $t_6 = 9.49$ and $t_7 = 7.0$ . On the basis of the t values we accept the null hypotheses (5), (6) and (7). From the t values of the estimated parameters $\alpha_2$, $\alpha_3$ and $\alpha_4$ it is seen that the differences between the effects of OF and (OF+IC), OF and (OF+NIC) and OF and (OF +IC+NIC) are all statistically significant at less than 1% probability level. All of them have positive influences on $Y_i$, the percentage of area under NF to the area under all crops. Further, it is observed that NIC has greater and statistically significant influence on $Y_i$ than IC and (IC+NIC) has much greater and statistically significant influence on $Y_i$ than IC or NIC taken separately.

Thus, from the above results of statistical analysis, the following observations can be made.

1. The credit, whether institutional or non-institutional, has played a significant role in changing the C.P. from foodgrains to more profit earning NF.
2. When (OF+NIC) has been supplemented by IC or (OF+IC) has been supplemented by NIC, their joint effect has been statistically significantly greater than that of (OF+IC) or (OF+NIC) alone.

## Credit and Cropping Pattern in Different Agro-Climatic Zones

Let us examine the impact of credit on the change in CP and other related issues in the individual sample blocks in the district under study. Each sample block has certain distinctly different agro-climatic features and, because of these differences, the blocks have specialized in the cultivation of those NF that suit the agro-climatic conditions obtaining in that block. For example, Sahid Matangini and Tamluk blocks are more suited for cultivation of betelvine, some parts of Panskura-I block have specialized in floriculture and some parts grow potato, jute and vegetable on large areas. Garbeta-I block produces potato as the main NF since the soil in the block is conducive to the production of potato. Some features common to these agro-climatically different zones need to be mentioned here. In case of the sample blocks, it is observed that for the sample households enjoying credit facilities the percentage of area under NF (viz., betelvine, flowers, potato, etc.) has increased in the post-loan period as compared to pre-loan period. In other words, it may be concluded that the credit (viz., IC, NIC or both) has a positive role in effecting change in CP in all the agro-climatic zones of Midnapore district (Table 9, 10, 11). On the other hand, in case of cultivators having no credit facilities, though the percentage of area under NF has increased in 1997 as compared to 1992, that increase is not so significant as in the case of cultivators with credit i.e., the rate of change in CP is lower for the households without credit than that for the households with credit in different agro-climatic zones of Midnapore district.

**Table 9**
**Frequency Distribution of Farm Households (FHs) by Percentage of Area Under Non-foodgrains in Saline Zone**

| | *FHs with Credit* | | *FHs without credit* | |
|---|---|---|---|---|
| *Percentage of Area under Non-foodgrains* | *Pre-loan period* | *Post-loan period* | *1992* | *1997* |
| 0 | 12 (60) | 0 (0) | 4 (20) | 4 (20) |
| 1-5 | 2 (10) | 9 (45) | 4 (20) | 3 (15) |
| 6-15 | 1 (5) | 3 (15) | 5 (25) | 7 (35) |
| 16-35 | 3 (15) | 4 (20) | 5 (25) | 4 (20) |
| 36-55 | 2 (10) | 4 (20) | 2 (10) | 2 (10) |
| 56-65 | 0 (0) | 0 (0) | 0 (0) | 0 (0) |
| Total | 20 (100) | 20 (100) | 20 (100) | 20 (100) |

*Note:* Figure in brackets indicate percentages to total.
*Source:* Field Level Survey.

In the following paragraphs we propose to test the hypotheses, outlined at the outset of this chapter, for individual blocks. For this purpose, the change in CP measured in terms of percentage of area under NF to GCA, has been regressed upon the amount of credit received by the farm households (Table-12) and also the correlation coefficients have been calculated and statistically tested, whenever required. We have also made tabular analysis in some cases.

**Table 10**

**Frequency Distribution of Farm Households (FHs) by Percentage of Area under Non-foodgrains in Laterite Zone**

| *Percentage of Area under Non-foodgrains* | *FHs with Credit* | | *FHs without credit* | |
|---|---|---|---|---|
| | *Pre-loan period* | *Post-loan period* | *1992* | *1997* |
| 0 | 5 (25) | 0 (0) | 4 (20) | 2 (10) |
| 1-5 | 3 (15) | 5 (25) | 2 (10) | 3 (15) |
| 6-15 | 8 (40) | 9 (45) | 15 (25) | 9 (45) |
| 16-35 | 3 (15) | 4 (20) | 5 (25) | 5 (25) |
| 36-55 | 1 (5) | 2 (10) | 2 (10) | 1 (5) |
| 56-65 | 0 (0) | 0 (0) | 2 (10) | 0 (0) |
| Total | 20 (100) | 20 (100) | 20 (100) | 20 (100) |

*Note:* Figure in brackets indicate percentages to total.
*Source:* Field Level Survey.

**Table 11**

**Frequency Distribution of Farm Households (FHs) by Percentage of Area under Non-foodgrains in Alluvial Zone**

| *Percentage of Area under Non-foodgrains* | *FHs with Credit* | | *FHs without credit* | |
|---|---|---|---|---|
| | *Pre-loan period* | *Post-loan period* | *1992* | *1997* |
| 0 | 7 (14) | 0 (0) | 0 (0) | 0 (0) |
| 1-5 | 17 (34) | 1 (2) | 6 (20) | 8 (26.27) |
| 6-15 | 10 (20) | 11 (22) | 10 (33.33) | 12 (40) |
| 16-35 | 9 (18) | 16 (32) | 5 (16.67) | 5 (16.67) |
| 36-55 | 3 (6) | 8 (16) | 6 (20) | 4 (13.33) |
| 56-65 | 4 (8) | 14 (28) | 3 (10) | 1 (3.33) |
| Total | 50 (100) | 50 (100) | 30 (100) | 30 (100) |

*Note:* Figure in brackets indicate percentage to total.
*Source:* Field Level Survey.

For Panskura-I block, a part of which produces flowers of different varieties and another part produces potato, jute and vegetable as NF, the following observations can be made.

1. All farmers with relatively small size of land holding allocate larger share of cultivable land for the NF cultivation (see Table 2).
2. The credit has been as important factor determining the change in CP away from foodgrains towards the NF, viz., flowers, vegetables, potato. (The regression coefficients of credit variable is statistically significant at 1% probability level.)
3. The expanding NF cultivation has made statistically significant impact on the profitablity of all categories of sample FHs in this block.
4. Side by side, the change in CP has also made significant impact on the employment generation (The regression coefficients of change in CP is statistically significant at 5% probability level.)

Similarly, for Tamluk and Sahid Matangini blocks where betelvine is the principal NF the relations between (1) size of land holdings and CP, (2) credit and CP, (3) credit and profit earned from cultivation of these crops and (4) employment and the area under NF have been found to be the same as in the case of Panskura-I.

For Garbeta-I block which grows potato as the main NF the findings on the above relations tally with those for Sahid Matangini and Tamluk and Panskura-I.

A common thing that emerges from the above analysis is that the change in CP largely explains higher levels of profit for the farm households enjoying credit facilities as compared to those without credit.

We have discussed in the above paragraphs how the availability of credit has played a catalytic role in effecting change in the CP by bringing more and more areas under cultivation of higher income generating NF.

## A Chain of Multiplier Effects of Credit–A Village Level Analysis

In the following paragraphs we discuss the secondary effect of use of credit on the cultivation of various foodgrains and NF that are grown by the farmer just after the crop or crops grown with the direct help of credit are harvested. For instance, in Maguri village under Panskura-I block, the cultivators used credit for potato cultivation or boro paddy cultivation. A part of net income earned from potato cultivation that was made possible with loan is reinvested for jute or vegetables or oilseeds cultivations. Again, a part of net income earned from the cultivation of jute, or vegetables or oilseeds (viz., mug) is further re-invested for the cultivation of paddy. Similarly, in Govindanagar and Kaya villages under Panskura-I block it is observed that the credit is primarily invested in the cultivation of flowers like rose or gladiolus and then a part of net income earned from floriculture is reinvested for the cultivation of either paddy or vegetables. In Lapuria village under Garbeta-I block the cultivators invest credit mainly for the cultivation of potato and then a part of net income earned from potato cultivation is reinvested for the cultivation of paddy. In Chitra and Harasankar villages it is found that the cultivators use credit mainly for betelvine cultivation and then a part of net income earned from betelvine cultivation is reinvested in the cultivation of paddy, particularly, boro paddy. Therefore, it may be concluded that, the provision of credit to the farmer has created a "chain of multiplier effects" which helps farmers to initiate multiple cropping round the year using HYV technology, wherever possible.

## Various Types of NIC and their Relative Effects on Economic Conditions of the Farmers

Let us now discuss the different types of NIC and their relative merits and/or demerits. The sample farmers obtain NIC in cash from rural money lenders, or mahajans, aratdars, commodity traders and in kind from fertiliser and pesticides traders, suppliers of water for irrigation, ploughing agencies, etc. for accelerating agricultural growth through change in CP. The loans from rural money lenders or mahajans generate interest cost, but the loans from others do not generate such interest cost since these authorities provide loans to farmers to serve their business interest. Therefore, the loans provided by aratdars, fertiliser and pesticides suppliers, etc. (except the loans from mahajans or rural money lenders) may reduce the cost of production, which, in turn, increases the profit per acre. Again, the loans from fertiliser and pesticide traders, aratdars, etc. benefits largely the very small or marginal farmers who have very limited scope of getting NIC from mahajans or money lenders and hence such type of loans creates benefits to the poorest of the poors. The timely availability and lower transaction cost of NIC also help farmers to maximise profit. Side by side, this process of provision of credit has created an interlocking of different markets, namely, product market, credit market and input market. In such cases some types of 'deprivation' of farmers (by aratdars ,fertiliser and pesticide traders, etc.) and perpetual debt condition of farmers are also noticed. For example, in case of betelvine, the aratdars or traders do not pay the full amount of sale value at one time to the farmers and so the farmers become unable to meet their cash requirements which is necessary to purchase either different agricultural inputs or their family consumption goods. As a result, the FHs are obliged to take loans either in cash or in kind from different sources like the traders of agricultural inputs, traders of consumer goods. This phenomenon forces the cultivators to fall into a perpetual debt condition. In the betelvine growing areas it has been found that most of the betelvine cultivators have been trapped into this type of perpetual debt conditions. Again, sometimes, it is also observed that the betelvine aratdars pay less than the contracted amount to the farmers showing the reason that the sold baskets of betel leaves contain many damaged betel leaves (which is often not true) and in this way these aratdars (who have supplied loans to the cultivators and to whom the cultivator is obliged to sell the product) deprive the cultivators of their dues. A perpetual debt condition is observed in the case of floriculture too, though a relatively lesser extent. The exorbitant interest rate and perpetual debt condition have

**Table 12**
**Estimated Regression Equations for Different Agroclimatic Zones in Midnapore District**

| *Alluvial Zone : Panskura-I : Govindanagar & Kaya villages (floriculture as main crop enterprise)* | *Fit* | | | $R^2$ |
|---|---|---|---|---|
| **FHs with credit** | | | | |
| | NFA | = 55.94 - | 5.47* LH (-2.31) | 0.29 |
| | NFA | = 10.07 + | 0.011** Tcr (4.95) | 0.65 |
| | GP | = 3204.08 + | 167.47** NFA (11.75) | 0.91 |
| | EMP | = 189.69 + | 4.56* NFA (2.24) | 0.28 |
| **FHs without credit** | | | | |
| | NFA | = 53.68 - | 4.35** LH (-1.84) | 0.21 |
| | GP | = 3296.32 + | 169.16** NFA (10.04) | 0.89 |
| | EMP | = 291.16 + | 458* NFA (2.34) | 0.29 |

| *Alluvial Zone : Panskura-I : Maguri Jagannathchak (Potato & Jute as main crop enterprise)* | *Fit* | | | $R^2$ |
|---|---|---|---|---|
| **FHs with credit** | | | | |
| | NFA | = 53.65 - | 4.20* LH (-2.68) | 0.18 |
| | NFA | = 10.45 + | 0.01** Tcr (4.46) | 0.60 |
| | GP | = 2926.39 + | 177.94** NFA (9.47) | 0.87 |
| | EMP | = 176.93 + | 4.95* NFA (2.64) | 0.38 |
| **FHs without credit** | | | | |
| | NFA | = 58.03 - | 5.21*** LH (-1.93) | 0.22 |
| | GP | = 2751.66 + | 184.77** NFA (8.49) | 0.47 |
| | EMP | = 162.44 + | 5.49* NFA (2.76) | 0.37 |

*Contd...*

| *Alluvial Zone : Tamluk-II : Chitra & Harasankar (Betelvine as main crop enterprise)* | *Fit* | | | $R^2$ |
|---|---|---|---|---|
| F.H.s with credit | NFA | = 53.65 - | 4.20* LH | 0.18 |
| | | | (-2.68) | |
| | NFA | = 16.39 + | 0.01** Tcr | 0.62 |
| | | | (4.57) | |
| | GP | = 58.85 + | 0.01** NFA | 0.64 |
| | | | (4.79) | |
| | EMP | = 189.69 + | 4.56* NFA | 0.28 |
| | | | (2.24) | |
| FHs without credit | | | | |
| | NFA | = 53.68 - | 4.35*** LH | 0.21 |
| | | | (-1.84) | |
| | GP | = 70.87 + | 0.008** NFA | 0.44 |
| | | | (3.22) | |
| | EMP | = 191.95 + | 4.52* NFA | 0.34 |
| | | | (2.36) | |

| *Laterite Zone : Garbeta-I : Lapuria (Potato as main crop enterprise)* | *Fit* | | | $R^2$ |
|---|---|---|---|---|
| FHs with credit | | | | |
| | NFA | = 53.65 - | 4.20* LH | 0.18 |
| | | | (-2.68) | |
| | NFA | = 11.29 + | 0.01** Tcr | 0.59 |
| | | | (4.31) | |
| | GP | = 34.19 + | 162.86** NFA | 0.93 |
| | | | (12.86) | |
| | EMP | = 191.95 + | 4.52** NFA | 0.30 |
| | | | (2.36) | |
| FHs without credit | | | | |
| | NFA | = 53.96 + | 4.81*** LH | 0.14 |
| | | | (-1.44) | |
| | GP | = 60.43 + | 0.01** NFA | 0.60 |
| | | | (4.68) | |
| | EMP | = 191.21 + | 4.55** NFA | 0.32 |
| | | | (2.46) | |

*Notes:* (i) NFA = % of area under non-foodgrains to GCA (NFA), LH = size of land holdings, Tcr = total credit per acre, GP=Gross Profit Per acre

(ii) figures in brackets indicate t ratios

(iii) *, **, *** indicate level of significance at 5, 1 & 10% level respectively.

*Source:* Field Level Survey.

created some detrimental effects on economic condition of sample FHs. Due to the interlocking of markets in the study area, the concentration of economic power is seen and as a result a new rural elite class (other than big farmers) who are not landlord, but lord of water, fertiliser, etc. is also found.

### A Summing-Up

The whole analysis reveals that the credit availability from both institutional and non-institutional sources has made a significant contribution on change in CP. But the impact of credit availability on CP change has been more significant in case of smaller size of LHs. Again, the profitability is also higher in case of small and marginal farmers. The closer supervision for cultivation, availability of credit, exclusion of cost of family inputs, viz., family labour, etc. are the factors behind this higher profitability. The profit per acre from non-foodgrains cultivation is larger than that from foodgrains. Despite this higher profitability, the factors like food security, higher cost of cultivation, non-availability of credit in right time and in right quantity, mixed crop-livestock farming system, etc. are hindering the wholesale shift of CP towards non-foodgrains cultivation. When OF+NIC has been supplemented by IC or OF+IC has been supplemented by NIC, their joint effect has been statistically significantly greater than that of OF+IC or OF+NIC alone. All farmers with relatively small size of LHs and credit allocate larger share of cultivable land for non-foodgrains cultivation. The expanding non-foodgrains cultivation has made statistically significant impact on the profitability of all categories of farmers. Side by side, the change in CP has also made significant impact on the employment generation. In the study area, NIC also plays an important role to change CP. It is observed that the provision of credit to the farmers has created a "chain of multiplier effects". The deprivation of farmers and a new rural elite class are also observed in the study area.

# CHAPTER–11

## SUMMARY AND CONCLUSION

In this concluding chapter, we summarise our work and state the main findings. The change in CP has an important role in accelerating agricultural growth. A rapid change in CP has occurred in different states of India during the post-green revolution period. This change in CP is influenced by several natural, historical, economic, social and policy factors. Among those factors, the finance is an important one. On the other hand, the nationalisation of commercial banks has made possible a massive inflow of agricultural credit to the farmers, particularly the poor, in the country. In this study, we have set out to examine the impact of credit, in particular, IC, on the change in CP influencing agricultural growth in India. To fulfill this objective, we have examined (i) the nature and pattern of agricultural growth and the relative role of CP on agricultural growth, (ii) the nature of change in CP, (iii) the trend of IC, (iv) the impact of credit, particularly IC, on the factors like irrigation, fertilizer, (v) the impact of several factors including credit, in particular IC, on the change in CP and also on agricultural growth and (vi) the economics of change in CP in relation to credit availability. To examine these issues we have applied different statistical techniques, viz., regression analysis, correlation analysis, hypothesis testing, CV analysis, growth rate analysis, Chow test, dummy variable model, etc. Both the primary and secondary data, used in this study covering India as a whole, twelve sample states, fifteen districts of W.B. and 54 blocks of Midnapore, have been collected from various published sources. Some villages in the district of Midnapore have also been selected for a household level study on the issues mentioned above. The present study emphasizes the relative change in CP highlighting the pro-superior cereals and pro-foodgrains CP change, pro-non-foodgrains CP change and pro-HYV CP change. Let us now summarise the major findings of our study in the following paragraphs.

First, it is found that the total agricultural output in India at constant 1990-93 prices has increased at a CAGR of 2.71% in the period 1962-65 to 1992-95. There were important changes in the pattern of agricultural development during the different sub-periods. The period from 1970-73 to 1980-83 is characterized by the extension of new seed fertilizer technology from wheat to rice and it spread from Punjab and Haryana to eastern U.P. as well as the rice-producing states in southern regions. The period from 1980-83 to 1990-95 marks a turning point in India's agricultural development with respect to the growth of agricultural output. The noticeable feature of the 1980s was that agricultural growth penetrated to all the regions in India. The most remarkable development was the notable acceleration of growth in the eastern region. Specifically creditable was the performance of W.B. where the growth rate increased to an unprecedented rate of 5.39% per annum. The CAGRs of agricultural production have been observed to be significant in India and the sample states except in H.P. and Gujarat during 1972-73 to 1994-95. Turning our attention to the growth of agriculture in districts of W.B. we find that the agricultural growth is significant for all the districts during 1972-73 to 1996-97. The acceleration in agricultural growth is observed in the districts like Midnapore, Bankura, Purulia, 24-Parganas, Hooghly and West Dinajpur. The growth rates of production and yield of foodgrains have been significant for most of the blocks in Midnapore. Side by side, the growth rates of production and yield of non-foodgrains have been significant for a good number of blocks in the district.

As regards the relative role of change in CP on agricultural growth it is observed that there exists a significant role of CP change on agricultural growth in the sample states like Tamil Nadu, Orissa, Haryana, Maharashtra, Rajasthan, U.P. and H.P. either in the whole period (1972-73 to 1995-96) or in the sub-periods. The Chow test reveals that there also exists differential influences of CP change on agricultural growth between two sub-periods, 1972-73 to 1982-83 and 1983-84 to 1995-96 in most of the sample states and India.

The similar analysis for the districts of W.B. shows that the relative contribution of CP change on agricultural growth is not statistically significant for most of the districts during the whole period from 1972-73 to 1995-96 or during either both or one of the sub-periods. It is significant only for six districts in the state, namely Coochbehar, Nadia, West Dinajpur, Purulia, Malda and Bankura. On the other hand, the Chow test reveals the differential influences of CP change on agricultural growth between twosub-periods in some of the districts under study.

An attempt has been made to measure the impact of change in CP, as measured by the percentage of area under HYV crops in the GCA, or the percentage of area under non-foodgrains in the GCA, on the agricultural production in the twelve sample states. For this purpose we have regressed agricultural production index on the change in CP. The results show that the CP change in terms of the percentage of area under HYV crops in the GCA has significant effect on agricultural production in the states of A.P., Rajasthan, Haryana, Orissa, Tamil Nadu, Punjab, U.P., Maharashtra and India. On the other hand, the change in CP in terms of percentage area under non-foodgrains in the GCA has significant effect on agricultural production in the states like A.P. and U.P., and in India. The similar analysis has also been done for the districts of W.B. It is observed that the CP change in terms of the percentage of area under HYV crops in the GCA has significant effect on agricultural production in the districts like Burdwan, Murshidabad, Malda, Nadia, Bankura, Howrah, Hooghly, 24-Parganas, Birbhum and Midnapore during the period from 1976-77 to 1995-96. Again, the impact of change in CP in terms of percentage of area under non-foodgrains in the GCA on the agricultural production has been statistically significant for the districts like Burdwan, Murshidabad, Nadia, Bankura, Birbhum, Coochbehar. Darjeeling and Midnapore during the period from 1972-73 to 1995-96.

Focussing on the nature of change in CP it is observed that there took place pro-superior cereals and pro-foodgrains CP changes in 8 of our 12 sample states, viz., Punjab, H.P., A.P. Gujarat, Haryana, U.P. and W.B. But the pro-non-foodgrains CP change has occurred in other 4 sample states and India. The pro-HYV CP change has occurred in all the sample states and districts of W.B. as well as in India. In W.B. the pro-superior cereals and pro-foodgrains CP changes have taken place in five sample districts, viz., West Dinajpur, Howrah, Coochbehar, Purulia and Darjeeling, while the pro-non-food grains CP change has been found to have occurred in the other ten sample districts. The change in CP by size classes of holdings shows that the CP has changed in favour of foodgrains in India and in favour of non-foodgrains in W.B. The CP has moved in favour of non-foodgrains cultivation in most of the blocks in Midnapore.

Focussing on the trend of IC it is observed that the amount of credit from various institutional sources to rural people has increased considerably during 1972-73 to 1996-97. It is observed that the CAGR of IC per hectare is statistically significant in the period from 1972-73 to 1996-97 for all the sample states and districts in W.B. as well as in India. The trend of commercial bank credit across the different size classes of land holdings shows that the percentage share of a size class in the total commercial bank credit has risen ,at relatively higher rates for the smaller size of land holdings, during 1980-95 for all the sample states as well as for India. The same holds true for that of PACs credit by size classes of land holdings in the states in India as well as in the districts of W.B. There exists a rising trend of crop wise PACS credit for all the sample states as well as for India as a whole. It is observed that the PLDBC has no definite trend for the size classes of land holdings in the districts of W.B. The IC per hectare has shown a rising trend for most of the blocks in Midnapore.

As regards the relationship between the credit availability and change in CP it is observed that the bank loans (after the nationalisation of commercial banks) have created a significant impact on the change in CP in India and most of the sample states. On the other hand, the larger provision of credit, particularly, IC, larger irrigational facilities, etc. on the supply side have contributed to change in CP in India, the sample states , the districts of W.B. and the blocks of Midnapore. Side by side, the significant growth of PCI or PCSDP on the demand side has made significant contribution to change in CP in the study area. Finally, the IC has played a catalytic role in accelerating agricultural growth in India, sample states, districts of W.B. as well as in the blocks of Midnapore. The micro level study based on the village level data shows that the credit availability (IC or non–IC or both) has made statistically significant effect on the change in CP across the sample cultivators in three agro-climatic zones of Midnapore district. On the other hand, the change in CP has also made statistically significant effect on the profitability and employment generation of sample cultivator households. The study on the behaviour of the households belonging to different size classes of landholdings with respect to cultivation of crops shows that the smaller is the size of land holdings, larger is the area under non-foodgrains cultivation.

Again, the provision of credit to the sample farmers has created a 'chain of multiplier effects' which helps farmers to initiate multiple cropping with the aid of capital intensive HYV technology. The non–IC has made larger benefit to the relatively small and marginal farmers. Again, the process of payment of value of sold product and other related factors have created a 'perpetual debt condition'. The provision of non–IC has accelerated agricultural growth to a large extent. Credit has helped the cultivators in building and/or acquiring fixed capital assets necessary for the expansion of circulating capital necessary for financing purchase of raw materials, labour power and other raw materials.

From the above observations, the following policy prescriptions can be made.

First, since the credit plays a substantial role in effecting a significant change in CP in favour of more remunerative crops or HYV crops or non-foodgrains needed to satisfy the rising demand/diversified demand in the context of rapid industrialization, urbanisation and population growth, so the Government of India (GOI) must put greater emphasis on increasing the availability of IC to the farmers, particularly, the poorer ones. If this is done, the country will experience not only the higher rate of agricultural growth through change in CP but also a remarkable acceleration in the pace of industrialisation which can, in turn, help the process of economic reforms initiated in the country.

Secondly, the IC should be made available on time and in required amount. Transaction cost of this credit should be reduced so that it can adequately compete with non -IC.

Thirdly, the credit policy of the Government should be designed in such a manner that the poorer sections of farmers (in relation to their size of land holdings and other economic powers) can get more opportunity of obtaining IC. As a result, the small and marginal farmers will be in a position to increase agricultural production and thereby their economic strength. This step may therefore go a long way in reducing income inequality among the people in the countryside. It has been found in our study that the small and marginal farmers have the higher tendency to go for high-yielding and high-remunerative crops, when provided with credit at reasonable rates.

Fourthly, the Government must make a policy for the larger provision of IC to the farmers for the following reasons: (i) to protect or rescue them from the perpetual debt condition prevailing in rural area, (ii) to bring about a 'chain of multiplier effect' which help the farmers to initiate multiple cropping with the aid of capital-intensive HYV technology, (iii) to help the farmers for the larger acquisition and building of fixed capital assets and possession of larger volume of circulating capital.

Fifthly, the Government must design an appropriate export policy by which the country can earn more foreign currency by exporting those agricultural commodities having higher demand abroad. If this is done, there will be an acceleration of agricultural growth through the rapid change in CP towards the cultivation of those crops having high demand abroad. In this respect, the government must design a suitable credit policy by which the cultivation of those crops can increase.

Sixthly, since the change in CP has a positive impact on the labour employment in agriculture, the change in CP may increase the employment of labour in agricultural production, so more IC must be supplied for the large pace of change in CP which can ultimately solve the unemployment problem to a large extent by increasing labour employment opportunities in agricultural production.

Seventhly, since the IC has a significant effect on irrigation and fertilizer consumption which also have a significant impact on the change in CP, the government should design the credit policy in such a manner that the farmers may avail the larger irrigational facilities and use larger volume of chemical fertilizers and pesticides which, in turn, can accelerate agricultural growth through the change in CP.

Lastly, the change in CP particularly in favour of commercial crops cultivation has jeoparadised the food security of the country in recent years. So, the government must design an appropriate cropwise credit policy which can ensure the food security of a populous country like India.

To conclude, the present note has excluded some factors, viz., prices of crops, natural factors like rainfall, temperature, etc. to examine the impact of different factors on the change in CP.

Again, there is a scope to examine the relationship between (i) the resource base and CP of households and credit, and (ii) liberalisation policy, GATT agreement and changes in CP and agricultural growth.

# BIBLIOGRAPHY


Acharya, S.S. and Agarwal, N.L. (1974), "Factors Affecting Supply and Growth Rates of Commercial Crops in Rajasthan" (Summary), *Indian Journal of Agricultural Economics ( IJAE)*, Vol. XXIX, No. 3, July–September.

Azad, M.P., Prasad, M., Yadav, R.N. and Singh, C.P. (1985), "Why Mixed Farming for Marginal and Small Farmers ? (A Case Study)" (Summary), *IJAE*, Vol. XL, No. 3, July–September.

Agricultural Credit Schemes of Commercial Banks : Report of Expert Group, *Op. Cit.* p. 50.

Athavale, M.C., Yadav, K.S. and Mishra, J.P. (1971), "Green Revolution and Short Term Co-operative Credit–A Study in Two Districts of Madhya Pradesh", *IJAE*, Vol. XXVI , No. 4, Oct.- Dece.

Alagh, Y. K. and Sharma, P.S. (1985), "Growth of Crop Production: 1960-61 to 1978-79, Is it Decelerating", *EPW.*

Agrawal, A. P., Puhazhendhi, K. P. and Satyasai, K.J.S. (1997), "Gearing Rural Credit for the Twenty-First Century", *EPW*, October 18.

Adhikari, M.K. and Sau, S. (2003), "Watershed Development Planning For Subtainable Agriculture in West Bengal: Some Issues and Problems", *IJRS*, Vol.XXXV, No. 2,

Anant, T.C.A. (2002), "Institutional Reform for Agricultural Growth", *IJAE*, Vol. 57, No. 3, July–Sept.

Ahluwalia, I. J. and Little, I.M.D. (1998), "India's Economic Reforms And Development, Essays For Manmohan Singh", OUP.

Bagchi, A. K. (1998), "Studies on the Economy of West Bengal since Independence", *EPW*, Nov. 21.

Balakrishnan, P. (2000), "Agriculture and Economic Reforms, Growth and Welfare", *EPW,* March 18.

Biswas, B.C. and Khambele, N.N. (1980), "Re-orientation of the Cropping Pattern on the Basis of Probabilistic Moisture Availability Index", *IJAE,* Vol. XXXV, No. 2, April–June.

Bhanja, P. K. (1971), "Institutional Credit and HYV Programme: An Analysis of Credit Requirement and Economic Feasibility (A Case Study in Birbhum, W.B.)", *IJAE,* Vol. XXVI, No. 4, October - December.

Bhalla, G. S. (2001), "Political Economy of Indian Development in the 20th Century, Indian Road to Freedom and Growth", Presidential Address, 83rd Annual Conference, Indian Economic Association, University of Jammu.

Bhalla, G. S. (2004), "State of the Indian Farmer: A Millennium Study, Globalisation and Indian Agriculture", Vol. 19, Academic Foundation, New Delhi and Ministry of Agriculture, Govt. of India.

Banerjee, H.K. and Kundu, S. (2001), "Agriculture in West Bengal- An Analysis of Recent Trends", *Artha Beekshan,* Vol. 10, No. 1, June.

Banerjee, H.K. and Sen, I. (1995), "Influence of Farm-Size on Marketable Surplus of Food Crops-A Quantitative Approach", *Artha Beekshan,* Vol. 4, No. 1, June.

Bhaumick, S. K. and Rahim, A. (1998), "Structure and Operation of Rural Credit Markets: Some Results Based on Field Level Surveys in West Bengal", Paper presented at the International Conference on Decentralisation: Devolution and Participation, Organised by State Planning Board, Govt. of West Bengal, Dec. 28-30, Calcutta.

Bhaumick, S. K. and Rahim, A. (2000), "Interlinked Credit Transactions in Rural West Bengal", *IJAE,* Vol. 54, No. 2.

Bera, S. K. and Ghoshal, R. K. (2003), "Institutional Finance and Agriculture Development in India: A Study of Rural Credit Co-Operative Institutions", *Artha Beekshan,* Vol. 12, No. 2, September.

Bezbaruah, M.P. and Roy, N. (2002), "Factors Affecting Cropping Intensity and Use of Fertilisers and High-Yielding Seeds in Barak Valley", *IJAE,* Vol. 57, No. 2, April- June.

Bhattacharya, R. and Chatterjee, B. (2005 & 2006), "Regional Disparities in Boro Paddy Cultivation in West Bengal", A Case Study of Six Districts", *Artha Beekshan,* Vol. 14, Nos. 3 & 4 , Dec. & March.

Bhattacharya, R. and Chatterjee, B. (2007), "Diversification in Agriculture in West Bengal During 1997–2005", *Artha Beekshan*, Vol. 15, No. 4. March.

Bhosale, S.S. and Gawade, B.B. (1991), "Relative Economics of Different Farming Systems in Western Maharashtra", (Summary), *IJAE*, Vol. 46, No. 3, July–September.

Bhardwaj, S.P., Dixit, V.N. and Gupta, H.C. (1991), "Impact of Risk and Resource Allocation on Cropping Pattern–A Case Study", (Summary), *IJAE* Vol. 46, No. 3 , July-September.

Balishter, Singh, R.K. and Singh, Roshan (1985), "A Study of Diversification of Agriculture in Agra District of Uttar Pradesh", (Summary), *IJAE* ,Vol. XL, No. 3, July–September.

Belshaw, H. (1959) "Agricultural Credit in Economically Under Developed Countries", FAO Agricultural Studies, No. 26, Rome, pp. 199-210.

Basu, S.K. (1979), "Commercial Banks and Agricultural Credit: A Study in Regional Disparity in India", Bombay, 1979.

Basu, J.P. and Dinda, S. (2003), " Market Integration: An Application of Error Correction Model to Potato Market in Hoogly District, West Bengal", *IJAE*, Vol. 58, No. 4, October- December.

Bansil, P.C. (1971), "Short Term Credit Requirements at the End of the Fourth Plan Period 1973–74", *IJAE*, Conference Number , Vol. XXVI , No.4, Oct.-Dec.

Borbora, S. and Mahanta, R. (2001), " District-wise Study of Agricultural Production in Assam Using Cobb-Douglas Production Function", *IJRS*, Vol. XXXIII, No. 1.

Bhalla, S. (1987), "Trends in Employment in Indian Agriculture, Land and Asset Distribution", *IJAE*, Vol. 42, No. 4, October–December.

Bhalla, G.S. and Singh, G. (1997), "Recent Developments in Indian Agriculture: A State Level Analysis", *EPW*, March 29.

Basu, K. (1998), "Analytical Development Economics, The Less Developed Economy Revisited", OUP.

Basu, K. (1984), "Implicit Interest Rates, Usury and Isolation in Backward Agriculture", *Cambridge Journal of Economics*, Vol. 8, No. 2.

Boyce, J. K. (1987), "Agrarian Impasse in Bengal: Agricultural Growth in Bangladesh and West Bengal, 1949–1980", Oxford University Press (O.U.P.).

Bhalla, G.S. and Alagh, Y.K. (1979), "Performance of Indian Agriculture–A District wise Study", Starling Publishers.

Bharadwaj, Krishna (1979), "A Note on Commercialisation of Indian Agriculture", O.U.P.

Bandyopadhya, N. (1975), "Changing Forms of Agricultural Enterprise in West Bengal", *EPW*, April 26.

Bhattacharya, S., (1996), "Evolution of the Agricultural Credit Market in West Bengal Since 1977: A Journey from Peasant Class Differentiation to Agriculture–Led Development", in "Economy of West Bengal" edited by A. Roychaudhuri, Allied Publishers, New Delhi.

Bhattacharya, S. (2003), "Rural Credit and Class differentiation: West Bengal under Left Interventionist Regime", KP Bagchi & Company, Kolkata.

Bhattacharya, S. (2005), "Interest Rates, Collateral and (De-) Interlinkage: A Micro-Study of Rural Credit In West Bengal", *Cambridge Journal of Economics*, Vol. 29, No. 3, May.

Bhattacharya, S. (2001), "Capitalist Development, Peasant Differentiation and The State: Survey Findings From West Bengal", *Journal of Peasants Studies*, Vol. 28, No. 4, July.

Breman, J and Mundle, S. (1991), "Rural Transformation in Asia", p. IX

Black, John D. (1955), "Contemporary Readings in Agricultural Economics", p. 332.

Birthal, P.S., Jha, A.K., Joshi, P.K. and Singh, B. K. (2006), "Agricultural Diversification in North Eastern Region of India Implications for Growth and Equity", *IJAE*, Vol. 61, No. 3, July- September.

Bhowmik, D. (2002), "Behavior of Indian Commercial Banks Profit During 1950-97", *Artha Beekshan*, Vol. 11, No. 3, December.

Bagchi, J.(2003), "Agriculture and WTO Opportunity for India", *SAMSKRITI*, New Delhi.

Chadha, G.K. (2003), "Indian Agriculture in the New Millennium: Human Response to Technology Challenges", *IJAE*, Vol. 58, No.1, Jan-March.

Chattopadhyay, A and Das, P. S. (1997), "Trends in Crop Output in West Bengal", *Artha Beekshan*, Vol. 6, No. 1, June.

Chottopadhyaya, A. (2001), "Growth, Stability and Equity in Agriculture: A Quantitative Analysis", Concept Publishing House, New Delhi.

Chattopadhyay, A and Das, P.S. (1997), "Estimation of Growth Rates: A Fresh Look", *Vidyasagar University Journal of Commerce,* Vol. 2, March.

Chottopadhyaya, A.K. and Das P. S. (2000), "Estimation of Growth Rate: A Critical Analysis with Reference to West Bengal Agriculture", *IJAE,* Vol. 55, No. 2, April-June.

Chattopadhyay, M. and Bhattacharya, C. (1987), "Growth of Indian Agriculture: A Reappraisal", *E.P.W.* , Vol. 42, No. 1.

Chattopadhyay, A. K. (2005), "Distributive Impact of Agriculture Growth In Rural West Bengal", *EPW,* Vol. XL, No. 53, December 31 to January 6.

Chatterji, A. (1966), "A Study of Agricultural Growth During 1950-1963 in India", *IJAE,* Vol. XXI, No. 4, October-December.

Chauhan, K. K. S.,Verma, R. C. and Shukla, G.C. (1974), "Impact of Commercial Crops on Farm Incomes and Resources Use in Jaipur District of Rajasthan", *IJAE,* Vol. XXIX, No. 3, July-September.

Chaudhury, M.K. and Aneja, D.R. (1991), "Impact of Green Revolution on Long-Term Sustainability of Land and Water Resoruces in Haryana", *IJAE,* Vol. 46, No. 3, July-September.

Chandrakar, M.R., Koshta, A.K. and Agrawal, K.G. (1991), "Changes in Crop-Mix in Sub-Region of Raipur District in Madhya Pradesh", (Summary), *IJAE,* Vol. 46, No. 3, July-September.

Chawla, J.S. and Chahal, T.S. (1991), "Employment and Wages of Agricultural Workers in the Post-Green Revolution Period in Punjab", (Summary), *IJAE,* Vol. 46, No. 3, July-September.

Chand, Kanwar Prakash and Sing, Ranveer (1985), "Diversification of Agriculture in Himachal Pradesh: A Spatio-Temporal Analysis", (Summary), *IJAE,* Vol. XL, No. 3, July-September.

Chand, R. and Kumar, P. (2002), " Long Term Changes in Coarse Cereals Consumption in India: Causes and Implications", *IJAE,* Vol. 57, No. 3, July-September.

Chawla J.S. and Chahal, T.S. (1985), "Towards Diversification in Punjab Agriculture-A Case Study", (Summary), *IJAE,* Vol. XL, No. 3, July-September.

Choubey, B.N. (1977), "Institutional Finance for Agricultural Development", Pune, p. 37.

Chand, Ramesh, Sidhu, D.A. and Kaul, J.L. (1985), "Impact of Agricultural Modernisation on Labour Use Pattern in Punjab with Special Reference to Women Labour"' *IJAE,* Vol. XL, No. 2, April-June.

Chakraborty, S. (1997), "Writing on Development", *OUP*, Delhi, P.227-228

Chakarborty, D. and Ghosh, S. (2003), "Improving Agricultural Productivity in India: An Urgent Need for Systematic Reform", *Artha Beekshan*, Vol. 12, No. 3, December.

Dharma Narain (1977), "Growth of Productivity in Indian Agriculture", *IJAE*, Vol. XXXII, No. 2, January-March.

Dasgupta, R. (2002), "Priority Sector Lending: Yesterday, Today and Tommorow", *EPW*, Vol. XXXVII, No. 41, Oct. 12.

Dhar, A., and Pal, D. (2005 & 2006), "Inter District Disparity of Foodgrains Production in West Bengal: Growth, Instability and Determinants of Yield", *Artha Beekshan*, Vol. 14, No. 3&4, Dec.-March.

Dayal, R. (1966), "Agricultural Growth Rates and Their Components", *IJAE*, Vol. XXI, No. 4, October-December.

Deoghare, P.G., Sharma, B.M. and Goel, S. K. (1991), "Impact of Mixed Farming System on Income and Employment on Small Farms in Karnal District of Haryana", *Agricultural Situation in India* (ASI), Vol. XIV, No. 10, January.

Dhindsa, K.S. and Sharma, A. (2001), *"Policy Planning and Liberalisation"*, Vol. I, II, and III, Concept Publishing Co., New Delhi.

Dangost, S.B. and Patil, R.G. (1985), "Potential for Increasing Farm Income in Dry Farming Areas Through Diversification" (Summary) *IJAE*, Vol. XL, No. 3, July-September.

Desai, D. K. (1988), "Institutional Credit Requirements for Agricultural Production-2000 A.D.", *IJAE*, Vol. 43, No. 3, July-September.

Desai, B. M., Gupta, V.K. and Singh, Gurdev (1988), "Institutional Credit for Green Revolution Areas under Semi-Arid Tropics in India", *IJAE*, Vol. 43, No. 1, January-March.

Dangat, S.B., Sale, D.L. and Nawadkar, D.A. (1991), "Labour Employment and Income Pattern under Different Farming Systems in Scarcity Zone of Maharashtra State", (Summary), *IJAE*, Vol. 46, No. 3, July-September.

Deshmukh. P.D. , Hange, D. S., Kale, N. K. and Pawar, J.R. (1991), "Resource Costs, Return and Resource Use Efficiency Under Different Farming Systems in Drought-Prone Area of Western Maharashtra", (Summary), *IJAE*, Vol. 46, No. 3, July-September.

Desai, B. M. (1977), "Analysis of Cropping Pattern of Farm Families, Surat District", *IJAE*, Vol. XXXIII, No. 1, January-March.

De, U.K. (2003), "Changing Copping Pattern System in Theory & Practice: An Ecónomic Insight into Agrarian West Bengal", *IJAE*, Vol. 58, No. 1, January-March.

De, U.K. (2000), "Diversification of Crop in West Bengal: A Spatio-Temporal Analysis", *Artha Vijnana*, Vol. XLII, No. 2, June.

Dev, S. Mahendra (1989), "Towards an Explanation of Spatial and Temporal Variations in Cropping Intensity in India", *IJAE*, Vol. 44, No.1, January-March.

Desai, D. Mahandra and Naik, D. Bharat (1971), "Prospects of Demand for Short Term Institutional Credit for High Yielding Varieties", *IJAE*, Vol. XXVI, No. 4, October-December.

Dikshit, S.P. and Singh, A.K. (1974), "Impact of Green Revolution on Agricultural Production Structure-A Case Study of the Changes in the Cropping Pattern in Uttar Pradesh", ASI, Vol. XXIX, May, No. 2.

Dadhichi, C.L. (1971), "Willful Default of Co-operative Credit in Rajasthan: Some Issues", (Summary), I.J.A.E., Conference Number, Vol. XXVI, No. 4, Oct.-Dec.

Debnath, J. and Majumdar, S. (2006), " The Changing Pattern of Food Consumption Expenditure in India During 1950-2001: An Analytical Study", *Artha Beekshan*, Vol. 15, No. 3.

Dadibhavi, R.V. (1988), "Dimensions of Regional Disparities in Institutional Credit to Agriculture", *IJAE*.

D'Souza, E. (2000), " Prudential Regulation in Indian Banking", *EPW*, January 29.

Dandekar, V.M. (1980), "Introduction to Seminar on Data Base and Methodology for the study of Growth Rates in Agriculture", *IJAE*, Vol. 35, No. 2, April-June.

Das, P. S. (1978), "Growth and Instability in Crop Output in Eastern India", *EPW*, Vol. 13, No. 41, October.14.

Dasgupta, B. (1987), "Agrarian Reform Programme in West Bengal", Calcutta, December.

Dasgupta, B. (1998), "Structural Adjustment, Global Trade and the New Political Economy of Development", Vistaar Publications, New Delhi.

Datt, R. and Sundharam, K.P.M. (1995), "Indian Economy", 32nd Revised Edition, S. Chand and Company Ltd., New Delhi.

Eighth Five Year Plan (1992-97), Govt. of India, Planning Commission.

Feroze, S.M., Chauhan, A.K. & Bhattacharya, B. (2007), "Diversification of Agriculture in West Bengal", *Artha Beekshan,* Vol. 15, No. 4, March.

Fourth Five Year Plan (1969-74), Govt. Of India, Planning Commission.

Giri, R., Sastri, A.V.K. and Samayajulu, D.S. (1966), "Components of Crop Output Growth in India", *IJAE,* Vol. XXI, No. 4, October-December.

Gupta, A.K. (1963), "Inter-State Differences in Cropping Pattern and Productivity", *IJAE,* 18(1), January-March.

Gupta, R.P. and Tewari, S.K. (1985), "Factors Affecting Crop Diversification : An Empirical Analysis", *IJAE,* Vol. XI, No. 3, July-September.

Gill, G.S. and Patel, R.K. (1985), "Diversification of Farm Production Patterns in the Rural Economy of Cotton Belt of the Punjab", (Summary), *IJAE,* Vol.XI, No.3, July-September.

Ghosal, R.N. (1997), " Regional Variation in The Growth of Crop Output in West Bengal: An Inter District Analysis", *Artha Beekshan,* Vol. 6, No. 2.

Ghosh, M. (2002), "Trends, Random Walks and Structural Breaks in Indian Agriculture", *IJAE,* Vol. 57, No. 3, July-Sept.

Ghose (Nee Dhar), A. and Pal, D. (2007), "Growth of Foodgrains Production: An Inter District Comparison of West Bengal", *Artha Beekshan,* Vol. 15, No. 4 , March.

Ghosh, S.K. (2007), " Food Security and Government Action in West Bengal", *Artha Beekshan,* Vol. 15, No. 4, March.

Ghosh, M. (2006), "Regional Convergence In Indian Agriculture", *IJAE,* Vol. 61, No.4, October-December".

Gadgilkar, V. D. and Gadve, N.A. (1988), "Structure of Rural Credit in Akola District of Maharashtra State", (Summary), *IJAE,* Vol. XXXIII, No. 4, October-December.

Gupta, S.B.L., Balera, M.M. and Venkatessardu, U. (1978), "Regional Inequality in the Supply of Co-operative Credit in India", (Summary), *IJAE,* Vol. XXXIII, No. 4, October-December.

Gupta, M. (2004), Changing Profile of Public Sector Banks", *Artha Beekshan,* Vol. 12, No. 4. March.

Gadgil, M.V. (1986), "Agricultural Credit in India: A Review of Preference and Policies", *IJAE,* Vol. 41, No. 3, July-September.

Gulati, A., Joshi, P.K. and JR. Cumminings, R. (2007), "Agricultural Diversification and Small Holders In South Asia", Academic Foundation, New Delhi.

Golait, R. and Pradhan, N.C. (2006), " Changing Food Consumption Pattern in Rural India: Implication on Food and Nutrition Security", *IJAE*, Vol. 61, No. 3, July-September.

Government of India, Ministry of Commerce, "Civil Supplies and Co-operation Report of the Working Group on Rural Credit and Co-operation, Medium Term Plan 1978–1983 (Mixed)", New Delhi, 1979, p. 5.

Gokhale Institute of Politics and Economics, Writing and Speeches of Prof. D.R. Gadgil on Co-operation, Pune, 1975, p. 113.

Government of India, Programme Evaluation Organization, Planning Commission, "Utilisation of Co-operative Credit", New Delhi, 1965, p. 69.

Giri, A. K. and Gangopadhyay, S. (1985), "Forms and Factors of Diversification", (Summary), *IJAE*, Vol. XI, No. 3, July– September.

Gharei, S.K. (1998), "Pace & Pattern of Commercialisation of Agriculture in West Bengal-A Study on the Role of Infrastructural Factors", A Seminar Paper Presented in 'National Seminar on 50 years of Rural Development in India' Organised by Dept. of Economics with Rural Development (R.D), Vidyasagar University (V.U.), West Bengal (W.B.).

Ghosh, M. (1998), "Agricultural Development, Agrarian Structure and Rural Poverty in West Bengal", *EPW*, Nov. 21.

Gujarati, D.N.(1998), "Basic Econometrics", Second Edition, McGraw-Hill International Editions.

Ghosh, D.N. (2000), " Weak Banks: A Strategy for Self-Renewal", *EPW*, January 29.

Ghosh, M. (1997), " Some Recent Trend in Development of Agriculture", *Artha Beekshan*, Vol. 6, No. 1, June.

Hussain, M. (1991), "Changing Cropping Patterns in India and Ecological Crisis", in 'Agriculture Planning and Development' edited by Reddy, M.V. Srinivasanan, S. S. Ramanaiah, Y.V. and Reddy, K.R.

Iyyampillai, S. and Swamikannan, D. (1991), "Inter-Regional Differences in the Farming Systems in Madurai District", (Summary), *IJAE*, Vol. 46, No. 3, July–September.

Indian Council of Social Science Research (1975) , "A Survey of Research in Economics", Vol. III, Agriculture(i), Bombay, p. 83.

Jasdanwalla, Z.Y. (1966), "Marketing Effciency in Indian Agriculture", Allied Pub. Pvt. Ltd., Bombay.

Johan, P.V. (1967), "Impact of Cropping Pattern on Agricultural Output : A Measurement", *IJAE*, Vol. XXII, No. 3, July-September.

Joshi, P.K. and Tyagi, N.K. (1991), "Sustainability of Existing Farming System in Punjab and Haryana-Some Issues on Ground Water Use", *I.J.A.E.*, Vol. 46, No. 3, July-September.

Joshi, P.K., Gulati, A., Birthal, P.S. and Tewari, L. (2004), "Agricultural Diversification in South Asia: Patterns, Determinants, and Policy Implication", *EPW*, June.12.

Joshi, P.K. and Saxena, R. (2002), "A Profile of Pulses Production in India: Facts Trends and Opportunities", *IJAE*, Vol. 57, No. 3, July-September.

Joshi, B.H. (1991), "Empioyment, Wage and Income of Agricultural Labourers in the Post-Green Revolution Belt of Saurashtra Region with Special Reference to Dhoraji Taluka", (Summary), *IJAE*, Vol. 46, No. 3, July-September.

Joshi, C.K. and Alshi, M. R. (1985), "Impact of High-Yielding Varieties on Employment Potential of Female Labour-A Study in Akola District (Maharashtra)", *IJAE*, Vol. XL, No. 2, April-June.

Johnston and Mellor (1961), " The Role of Agriculture in Economic Development", American Economic Review, Sept., p. 571-81 Reprinted in Meir (ed.), "Leading Issues in Economic Development", Oxford.

Jugale,V.B. and Jugale, Y.V. (2006), "*The State of Indian Economy*", Vol. 1, 2, 3, and 4, Serials Publications, New Delhi.

Kumari, A.K., Reddy, K.R. and Naik, H.(2004), "Agricultural Modernisation among Farmers in Nellore District, Andhra Pradesh", *IJRS*, Vol. XXXVI, No.1.

Kar, S.C., & Roy , S. N. (2007), "Rural Employment : A Case study of A Depressed but Resource-Rich Area of West Bengal Economy", *Artha Beekshan*, Vol. 15, No. 4, March.

Kaul, J.L. (1966), "Value Productivity Growth of Some Important Crops in the Punjab", *IJAE*, Vol. XXI, No. 4, October-December.

Kohli, R. (1992), "Rural Bank Branches and Financial Reform", *EPW*, January 16-23.

Kumar, A. (1989), "Spatio-Temporal Variations and Determinants of Rice Cultivation in Bihar", ASI, Vol. XLIV, No. 3, June.

Kumar, R., Chahal, S, S. and Singh, I. (2004), " Factors Affecting Capital Investment Pattern On the Farms of a Developed Village Economy in Punjab-A Micro Approach", *Indian Journal of Regional Science (IJRS)*, Vol. XXXVI, No. 2.

Kymal, P.K. and Ramakrishna, M. (1973), "Change in Food Habits in Relation to Food Production in India", *ASI*, Vol. XXVIII, No. 5, August.

Khatkar, R.K., Chamola, S.D. and Manocha, V. (1991), "Post-Green Revolution Scenario in Haryana Agriculture", (Summary), *IJAE*, Vol. 46, No. 3, July-September.

Kadian, R.S., Kaushik, C.R. and Kumar, R. (1991), "Impact of Diversified Farming System on the Rural Poor in Haryana", (Summary), *IJAE*, Vol. 46, No. 3, July-September.

Kalita, D. R. (1985), "Potentialities of Mixed Farming on the Rainfed Medium Sized Farms in Jorhat Subdivision of Assam", (Summary), *IJAE*, Vol. XL, No. 3, July-September.

Kharinta, S., Aggarwal, K. and Kaur, M. (1985), "Diversification in Agriculture Through Grape Cultivation-A Case Study", (Summary), *IJAE*, Vol. XL, No. 3, July-September.

Krishnaji, N. (1980), "Measuring Agricultural Growth", *IJAE*, Vol. 35, No. 2, April-June.

Kuznets, S., "Economic Growth and the Contribution of Agriculture" in Economic Growth and Structure, pp. 236-256. Reprinted in "Agriculture in Economic Development" Edited by Eicher and Witt.

Konar, D.N. (1997), "Trend of Indian Agriculture in the Post-Independence Period", *Artha Beekshan*, Vol. 6, No. 2.

Kumar, D. and Kahlon, A.S. (1978), "A Critical Study of Farmers Indebtedness in Ludhiana District", *Financing Agriculture*, Vol. X, No. 3, October-December.

Kumar, P. and Mathur, V.C. (1996), " Structural Changes in the Demand for Food in India", *IJAE*, Vol. 51, No. 4, October-December.

Khundrakpam, J.K. (2000), "Priority Sector Bank Lending, Some Important Issues", *EPW*, April 1.

Kamboj, S. (2003), "Bengal: From Rice Bowl to Begging Bowl? Planning Commission Study Says Paddy Production will be less than Demand. Rice Shortfall by 2007 Predicted", The Sunday Statesman, May 18, Sunday.

Kanwar, U. (1981), "Banking and Agricultural Growth-A Study of Land Development Bank", Deep and Deep Publications, New Delhi.

Kalirajan, K.P.and Shand, R.T.(1997), "Sources of Output Growthin Indian Agriculture", *IJAE*, Vol. 52, No. 4.

Kapila, R. and Kapila, U.(2002), *"Indian Agriculture in the Changing Environment"*, Vol. 1 & 2, Academic Foundation, New Delhi.

Minhas, B.S. and Vaidyanathan, A. (1965), "Growth of Crop Output in India : 1951-1954 to 1956-1961, An Analysis by Component Elements", *Journal of Indian Society of Agricultural Statistics. (J.I.S.A.S.)*, Vol. XVII, No. 2, December.

Mihas, B.S. (1966), "Rapporteur's Report on Measurement of Agricultural Growth", *IJAE*, Vol.XXI, No.4, October-December.

Mathur, R. N. (1963), "Cropping Pattern and Employment in Vidarbha", *IJAE*, 18(1), January-March.

Majid, A (1963), "Cropping Pattern and Size of Cultivated Holdings" *IJAE*, 18(1), January-March.

Malya, M.M. (1963), "Urbanization and Cropping Pattern", *IJAE*, 18(1), January-March.

Metha, P., Sharma, R.K. and Singh, J.N. (1982), "Impact of Farming on Economic Efficiency of Different Farm Sizes", *ASI*, Vol. 38, No. 8, April.

Madhava Swamy, G. (1991), "Role of Sericulture Industry for Development of Diversified Farming", (Summary), *IJAE*, Vol. 46, No. 3, July-September.

Madalia, V.K. (1985), "Diversification in Farming and its Impact on Farm Income and Employment-A Study of Mehsana District in Gujarat State" (Summary), *IJAE*, Vol. XL, No. 3, July-September.

Mukundan, K. (1985), "Diversified Farming for Augmenting Farm Income : A Study of Typical Farming Situation in Trichur District, Kerala State", (Summary), *IJAE*, Vol. XL, No. 3, July-September.

Mani, K. and Varadarajan, S. (1985), "Diversification of Farms", (Summary), *IJAE*, Vol. XL, No. 3, July-September.

Majumdar, N.A., (1989), "Rapporteur's Report on Institutional Credit-Rural", *IJAE*, Vol. 44, No. 2, January.

Madhava Swamy, G. (1985), "Effect of Diversification of Agriculture on Income and Employment", (Summary), *IJAE*, July - September.

Mahendradev, S. (1987), "Regional Variations in Agricultural Labour Productivity and Rural Poverty in India" *EPW*, September 26.

Mehra, S. (1981), "Instability in Indian Agriculture in the Context of the New Technology" Inernational Food Policy Research Institute, Research Paper.

Meier, G. (1980), "Population and Poverty: A Note" in G. Meier (ed.), "Leading Issues in Economic Development", OUP.

Maity, B. and Chatterjee, B. (2006), " Impact of Modern Technology on Foodgrain Production in West Bengal: An Econometric Analysis", *IJRS*, Vol. XXXVIII , No. 2.

Mitra, A. (1968), "Bumper Harvests Created Some Dangerous Illusions", The Statesman, 14-15 October.

Mukhopadhyay, D. (2002), " Changing Profile of Public Sector Banks in the Post-Reform Era", *Artha Beekshan*, Vol. 11, No. 3, December.

Mitra, S. (1987), "Agricultural Growth in the District of Midnapore, 1960–1986", M.Sc. Dissertation (Unpublished), Vidyasagar University, Midnapore.

Mukhopadhya, S. (1982), "Agricultural Growth and Potential in West Bengal", Report of the Study Sponsored by the Planning Commission, Government of India.

Mahandradev, S. (1987), "Growth and Instability in Foodgrains Production–An Inter State Analysis", *E.P.W.*, September, 26.

Mallik, J.K. (1997), "Growth of Agriculture in Independent India: 50 years and After", Reserve Bank of India (RBI) Occassional Papers, Vol. 18, No. 2 & 3, Special Issue, June & Sept.

Mallik, M. (2004), "Role of Public Sector Banks in Rural Development of India since 1970 s", *Artha Beekshan,* Vol. 13, No. 3, December.

Mishra, S. N. (1997), "Agricultural Liberalisation and Development Strategy in Ninth Plan", *EPW*, March.

Mruthyunjoy & Kumar (1989), "Crop Economics and Cropping Pattern", *EPW*, December. 23–30.

Mruthyunjay, Pal, S., Joshi, P.K. and Saxena, R. (2002), "Institutional Reforms & Agricultural Development under New Economic and Technological Environment", *IJAE*, Vol. 57, No.3, July–Sept.

Mishra, V.N.(2004),*"State of the Indian Farmer: A Millennium Study, Terms of Trade"*, Vol. 15, Academic Foundation, New Delhi, and Ministry of Agriculture, Govt. of India.

Mohan, R.(2006), "Agricultural Credit in India: Status, Issues and Future Agenda", *EPW,* Vol. XLI, No.11, March 18.

Majumdar, N.A. and Kapiala, U. (2007), *"Indian Agriculture in the New Millennium"*, Vol. 1 & 2, Academic Foundation, New Delhi.

Mallik, A.(2006), " The Impact of Economic Reforms on Indian Agriculture with reference to Food Security: A Note", *Artha Beekshan,* Vol. 15, No. 3, December.

Nadkarni, M.V. (1996), "Accelerating Commercialisation of Agriculture, Dynamic Agriculture and Stagnating Peasants?" *EPW,* June 29.

Naidu, P. Devasina (1989), "Impact of Cropping Pattern on Agricultural Production at the Disaggregation Level in Andhra Pradesh", *ASI,* Vol. XLIV.

Nair, T.S. (2000), 'Rural Financial Intermediation and Commercial Banks, Review of Recent Trends", *EPW,* January 29.

Nadda, A.L., Sharma, L.R. and Bhaduria, N.S. Swarup (1978), "Optimizing Crop Production Patterns in Different Agro-Climatic Zones of Himachal Pradesh", *IJAE,* Vol. XXXIII, No.4, October-December.

Nandi, A.K. & Chatterjee, P. K. (2007), "Resource Use Efficiency in West Bengal Agriculture", *Artha Beekshan,* Vol. 15, No. 4, March.

Narayanamoorthy, A. and Hanjra, M.A. (2006), " Rural Infrastructure and Agricultural Output Linkages: A Study of 256 Indian Districts", *IJAE,* Vol. 61, No. 3, July- September.

National Council of Applied Economic Research (1974), "Credit Requirements for Agriculture", New Delhi , p. 80, Table 32.

Nambiar, P.C.D. (1978), "Bank Finance for Agriculture", Commerce, Vol. 137, No. 3507, August 26.

Nanda, P. and Raikhy, P.S. (2000), "Exports from Punjab-Performance and Potentials", in 'Punjab Economy: Emerging Issues' edited by R. S. Bawa and P.S. Raikhy (Guru Nanak Dev University, Amritsar).

Nancharaih, G. (2003), "Economic Reforms and Rural Development In India", Academic Foundation, New Delhi.

Ninth Five Year Plan, Govt. of India.

Organisational Framework for the Implementation of Social Objectives, Report of the Study Group of the National Credit Council, *Op.Cit,* pp. S 79–81.

Pal, G.P. (1999), "Agricultural Structure & Growth in West Bengal", *Artha Beekshan,* Vol. 8, No. 1, June.

Prihar, R.S. and Singh, N. (1988), "A Study into Institutional Finance for the Agricultural Sector in Punjab", *IJAE,* Vol. 43, No. 3, July–September.

Pal Singh, Inder and Sankhayan, P.L. (1991), "Sustainability of Water Resources during the Post-Green Revolution Period in Punjab". *IJAE,* Vol. 46, No. 3, July–September.

Paul Singh, Invinder, Verma, S. and Gangwar, A.C. (1991), "Impact of Technological Change on Farm Investment and Farm Income Distribution in Haryana Agriculture", (Summary), *IJAE,* Vol. 46, No. 3, July–September.

Pal, D.P. and Pal, G.P., (1985), "Crops Diversification : Its Causes and Impacts on the Farm Economy" , (Summary), *IJAE,* Vol. XL. No. 3, July–September.

Prahaladachar, M and Mohansundaram, V. (1985), "Impact of Diversification of Rural Economy on Income and Employment of Small farmers and Agricultural Labourers: A Case Study of Dairy Enterprise in Rural Tamil Nadu" (Summary), *IJAE,* Vol. XL, No. 3, July–September.

Patel, A.R. (1975), "Growth of Agricultural Advances: A Review", Financing Agriculture, Annual No., pp. 61–66.

Pal, B.K. (1973), "Agricultural Finance in West Bengal", Calcutta.

Pal, P. (2003), "Agricultural Production in India: Growth and Instability", *Artha Beekshan,* Vol. 12, No. 3, December.

Pandey, R.N., Prasad, B. , Rawat, D.S. and Kumar, P.S. (1978), "Changes in Credit Use with Advancement in Agriculture", (Summary), *IJAE,* Vol. XXXIII, No. 4, October–December.

Pandey, V.K. and Sharma, K.C. (1996), " Crop Diversification and Self-Sufficiency in Foodgrains", *IJAE,* Vol. 51, No. 4, October-December.

Pandey, U.K., and Muralidharan, M.A. (1980), "Overdues and the Size of Holdings of Defaulters", *Financing Agriculture,* Vol. XL, No. 3 & 4, October–December.

Puhazhendhi, V. and Jayaraman, B. (1999), "Rural Credit Delivery: Performances and Challenges Before Banks", *EPW,* January 16-23.

Pany, R.K. (1982), "Institutional Credit for Agrriculture in India" Ashish Publishing Home, New Delhi.

Pillai, R. (2001), "An Analysis of Paddy Productivity Growth in West Bengal and Orisa", *IJAE*, Vol. 56 , No. 4, October-December.

Patnaik, U. (1997), "Agriculture and Rural Development", Alternative Economic Survey, 1996-97.

Patnaik, U.(1999), "The Long Transition, Essays on Political Economy", Tulika, New Delhi.

Pal, P.K.(2007), "Economic Growth and Development, Emerging Issues", Deep & Deep Publications, New Delhi.

Poddar, R.S., Kumar, H.S.V., Murthy, H.G.S. and Murthy, T.N.V.(1995), "Impact of Credit on Cropping Pattern and Cropping Intensity on Farms in Unirrigated and Irrigated Villages in Bijapur District of Karnataka-An Application of Linear Programming Technique", *AERR*, Vol. 8, No. 1.

Ramasamy, C. (2004), "Constraints to Growth in Indian Agriculture: Needed Technology, Resource Management, and Trade Strategies", *IJAE*, Vol. 59, No. 1, Jan.-March.

Ramasamy, C(2002), " Pulses, Oil Seeds and Coarse Cereals: Why They are Slow Growth Crops", *IJAE*, Vol. 57, No. 3, July-September.

Ranade, C.G. (1980), "Impact of Cropping Pattern on Agricultural Production", *IJAE*, Vol. XXXV, No. 2 , April-June.

Ravallion, M. (2000), " What is Needed for A More Pro-Poor Growth Process in India?" *EPW*, March 29.

Rao, C.H. Hanumantha (1985), "Growth of Agriculture in the Punjab during the Decade 1952-1962," *IJAE*, Vol. XX, No.3, July- September.

Rao, D. and Parwez, S. (2005), " Dynamics of Cropping Pattern in Sorghum Growing States of India", *IJAE*, Vol. 60, No. 4, October-December.

Ramalingam, C. (1963), "Some Economic Aspects of Cropping Patterns", *IJAE*, 18(1), January-March.

Rao, T . Ramkrishna, (1965), "Urbanisation and Efficiency in Farming", *IJAE*, Vol. XX, No. 4, October-December.

Rajagopalanan, V., and Varadrajan, S. (1978), "Impact of Risk and Uncertainty on Farm Production and Income in the Hills of the Nilgiris, Tamil Nadu", *IJAE*, Vol. XXXIII, No. 4, October-December.

Rajeev, M. and Dev, S. (1998), "Institutional and Non-Institutional

Credit in Agriculture: Case Study of Hugli District of West Bengal", *EPW*, Nov. 21.

Ramasubban, T.A. (1963), "Some Statistical Measures to Determine Changes in Cropping Pattern", *ASI*, Vol. XVII, No. 11, February.

Reddy, Y.V.R. (1985), "Different Sources of Cash Income and Expenditure in Dryland Areas", (Summary), *IJAE*, Vol. XL, No. 3, July–September.

R.B.I., Report of the Committee to Review Arrangements for Institutional Credit For Agriculture and Rural Development (CARFFICARD), *Op. Cit.*, p. 55.

R.B.I., Report of the Committee on Co-operative Land Development Banks, Bombay (1975), pp. 63–66.

R.B.I. Multi-Agency Approach in Agricultural Finance, Report of the Working Group, Bombay (1987), pp. 18-32.

Report of the All-India Rural Credit Review Committee, *Op. Cit*, pp. 81-82.

Rao, V.M. , Nadkarni, M.V. and Despande, R.S. (1980), "Measurement of Growth and Fluctuations in Crop Output :An Approach Based on the Concept of Non-systematic Component", *IJAE*, Vol. 35, No. 2

Raj, K.N. (1969), "Some Questions Concerning Growth, Transformation and Planning of Agriculture in Developing Countries," *Journal of Development Planning*, No. 1.

Raj, K.N. (1985), "In Production, Essays on the Commercialisation of Indian Agriculture", OUP.

Raj, K. N. (1990), "Organisational Issues in Indian Agriculture", OUP.

Ray, S.K. (1983), "Variation in Crop Output", Institute of Economic Growth, Delhi.

Roy, B.C. and S. Pal (2003), "Investment, Agricultural Productivity and Rural Poverty in India : A State Level Analysis", *IJAE*, Vol. 57, No. 4, Oct–Dec.

Rudra, A. (1992), "Political Economy of Indian Agriculture", K. P. Bagchi & Company, Calcutta.

Rath, N. (1980), "A Note on Agricultural Production in India During 1955-78", *IJAE*, Vol. 35, April.

Rajwade, A.V. (2000), " Nationalised Banks, No Cogent Plan", *EPW*, March 11.

Rao, R.M.M. (1991), "Commercialisation and Agricultural Growth in Asia : A Comparative Study of South Korea and Coastal

Andhra", In J. Breman, and S. Mundle (Eds.) , Rural Transformation in Asia, OUP, New Delhi.

Rao, K.S.R., Das, A. and Singh, A.K. (2006), "Commercial Bank Lending to Small - Scale Industry", *EPW*, Vol. XLI, No. 11, March 18.

Ray, S.K. and Panja, S. (1998), "Role of Institutional Credit on Changes in Cropping Pattern : A Study of Midnapore District in West Bengal", in S.N. Sau (ed.), 'Rural Development Theories and Experiences', Allied Publishers Ltd., New Delhi.

Ray, S.K. (1994), "Availability of Institutional Finance and Changes in Cropping Pattern: A Preliminary Study", *VU Journal of Economics*, Vol. II.

Ray, S.K, Sau, S.N. and Nayak, A. (1994), "Availability of Institutional Finance and Its Impact on Change in Cropping Pattern : A Study of Midnapore District of West Bengal", 11th Conference Volume ,Bangiya Arthaniti Parishad, W.B.

Ray, S.K. and Sau, S.N. (1998), "Diverse Changes in Cropping Pattern Across States in India : An Analysis", (Abstract), Published in Indian Economic Association (I.E.A.) Conference Volume.

Ray, S.K. and Sau, S.N. (1998), "Constrained Acceleration of Commercialisation of Agriculture in India : Stagnating Infrastructures but Dynamic Peasants", (Abstract), Published in I.E.A. Conference Volume.

Ray, S.K. and Sau, S.N. (1992), "Commercial Banks Financing Rural Development : Some Issues and Problems", Abstract of Papers, 12th Annual Conference of Bangiya Arthaniti Parishad, W.B.

Ray, S.K. (2000), "Progress of Commercial Banking and Change in Cropping Pattern : A Study With Reference to Eastern and North-Eastern India", 20th Annual Conference Volume, Bangiya Arthaniti Parishad, W.B.

Ray, S.K. (2006), "Bank Nationalisation, Institutional Credit, Cropping Pattern and Agricultural Growth in India: A State Level Analysis", *Artha Beekshan,* Vol. 15, No. 1.

Ray, S.K. (2005 & 2006), "The Changing Role of Financial Sector in the Provision of Credit for Agricultural Growth: A Study with Reference to West Bengal", *Artha Beekshan,* Vol. 14, Nos. 3 & 4,

Ray, S.K. (2007), "Decentralised Agricultural Development Planning, Change in Cropping Pattern and Agricultural Growth : A Block

Level Study in Medinipur District", (Abstract), *Artha Beekshan*, Vol. 15, No. 4.

Ray, S. K. (2001), "Privatisation of Commercial Banks and Growth of Priority Sectors: Analysis of Some Issues", BEA 21st Conference Volume, Bangiya Arthaniti Parishad, W.B.

Ray, S.K. (1999), "Diverse Change in Cropping Pattern Across States In India: An Analysis", *Vidyasagar University Journal of Economics*, Vol. VI.

Ray, S.K. (2005), "Impact of Change in Cropping Pattern on Agricultural Growth: A Study With Special Reference to West Bengal", *Indian Journal of Regional Science*, Vol. XXXVII, No. 2.

Ray, S.K. (2003), "Progress, Problems and Prospects of Floriculture: An Economic Analysis with Special Reference to West Bengal", In K.R. Gupta ( ed.), *'Rural Development in India'*, Vol. I, Atlantic Publishers & Distributors, New Delhi

Ray, S.K. (2006), "Bank Nationalisation, Availability of Institutional Credit, Change in Cropping Pattern, Commercialisation of Agriculture and Agricultural Growth in India: A State Level Analysis", in V.B. Jugale & Y.V. Jugale (ed.), 'The State of Indian Economy: Rural & Agricultural Development', Vol. 3, Serials Publications, New Delhi (India).

Ray S.K. & Sau, Sachinandan (2004), "Decentralised Credit Planning and Change in Cropping Pattern: A Block Level Study in Medinipur District", DRS Occasional Paper, No.17, Department of Economics with Rural Development Under DRS (SAP), UGC, Vidyasagar University, Midnapore, W.B.

Ray, S.K.(1998) "Growth of Non-food grains: An Analysis", A Seminar Paper Presented in a National Seminar On 50 years of Rural Development in India" Organised by Dept. of Economics with R.D., V.U., W.B.

Ray, S. K.(2001), " Role of Institutional Credit in Changes in Cropping Pattern in West Bengal : An Analysis", in S.N.Sau & P.S.Das (ed.), "Fifty Years of Rural Development in India", Firma KLM Private Limited, Kolkata.

Ray, S.K. (2007), "Economics of Change in Cropping Pattern in Relation to Credit: A Micro Level Study In West Bengal" (Forthcoming), *IJAE*, April-June.

Rao, C.H.H. (1998), "Agricultural Growth, Sustainability and Poverty Alleviation: Recent Trends and Major Issues of Reform", *E.P.W.*, Vol. XXXIII, November 29 and 30, July 18-24/25-31.

Rao, J.M. (1998), "Food , Agriculture and Reform: Change and Continuity", *E.P.W.*, Vol. XXXIII, No. 29 and 30, July 18-24/25-31.

Rahim, K.MB(1995), "Capital Requirement for Optimisation of Resource Use on Marginal and Small Farms in Old Alluvial Zone of West Bengal", Agricultural Economics Research Review (*AERR*), Vol. 8, No. 1.

Roy, M. (2006), "A Review of Bank Lending to Priority and Retail Sectors", *EPW*, Vol. XLI, No. 11, March 18.

Rogaly, B., Harris-White, B.and Bose, S. (1995), "Sonar Bangla? Agricultural Growth and Agrarian Change in West Bengal and Bangladesh", *EPW*, Vol. 30, No. 29, July 22.

Rogaly, B. *et. al.* edited (1999), "Sonar Bangla? Agricultural Growth and Agrarian Change in West Bengal and Bangladesh", Sage Publication, New Delhi.

Ray, D. (1998), "Development Economics", OUP.

Rao, C.H.H.(2007), "Agricultural Growth, Farm Size and Rural Poverty Alleviation in India" , Academic Foundation, New Delhi.

Ramachandran, V.K. and Swaminathan , M.(2002), "Agrarian Studies, Essays on Agrarian Relations in Less Developed Countries", Tulika Books, New Delhi.

Sengupta, R. and Sinha, A.K. (2003), "Challenge of Sustainable Development: The Indian Dynamics", First Edition, IIM, Calcutta and Manak Publications Pvt. Ltd., Delhi.

Sanyal, M.K., Biswas,P.K., and Bardhan, S. (1998), "Institutional Change and Output Growth in West Bengal Agriculture", *EPW*, Nov. 21.

Shiva, V. and Bedi, G. (2002), "Sustainable Agriculture and Food Security, The Impact of Globalisation", Sage Publications, New Delhi

Samal, B. (2002), " Institutional Credit Flow to West Bengal Agriculture: Revisited", *IJAE*, Vol. 57, No. 3, July-September.

Sharma, P.V. (1974), "Identification of Contribution of Each Element to the Growth Rate of Commercial Crop Output in the Districts of Andhra Pradesh", *IJAE*, Vol. XXIX, No. 3, July–September.

Singh, S. and Vidyasagar (2004), "State of the Indian Farmer: A Millennium Study, Agricultural Credit in India", Vol. 7, Academic Foundation, New Delhi, and Ministry of Agriculture, Govt. of India.

Shah, C.H. (1963), "Crop Variations in 1943-1954, A Study in Crop pattern Change for Farmers in Different Size Groups of Holdings", *IJAE*, 11(3), July-September.

Sharma, J.S. & Prasad, B. (1971), "An Assessment of Production Credit Needs in Developing Agriculture", *IJAE*, Vol. XXVI, No. 4, October-December.

Satish, P. and Gopalakrishana, C.K.(1997), "Viability of Rural Banking", *EPW*, October 18.

Shah, S.M. (1963), "Cropping Pattern in Relation to Irrigation", *IJAE*, 18(1) January-March.

Sridharan, B. and Radhakrishnan, S.A., (1978), "A Study on The Factors Affecting Changes in the Cropping Pattern in Nilgiris District, Tamil Nadu", *IJAE*, Vol. XXXIII, No. 4, October-December.

Sridharan, B. (1971), "A Micro Level Study on Factors Influencing Cropping Pattern in Coimbatore District", Ph.D. Dissertation submitted to Tamil Nadu Agricultural University (Unpublished).

Swaminathan, M. (2003), "Foodgrain Marketing in India: Is Privatisation the Way Ahead?" *Artha Beekshan*, Vol. 12, No. 3, December.

Shah, S.L. (1978), "Rapporteur's Report on Farming Systems in Hill Areas", *IJAE*, Vol. XXXIII, No. 4, October-December.

Sinha, A.K. (1978), "Impact of Lift Irrigation on Cropping Pattern and Crop Yields: Based on a Five Village Survey in Bhiwani (Haryana)", *I.J.A.E.*, Vol. XXXIII , No. 1, January-March.

Singh, K. and Rahim K.M.B. (1978), "Identification and Evaluation of Optimal Cropping Systems for a Typical Water Shed in Uttar Pradesh Hill", *I.J.A.E.*, Vol. XXXIII, No. 4, October-December.

Sharma, B.W. and Sirohi, A.S. (1980), "Labour Employment Implication on Farm Mechanisation and Multiple Cropping in Alipore Block (Delhi)", *ASI*, Vol. 35, No. 7, October.

Sau, S. (1998), "Role of Infrastructure in Rice and Foodgrains Productivity: A Regional Study" in S.Sau (ed.) "Rural Development: Theories & Experience", Allied Publishers Ltd., New Delhi.

Sasmal, J. (2001), "The Growth of West Bengal Agriculture in the Last Two Decades: A Probe into the Factors Behind the Growth", *Artha Beekshan*, Vol. 10, No. 2, December.

Sasmal, J. (1998), "Adoption of High-Yielding Variety Paddy-A Micro Level Study in India", in S. Sau (ed.) "Rural Development: Theories & Experience", Allied Publishers Ltd., New Delhi.

Sharma, A.K. and Pandy, R.N. (1988), "Capital Requirement for Modernisation of Agricultural Production in Haryana in 2000 A.D." *IJAE*, Vol. 43, No. 3, July-September.

Singh, R., Singh, R.K. and Balishter (1980) "Flow of Institutional Credit in Agriculture (With Special Reference to Commercial Bank's Finance)", *ASI*, Vol.35 , No.7 December.

Singh, A.J. and Dhawan, K.C. (1979), "Source, Utilisation and Productivity of Agricultural Credit in Ludhiana District of the Punjab State", *A.S.I.*, Vol. 34, No. 8, November.

Sharma, L.R., Bhato, J.P. and Singh, R. (1991), "Emerging Systems in Himachal Pradesh : Key Issues in Sustainability", *IJAE*, Vol. 46, No. 3, July-September.

Singh, J. and Grover, D.K. (1991), "The Impact of Technological Advance on Inter-Regional Disparities in Land Use and Farm Incomes in Punjab", *IJAE*, Vol. 46, No. 3, July-September.

Sharma, A.K. and Moorti, T.V. (1991), "Risk Efficient Farming System for Small Vegetable Farms in Kangra District of Himachal Pradesh (An Application of Parametric Risk Programming)", (Summary), *IJAE*, Vol. 46, No. 3, July-September.

Singh, A.K., Singh, L.R. and Sharma, J.S. (1991), "Income and Employment Increasing Potentiality in Various Farming Systems on Marginal Farms in Mid-Western Uttar Pradesh", (Summary), *IJAE*, Vol. 46, No. 3, July-September.

Sanyal, B., Bhattacharya, B. (2000), "Determinants of Profitability of Commercial Banks in India: Pre -Reform and Post-Reform Scenario", 20th Annual. Conference Volume, 2000, Bangiya Arthaniti Parishad, W.B.

Sain, I. (1991), "A Comparative Study on Wheat-Paddy versus Wheat-Cotton Farming Systems in the South-Western Punjab", (Summary), *IJAE*, Vol. 46, No. 3, July-September.

Singh, S.N., Singh, K.P., Gupta, R.L., Kadian, V.S. and Singh, N. (1991), "Comparative Studies on Income and Employment Potentials of Integrated Farming Systems on Small Holdings", (Summary), *IJAE*, Vol. 46, No. 3, July-September.

Singh, V.K. and Singh, H. (1991), "Post-Green Revolution Changes in the Cropping and Production Patterns in Haryana Agriculture" (Summary), *IJAE*, Vol. 46, No. 3, July-September.

Singh, A.J., Jain, K.K. and Sain, I. (1985), "Diversification of Punjab Agriculture: An Econometric Analysis", *IJAE,* Vol. XL, No. 3, July-September.

Saini, A.S. and Singh, R. (1985), "Impact of Diversification on Income, Employment and Credit Needs of Small Farmers in Punjab", *IJAE,* Vol. XL, No. 3, July-September.

Satheesh, A.V., Sharma, B.M. and Sharma, B.K. (1985), "Impact of Diversification and Liberal Credit Policy on Income and Employment of Non-Viable Farmers in Pithapuram Block of East Godavari District, A.P." *IJAE,* Vol. XL, No. 3, July-September.

Singh, B. ,Bal, H.S. and Kumar, N. (1985), "Diversification of Farming with Crop Cultivation and Dairying in Punjab-An Appraisal of Success and Constraints", (Summary), *IJAE,* Vol. XL, No. 3, July- September.

Sharma, A.C. and Singh, R. (1985), "Diversification of Small Unirrigated Farms: A Feasibility Study in the Arid Zone of the Sub-Mountainous Areas of Hoshiarpur District of the Punjab State", (Summary), *IJAE,* Vol. XL, No. 3, July-September.

Sharma, H.R. (2005), "Agricultural Development and Crop Diversification in Himachal Pradesh: Understanding the Patterns, Processes, Determinants and Lessons," Vol. 60, No. 1, January- March.

Sharma, R. (2002), "Public Sector Banks' Deposit Credit Ratio in Assam and Other N.E. States", *Artha Beekshan,* Vol. 11, No. 3, December.

Singh, R.I., Singh G.N. and Sengar, S.D.S. (1985), "Impact of Diversification in Agriculture on Level of Income and Employment of Rural Poor in District Kanpur, U.P.", (Summary), *I.J.A.E.,* Vol.XL, No.3, July - September.

Singh, R.P. and Pandey, R.K. (1991), "Economic Analysis of Marginal and Small Farms of Green Revolution Belt and Non-Green Revolution Belt–A Case Study", *I.J.A.E.,* Vol. 46, No. 3, July- September.

Srivastava, G.C. (1985), "Industrial Growth and Diversification of Farm Economy–An Empirical Investigation", (Summary), *IJAE,* Vol. XL, No. 3, July–September.

Singh, C.B. and Sharma, S.P. (1985), "Impact of Diversification on the Household Economy of Weaker Sections", (Summary), *IJAE,* Vol. XL, No. 3, July–September.

Sinha, R. (2002), "Public Sector Banks in the Reform Era", *Artha Beekshan*, Vol. 11, No. 3, December.

Sharma, R.K. (1985), "Mixed Farming on Tribal Farms–A Study of Bharmaur Tehsil of Chamba District, H.P.", (Summary) *IJAE*, Vol. XL, No. 3, July–September.

Singh, L.R. (1989), "Commercial Crops (Excluding Plantation Crops)", *IJAE*, Vol. 44, No.1, January–March.

Subrahmanyam, K.V. and Patel, R.K. (1975), "Impact of Capital Availability on Farm Income and Demand for Short Term Credit in West Godavari District, A.P.", *ASI*, Vol. XXVIII, No. 3, June.

Sau, S.N. (1990), "Productivity Differential in Rice and Foodgrains : A Regional Study", Indian Journal of Regional Science (IJRS), Vol. XXII, No. 2.

Subash, S. , Bino Paul, G.D. and Ramanathan, A. (2004), "Technical Efficiency in Rice Production: An Application of Stochastic Frontier Analysis", *IJRS*, Vol. XXXVI, No. 1.

Smith, A, "Wealth of Nations", p. 6.

Samuelson, P.A., and Solow R.M. (1975), "Balanced Growth under Constant Returns to Scale", p. 412.

Sen. S.R. (1967), "Growth and Instability in Indian Agriculture", *ASI*, Vol. 21 January.

Sahu, G.B. and Raja Sekhar (2005), " Banking Sector Reform and Credit Flow To Indian Agriculture", *EPW*, Vol. XL, No. 53, December 31-January 6.

Srinivasan, T.N. (1979), "Trends in Agriculture in India", *E.P.W*, Vol. 14, August.

Sawant, S.D. and Achuthan, C.V. (1995), "Agricultural Growth Across Crops and Regions: Emerging Trends and Patterns", *EPW*, March. 25.

Shajahan, K.M.(1998), "Priority Sector Bank Lending: Some Important Issues", *E.P.W.*, October 17-24.

Sawant, S.D. (1997), "Performance of Indian Agriculture with Special Reference to Regional Variations", *IJAE*, Vol. 52, No. 5, July-Sept,

Saha, A. (1996), "Adoption of Modern Agricultural Technology in Rice Cultivation in West Bengal", in 'Economy of West Bengal' edited by A. Roy Chaudhuri, Allied Publishers, New Delhi.

Saha, A., (2001), "Risk in HYV and Traditional Rice Cultivation: An Enquiry in West Bengal Agriculture", *IJAE*, Vol. 56, No. 1, January–March.

Schultz, W. (1964), " Transforming Traditional Agriculture", P.70, New Haven.

Sengupta, A. and Kundu, S. (2006), " Scale Efficiency Of Indian Farmers: A Non-Parametric Approach", *IJAE*, Vol. 61, No. 4 October-December.

Sen, R.K. and Dasgupta, A. (2007), "West Bengal Today, 25 Years of Economic Development", Deep and Deep Publications Pvt. Ltd. New Delhi.

Sohi, R.S. and Chahal, S.S. (2004), "Interlinked Credit Transactions In Rural Punjab", *IJAE*, Vol. 59, No. 1, January-March.

Shivamaggi, H.B.(2000), " Reforms in Rural Banking, Need for Bolder Approach", *EPW*, May 13.

Shanmugam, K.R. and Venkataramani, A. (2006), "Technical Efficiency in Agricultural Production and its Determinants: An Exploratory Study at the District Level", *IJAE*, Vol. 61, April-June.

Sarkar, D. and De, S. (2004), "High Technical Efficiency of Farms in Two Different Agricultural Lands: A Study Under Deterministic Production Frontier Approach", *IJAE*, Vol. 59, No. 2, April-June.

Sayulu, K. (1994), "Rural Marketing in India", Anmol Publications Pvt. Ltd., New Delhi.

Thomas, K. Jery and Devi, P. Indira (1990), "An Analysis of Cropping Pattern in Kerala", *ASI*, Vol. XLV, No. 3, June.

Tewari, S.K. and Sharma, J.S. (1980), "Impact of Credit as a Source of Liquidity upon Adoption of Modern Technology and Income Levels on Small Farms", *ASI*, Vol. 34, No. 11, February.

Thakur, D.S., Sharma, K.D. and Saini, A.S. (1991), "Optimum Farming System for Quantum Jump in Farm Production and Income in Post-Green Revolution Belt", (Summary), *IJAE*, Vol. 46, No. 3, July-September.

Thorve, P.V. and Gahgilikar, V.D. (1985), "Economics of Diversification of Farming with Dairy Enterprise", *IJAE*, Vol. XL, No. 3. July-September.

Thakur, D.S., Kapila and Moorti, T.V. (1985), "Vegetable Production for Diversification of Farm Economy", (Summary), *IJAE*, Vol. XL, No. 3 July-September.

Toor, M.S. and Singh, R.P. (1991), "Green Revolution and Inequality in India with Special Reference to Punjab", (Summary), *IJAE*, Vol. 46, No. 3. July-September.

Thingalay, N.K. (1978), "Rapporteur's Report on Rural Credit Structure and Flow", *IJAE*, Conference Number, Vol. XXXIII, No. 4, October–December, p. 173.

Tuteja, U. (2006), " Growth Performance and Acreage Response of Pulse Crops: A State Level Analysis", *IJAE*, Vol. 61, No. 2, April-June.

Todaro, M.P. (1987), "Agricultural Transformation and Rural Development–Economic Development in the Third World", Orient Longman Limited.

Todaro, M.P. (1985), "Economic Development in the Third World", p. 28.

Thirlwall, A.P.(1980), "Growth and Development with special reference to Developing Economies", ( Second Edition), ELBS and Macmillan.

Tyagi, B.P. (1952), "Agricultural Economics and Rural Development", Chhotery Lal Printing Press.

Thorat, Y.S.P.(2006), "Rural Credit in India: Issues and Concerns", *IJAE*, Vol. 61, No. 1, Jan.-March.

Thamarajakshi, R. (1999), "Agriculture and Economic Reform", *EPW*, August 14.

Upadhyay, S.R., Singh, B. and Upadhyay, R.G. (1985), "Pattern of Concentration and Diversification of Crop Enterprise in Uttar Pradesh (Plain)" (Summary), *IJAE*, Vol. XL, No. 3, July–September.

Venkataraman , L.S. and Prahaladachar, M. (1980), "Growth Rates and Cropping Pattern Changes in Agriculture in Six States: 1950– 1975", *IJAE*, Vol. XXV, No. 2, April–June.

Vidyasgar (1977), "A Component Analysis of the Growth of Productivity and Production in Rajasthan: 1956–1961 to 1960–1974", *IJAE*, Vol. XXXIII, No. 4, January–March.

Vyas, V.S. (1974), "LDBs Need for Reappraisal", Commerce, Annual Number.

Vyas, V. (2007), "Food Security in Asian Countries in the Context Of Millennium Goals", Academic Foundation, New Delhi.

Vidyasagar (1980), "Decomposition of Growth Rates and Certain Related Issues", *IJAE*, Vol. XXXV, No. 2, April–June.

Verma, R.C., Dhaka, J.M. and Gupta, S.S. (1991), "Impact of Green Revolution on Cropping Pattern–Agricultural Growth and Average Value Producitivity in Sriganganagar District of Rajasthan", (Summary), *IJAE,* Vol. 46, No. 3, July–September.

Vyas, V.S. (2000-01), "Agriculture: The Second Round of Economic Reforms", *The Indian Economic Journal,* Vol. 48, No. 3, January–March.

Vani, B. P. and Vysulu V. (1996), "Growth, Variability, and Instability of three Major Crops in Karnataka–A District Level Analysis from 1955-56 to 1989–90", *EPW,* June 29.

Vyas, V. S. (2006), "Our Agrarian Future: A Medium-Term Perspective on Asian Agriculture", Vol. XXXVII, No. 50, Dec. 14.

Vaidyanathan, A.(2000), "India's Agricultural Development Policy", *EPW,* May 13.

Waghmare, P.R., Tak, V.B. and Hedgire, D.N. (1991), Impact of Agro-Climatc Zones on Cropping Pattern, Production and Employment in Aurangabad District of Maharashtra", (Summary), *IJAE,* Vol. 46, No. 3, July–September.

Yadav, S.R., Rana, O.P., Singh, T.R. and Nigam, H.D. (1985), "Pattern of Income and Its Disposal Under Diversified Farming in District Kanpur", (Summary), *IJAE,* Vol. XL., No. 3, July–September.

# INDEX